T0328223

GEOPHYSICAL DATA ANALYSIS
Discrete Inverse Theory

GEOPHYSICAL DATA ANALYSIS
Discrete Inverse Theory

FOURTH EDITION

William Menke

ACADEMIC PRESS

An imprint of Elsevier

Academic Press is an imprint of Elsevier
125 London Wall, London EC2Y 5AS, United Kingdom
525 B Street, Suite 1800, San Diego, CA 92101-4495, United States
50 Hampshire Street, 5th Floor, Cambridge, MA 02139, United States
The Boulevard, Langford Lane, Kidlington, Oxford OX5 1GB, United Kingdom

© 2018 Elsevier Inc. All rights reserved.

No part of this publication may be reproduced or transmitted in any form or by any means, electronic or
mechanical, including photocopying, recording, or any information storage and retrieval system, without permission
in writing from the publisher. Details on how to seek permission, further information about the Publisher's
permissions policies and our arrangements with organizations such as the Copyright Clearance Center and the
Copyright Licensing Agency, can be found at our website: www.elsevier.com/permissions.

This book and the individual contributions contained in it are protected under copyright by the Publisher (other than
as may be noted herein).

Notices
Knowledge and best practice in this field are constantly changing. As new research and experience broaden our
understanding, changes in research methods, professional practices, or medical treatment may become necessary.

Practitioners and researchers must always rely on their own experience and knowledge in evaluating and using
any information, methods, compounds, or experiments described herein. In using such information or methods they
should be mindful of their own safety and the safety of others, including parties for whom they have a
professional responsibility.

To the fullest extent of the law, neither the Publisher nor the authors, contributors, or editors, assume any liability
for any injury and/or damage to persons or property as a matter of products liability, negligence or otherwise, or
from any use or operation of any methods, products, instructions, or ideas contained in the material herein.

Library of Congress Cataloging-in-Publication Data
A catalog record for this book is available from the Library of Congress

British Library Cataloguing-in-Publication Data
A catalogue record for this book is available from the British Library

ISBN 978-0-12-813555-6

For information on all Academic Press publications
visit our website at https://www.elsevier.com/books-and-journals

  Working together
to grow libraries in
developing countries

www.elsevier.com • www.bookaid.org

Publisher: Candice Janco
Acquisition Editor: Marisa LaFleur
Editorial Project Manager: Katerina Zaliva
Production Project Manager: Prem Kumar Kaliamoorthi
Cover Designer: Christian J. Bilbow

Typeset by SPi Global, India

Contents

12. Sample Inverse Problems

13. Applications of Inverse Theory to Solid Earth Geophysics

Preface

*For now we see through a glass, darkly, but then ... **Paul of Tarsus***

Every researcher in the applied sciences who has analyzed data has practiced inverse theory. Inverse theory is simply the set of methods used to extract useful inferences about the world from physical measurements. The fitting of a straight line to data involves a simple application of inverse theory. Tomography, popularized by the physician's CT and MRI scanners, uses it on a more sophisticated level.

The study of inverse theory, however, is more than the cataloging of methods of data analysis. It is an attempt to organize these techniques, to bring out their underlying similarities and pin down their differences, and to deal with the fundamental question of the limits of information that can be gleaned from any given data set.

Physical properties fall into two general classes: those that can be described by discrete parameters (e.g., the mass of the earth or the position of the atoms in a protein molecule) and those that must be described by continuous functions (e.g., temperature over the face of the earth or electric field intensity in a capacitor). Inverse theory employs different mathematical techniques for these two classes of parameters: the theory of matrix equations for discrete parameters and the theory of integral equations for continuous functions.

Being introductory in nature, this book deals mainly with "discrete inverse theory," that is, the part of the theory concerned with parameters that either are truly discrete or can be adequately approximated as discrete. By adhering to these limitations, inverse theory can be presented on a level that is accessible to most first-year graduate students and many college seniors in the applied sciences. The only mathematics that is presumed is a working knowledge of the calculus and linear algebra and some familiarity with general concepts from probability theory and statistics.

Nevertheless, the treatment in this book is in no sense simplified. Realistic examples, drawn from the scientific literature, are used to illustrate the various techniques. Since in practice the solutions to most inverse problems require substantial computational effort, attention is given to how realistic problems can be solved.

The treatment of inverse theory in this book is divided into four parts. Chapters 1 and 2 provide a general background, explaining what inverse problems are and what constitutes their solution as well as reviewing some of the basic concepts from linear algebra and probability theory that will be applied throughout the text. Chapters 3–7 discuss the solution of the canonical inverse problem: the linear problem with Gaussian statistics. This is the best understood of all inverse problems, and it is here that the fundamental notions of uncertainty, uniqueness, and resolution can be most clearly developed. Chapters 8–11 extend the discussion to problems that are non-Gaussian, nonlinear, and continuous. Chapter 11 devotes special attention to the so-called *adjoint method*, a mathematical technique that, over the past two decades, has become an increasingly important tool for solving inverse problems in seismology and climate science. Chapters 12–13 provide examples of the use of inverse theory and a discussion of the steps that must be taken to solve inverse problems on a computer.

MatLab scripts are used throughout the book as a means of communicating how the formulas of inverse theory can be used in computer-based data processing scenarios. *MatLab* is a commercial software product of *The MathWorks, Inc.* and is widely used in university settings as an environment for scientific computing. All of the book's examples, its recommended homework problems, and the case studies of Chapter 12 use *MatLab* extensively. Further, all the *MatLab* scripts used in the book are made available to readers through the book's website. The book is self-contained; it can be read straight through, and profitably, even by someone with no access to *MatLab*. But it is meant to be used in a setting where students are actively using *MatLab* both as an aid to studying (that is, by reproducing the examples and case studies described in the book) and as a tool for completing the recommended homework problems.

Many people helped me write this book. I am very grateful to my students at Columbia University and at Oregon State University for the helpful comments they gave me during the courses I have taught on inverse theory. Mike West, of the Alaska Volcano Observatory, did much to inspire recent revisions of the book, by inviting me to teach a mini-course on the subject in the fall of 2009. The use of *MatLab* in this book parallels the usage in *Environmental Data Analysis with MatLab* (Menke and Menke, 2011), a data analysis textbook that I wrote with my son Joshua Menke in 2011. The many hours we spent working together on its tutorials taught us both a tremendous amount about how to use that software in a pedagogical setting. Finally, I thank the many hundreds of scientists and mathematicians whose ideas I drew upon in writing this book.

Reference

Menke, W., Menke, J., 2011. Environmental Data Analysis With MatLab. Academic Press, Elsevier Inc., Oxford. 263 pp.

Companion Web Site

https://www.elsevier.com/books-and-journals/book-companion/9780128135556.

Introduction

I.1 FORWARD AND INVERSE THEORIES

Inverse theory is an organized set of mathematical techniques for reducing data to obtain *knowledge* about the physical world on the basis of inferences drawn from observations. Inverse theory, as we shall consider it in this book, is limited to observations and questions that can be represented numerically. The observations of the world will consist of a tabulation of measurements, or *data*. The questions we want to answer will be stated in terms of the numerical values (and statistics) of specific (but not necessarily directly measurable) properties of the world. These properties will be called model *parameters* for reasons that will become apparent. We shall assume that there is some specific method (usually a mathematical theory or model) for relating the model parameters to the data.

The question, "what causes the motion of the planets?," for example, is not one to which inverse theory can be applied. Even though it is perfectly scientific and historically important, its answer is not numerical in nature. On the other hand, inverse theory can be applied to the question, assuming that Newtonian mechanics applies, determine the number and orbits of the planets on the basis of the observed orbit of Halley's comet. The number of planets and their orbital ephemerides are numerical in nature. Another important difference between these two problems is that the first asks us to determine the reason for the orbital motions, and the second presupposes the reason and asks us only to determine certain details. Inverse theory rarely supplies the kind of insight demanded by the first question; it always requires that the physical model or theory be specified beforehand.

The term *inverse theory* is used in contrast to *forward theory*, which is defined as the process of predicting the results of measurements (predicting data) on the basis of some general principle or model and a set of specific conditions relevant to the problem at hand. Inverse theory, roughly speaking, addresses the reverse problem: starting with data and a general principle, theory, or quantitative model, it determines estimates of the model parameters. In the earlier example, predicting the orbit of Halley's comet from the presumably well-known orbital ephemerides of the planets is a problem for forward theory.

Another comparison of forward and inverse problems is provided by the phenomenon of temperature variation as a function of depth beneath the earth's surface. Let us assume that the temperature increases linearly with depth in the earth, that is, temperature T is related to depth z by the rule $T(z) = az + b$, where a and b are numerical constants that we will refer to as *model parameters*. If one knows that $a = 0.1$ and $b = 25$, then one can solve the forward problem simply by evaluating the formula for any desired depth. The inverse problem would be to determine a and b on the basis of a suite of temperature measurements made at different depths in, say, a bore hole. One may recognize that this is the problem of fitting a straight line to data, which is a substantially harder problem than the forward problem of evaluating a first-degree polynomial. This brings out a property of most inverse problems: that they are substantially harder to solve than their corresponding forward problems.

Forward problem:
  estimates of model parameters → quantitative model → predictions of data

Inverse problem:
  observations of data → quantitative model → estimates of model parameters

Note that the role of inverse theory is to provide information about unknown numerical parameters that go into the model, not to provide the model itself. Nevertheless, inverse theory can often provide a means for assessing the correctness of a given model or of discriminating between several possible models.

The model parameters one encounters in inverse theory vary from discrete numerical quantities to continuous functions of one or more variables. The intercept and slope of the straight line mentioned earlier are examples of discrete parameters. Temperature, which varies continuously with position, is an example of a continuous function. This book deals mainly with discrete inverse theory, in which the model parameters are represented as a set of a finite number of numerical values. This limitation does not, in practice, exclude the study of continuous functions, since they can usually be adequately approximated by a finite number of discrete parameters. Temperature, for example, might be represented by its value at a finite number of closely spaced points or by a set of splines with a finite number of coefficients. This approach does, however, limit the rigor with which continuous functions can be studied. Parameterizations of continuous functions are always both approximate and, to some degree, arbitrary properties, which cast a certain amount of imprecision into the theory. Nevertheless, discrete inverse theory is a good starting place for the study of inverse theory, in general, since it relies mainly on the theory of vectors and matrices rather than on the somewhat more complicated theory of continuous functions and operators. Furthermore, careful application of discrete inverse theory can often yield considerable insight, even when applied to problems involving continuous parameters.

Although the main purpose of inverse theory is to provide estimates of model parameters, the theory has a considerably larger scope. Even in cases in which the model parameters are the only desired results, there is a plethora of related information that can be extracted to help determine the "goodness" of the solution to the inverse problem. The actual values of the model parameters are indeed irrelevant in cases when we are mainly interested in using inverse theory as a tool in experimental design or in summarizing the data. Some of the questions inverse theory can help answer are the following:

1. What are the underlying similarities among inverse problems?
2. How are estimates of model parameters made?
3. How much of the error in the measurements shows up as error in the estimates of the model parameters?
4. Given a particular experimental design, can a certain set of model parameters really be determined?

These questions emphasize that there are many different kinds of answers to inverse problems and many different criteria by which the goodness of those answers can be judged.

Much of the subject of inverse theory is concerned with recognizing when certain criteria are more applicable than others, as well as detecting and avoiding (if possible) the various pitfalls that can arise.

Inverse problems arise in many branches of the physical sciences. An incomplete list might include such entries as

1. medical and seismic tomography,
2. image enhancement,
3. curve fitting,
4. earthquake location,
5. oceanographic and meteorological data assimilation,
6. factor analysis,
7. determination of earth structure from geophysical data,
8. satellite navigation,
9. mapping of celestial radio sources with interferometry, and
10. analysis of molecular structure by X-ray diffraction.

Inverse theory was developed by scientists and mathematicians having various backgrounds and goals. Thus, although the resulting versions of the theory possess strong and fundamental similarities, they have tended to look, superficially, very different. One of the goals of this book is to present the various aspects of discrete inverse theory in such a way that both the individual viewpoints and the "big picture" can be clearly understood.

There are perhaps three major viewpoints from which inverse theory can be approached. The first and oldest sprang from probability theory—a natural starting place for such "noisy" quantities as observations of the real world. In this version of inverse theory, the data and model parameters are treated as random variables, and a great deal of emphasis is placed on determining the probability density functions that they follow. This viewpoint leads very naturally to the analysis of error and to tests of the significance of answers.

The second viewpoint developed from that part of the physical sciences that retains a deterministic stance and avoids the explicit use of probability theory. This approach has tended to deal only with estimates of model parameters (and perhaps with their error bars) rather than with probability density functions per se. Yet what one means by an estimate is often nothing more than the expected value of a probability density function; the difference is only one of emphasis.

The third viewpoint arose from a consideration of model parameters that are inherently continuous functions. Whereas the other two viewpoints handled this problem by approximating continuous functions with a finite number of discrete parameters, the third developed methods for handling continuous functions explicitly. Although continuous inverse theory is not the primary focus of this book, many of the concepts originally developed for it have application to discrete inverse theory, especially when it is used with discretized continuous functions.

I.2 MATLAB AS A TOOL FOR LEARNING INVERSE THEORY

The practice of inverse theory requires computer-based computation. A person can learn many of the *concepts* of inverse theory by working through short pencil-and-paper examples and by examining precomputed figures and graphs. But he or she cannot become proficient in the *practice* of inverse theory that way because it requires skills that can only be obtained through the experience of working with large data sets. Three goals are paramount: to develop the judgment needed to select the best solution method among many alternatives; to build confidence that the solution can be obtained even though it requires many steps; and to strengthen the critical faculties needed to assess the quality of the results. This book devotes considerable space to case studies and homework problems that provide the practical problem-solving experience needed to gain proficiency in inverse theory.

Computer-based computation requires software. Many different software environments are available for the type of scientific computation that underpins data analysis. Some are more applicable and others less applicable to inverse theory problems, but among the applicable ones, none has a decisive advantage. Nevertheless, we have chosen *MatLab*, a commercial software product of *The MathWorks, Inc.* as the book's software environment for several reasons, some having to do with its designs and other more practical. The most persuasive design reason is that its syntax fully supports linear algebra, which is needed by almost every inverse theory method. Furthermore, it supports *scripts*, that is, sequences of data analysis commands that are communicated in written form and which serve to document the data analysis process. Practical considerations include the following: it is a long-lived and stable product, available since the mid-1980s; implementations are available for most commonly used types of computers; its price, especially for students, is fairly modest; and it is widely used, at least, in university settings.

In *MatLab*'s scripting language, data are presented as one or more *named variables* (in the same sense that c and d in the formula, $c = \pi d$, are named variables). Data are manipulated by typing formula that create new variables from old ones and by running *scripts*, that is, sequences of formulas stored in a file. Much of inverse theory is simply the application of well-known formulas to novel data, so the great advantage of this approach is that the formulas that are typed usually have a strong similarity to those printed in a textbook. Furthermore, scripts provide both a way of documenting the sequence of a formula used to analyze a particular data set and a way to transfer the overall data analysis procedure from one data set to another. However, one disadvantage is that the parallel between the syntax of the scripting language and the syntax of standard mathematical notation is nowhere near perfect. A person needs to learn to translate one into the other.

I.3 A VERY QUICK MATLAB TUTORIAL

Unfortunately, this book must avoid discussion of the installation of *MatLab* and the appearance of *MatLab* on your computer screen, for procedures and appearances vary from computer to computer and quickly become outdated, anyway. We will assume that you have successfully installed it and that you can identify the Command Window, the place where *MatLab* formula and commands are typed. Once you have identified the Command Window, try typing:

```
date
```

MatLab should respond by displaying today's date. All the *MatLab* commands that are used in this book are in freely available *MatLab* scripts. This one is named `gda01_01` and is in a

MatLab script file (*m-file*, for short) named `gda01_01.m` (conventionally, m-files have filenames that end with ".m"). In this case, the script is very short, since it just contains this one command, `date`, together with a couple of comment lines (which start with the character "%"):

```
% gda00_01
% displays the current date
date
```

(*MatLab* script gda00_01)

All the scripts are in a folder named `gda`. You should copy it to a convenient and easy-to-remember place in your computer's file system. The *MatLab* command window supports a number of commands that enable you to navigate from folder to folder, list the contents of folders, etc. For example, when you type:

```
pwd
```

(for "print working directory") in the Command Window, *MatLab* responds by displaying the name of the current folder. Initially, this is almost invariably the wrong folder, so you will need to `cd` (for "change directory") to the folder where you want to be—the `ch00` folder in this case. The pathname will, of course, depend upon where you copied the `gda` folder but will end in `gda/ch00`. On the author's computer, typing:

```
cd c:/menke/docs/gda/ch00
```

does the trick. If you have spaces in your pathname, just surround it with single quotes:

```
cd 'c:/menke/my docs/gda/ch00'
```

You can check if you are in the right folder by typing `pwd` again. Once in the `ch00` folder, typing:

```
gda00_01
```

will run the `gda00_01` m-script, which displays the current date. You can move to the folder above the current one by typing:

```
cd ..
```

and to one below it by giving just the folder name. For example, if you are in the `gda` folder you can move to the `ch00` folder by typing:

```
cd ch00
```

Finally, the command `dir` (for "directory") lists the files and subfolders in the current directory.

```
dir
```

(*MatLab* script gda00_02)

The *MatLab* commands for simple arithmetic and algebra closely parallel standard mathematical notation. For instance, the command sequence

```
a=4.5;
b=5.1;
c=a+b;
c
```

(*MatLab* script gda00_03)

evaluates the formula $c = a + b$ for the case $a = 4.5$ and $b = 5.1$ to obtain $c = 9.6$. Only the semicolons require explanation. By default, *MatLab* displays the value of every formula typed into the Command Window. A semicolon at the end of the formula suppresses the display. Hence, the first three lines, which end with semicolons, are evaluated but not displayed. Only the final line, which lacks the semicolon, causes *MatLab* to print the final result, c.

Note that *MatLab* variables are *static*, meaning that they persist in *MatLab*'s *Workspace* until you explicitly delete them or exit the program. Variables created by one script can be used by subsequent scripts. At any time, the value of a variable can be examined, either by displaying it in the Command Window (as we have done above) or by using the spreadsheet-like display tools available through *MatLab*'s Workspace Window. The persistence of *MatLab* variables can sometimes lead to scripting errors, such as when the definition of a variable in a script is inadvertently omitted, but *MatLab* used the value defined in a previous script. The command clear all deletes all previously defined variables in the Workspace. Placing this command at the beginning of a script causes it to delete any previously defined variables every time that it is run, ensuring that it cannot be affected by them.

The four commands discussed earlier can be run as a unit by typing gda00_03. Now open the m-file gda01_03 in *MatLab*, using the File/Open menu. *MatLab* will bring up a text-editor type window. First, save it as a new file, say, mygda01_03; edit it in some simple way, say, by changing the 3.5 to 4.5; save the edited file; and run it by typing mygda01_03 in the Command Window. The value of c that is displayed will have changed appropriately.

A somewhat more complicated formula is

$$c = \sqrt{a^2 + b^2} \text{ with } a = 6 \text{ and } b = 8$$

```
a=6;
b=8;
c = sqrt(a^2 + b^2);
c
```

(*MatLab* gda00_04)

Note that the *MatLab* syntax for a^2 is a^2 and that the square root is computed using the function, sqrt(). This is an example of *MatLab*'s syntax differing from standard mathematical notation.

A final example is

$$c = \sin\frac{n\pi(x - x_0)}{L} \text{ with } n = 3, \ x = 4, \ x_0 = 1, \ L = 6$$

```
n=3; x=4; x0=1; L=6;
c = sin(n*pi*(x-x0)/L);
c
```

(*MatLab* gda00_05)

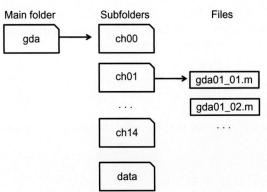

Main folder   Subfolders   Files

gda → ch00

ch01 → gda01_01.m

... gda01_02.m

ch14 ...

data

FIG. I.1 Folder (directory) structure used for the files accompanying this book.

Note that several formulas, separated by semicolons, can be typed on the same line. Variables, such as x0 and pi, above, can have names consisting of more than one character and can contain numerals as well as letters (though they must start with a letter). *MatLab* has a variety of predefined mathematical constants, including pi, which is the usual mathematical constant, π.

Files proliferate at an astonishing rate, even in the most trivial data analysis project. Data, notes, m-scripts, intermediate and final results, and graphics will all be contained in files, and their numbers will grow during the project. These files need to be organized through a system of folders (directories), subfolders (subdirectories), and filenames that are sufficiently systematic that files can be located easily and so that they are not confused with one another. Predictability both in the pattern of filenames and in the arrangement of folders and subfolders is an extremely important part of the design.

The files associated with this book are in a two-tiered folder/subfolder structure modeled on the chapter format of the book itself (Fig. I.1). Folder and filenames are systematic. The chapter folder names are always of the form chNN, where NN is the chapter number. The m-script filenames are always of the form gdaNN_MM.m, where NN is the chapter number and MM are sequential integers. We have chosen to use leading zeros in the naming scheme (e.g., ch00) so that filenames appear in the correct order when they are sorted alphabetically (as when listing the contents of a folder).

Many *MatLab* manuals, guides, and tutorials are available, both in the printed form (e.g., Menke and Menke, 2011; Part-Enander et al., 1996; Pratap, 2009) and on the Web (e.g., at www.mathworks.com). The reader may find that they complement this book by providing more detailed information about *MatLab* as a scientific computing environment.

I.4 REVIEW OF VECTORS AND MATRICES AND THEIR REPRESENTATION IN *MATLAB*

Vectors and matrices are fundamental to inverse theory both because they provide a convenient way to organize data and because many important operations on data can be very succinctly expressed using *linear algebra* (i.e., the algebra of vectors and matrices).

In the simplest interpretation, a *vector* is just a list of numbers that are treated as unit and given a symbolic name. The list can be organized horizontally, as a row, in which case it is called a *row vector*. Alternately, the list can be organized vertically, as a column, in which case it is called a *column vector*. We will use lower-case bold letters to represent both kinds of vector. An exemplary 1×3 row vector $\mathbf{r}$ and a 3×1 column vector $\mathbf{c}$ are

$$\mathbf{r} = [2\ 4\ 6] \quad \text{and} \quad \mathbf{c} = \begin{bmatrix} 1 \\ 3 \\ 5 \end{bmatrix}$$

A row vector can be turned into a column vector, and vice versa, by the *transpose* operation, denoted by the superscript "T." Thus,

$$\mathbf{r}^T = \begin{bmatrix} 2 \\ 4 \\ 6 \end{bmatrix} \quad \text{and} \quad \mathbf{c}^T = [1\ \ 3\ \ 5]$$

In *MatLab*, row vector and column vectors are defined by

```
r = [2, 4, 6];
c = [1, 3, 5]';
```

(*MatLab* gda00_06)

In *MatLab*, the transpose is denoted by the single quote, so the column vector c in the script above is created by transposing a row vector. Although both column vectors and row vectors are useful, our experience is that defining both in the same script creates serious opportunities for error. A formula that requires a column vector will usually yield incorrect results if a row vector is substituted into it, and vice versa. Consequently, we will adhere to a protocol where all vectors defined in this book are column vectors. Row vectors will be created when needed—and as close as possible to where they are used in the script—by transposing the equivalent column vector.

An individual number within a vector is called an *element* (or, sometimes, *component*) and is denoted with an integer index, written as a subscript, with the index 1 in the left of the row vector and top of the column vector. Thus, $r_2 = 4$ and $c_3 = 5$ in the example above. In *MatLab*, the index is written inside parentheses, as in

```
a = r(2);
b = c(3);
```

(*MatLab* gda00_06)

Sometimes, we will wish to indicate a generic element of the vector, in which case we will give it a variable index, as in r_i and c_j, with the understanding that i and j are integer variables.

In the simplest interpretation, a *matrix* is just a rectangular table of numbers. We will use bold uppercase names to denote matrices, as in

$$\mathbf{M} = \begin{bmatrix} 1 & 2 & 3 \\ 4 & 5 & 6 \\ 7 & 8 & 9 \end{bmatrix}$$

In the earlier example, the number of rows and number of columns are equal, but this property is not required; matrices can also be rectangular. Thus, row vectors and column vectors are just special cases of rectangular matrices. The transposition operation can also be performed on a matrix, in which case its rows and columns are interchanged.

$$\mathbf{M}^T = \begin{bmatrix} 1 & 4 & 7 \\ 2 & 5 & 8 \\ 3 & 6 & 9 \end{bmatrix}$$

A square matrix $\mathbf{M}$ is said to be *symmetric* if $\mathbf{M} = \mathbf{M}^T$. In *MatLab*, a matrix is defined by

```
M=[ [1,4,7]',[2,5,8]',[3,6,9]'];
```

(*MatLab* gda00_06)

or, alternately, by

```
M2 = [1, 2, 3;4, 5, 6;
7, 8, 9];
```

(*MatLab* gda00_06)

The first case, the matrix, $\mathbf{M}$, is constructed from a "row vector of column vectors" and in the second case, as a "column vector of row vectors." The individual elements of a matrix are denoted with two integer indices, the first indicating the row and the second the column, starting in the upper left. Thus, in the earlier example, $M_{31} = 3$. Note that transposition swaps indices; that is, M_{ji} is the transpose of M_{ij}. In *MatLab*, the indices are written inside parentheses, as in

```
c = M(1,3);
```

(*MatLab* gda01_06)

One of the key properties of vectors and matrices is that they can be manipulated symbolically—as entities—according to specific rules that are similar to normal arithmetic. This allows tremendous simplification of data processing formulas, since all the details of what happens to individual elements within those entities are hidden from view and automatically performed.

In order to be added, two matrices (or vectors, viewing them as a special case of a rectangular matrix) must have the same number of rows and columns. Their sum is then just the matrix that results from summing corresponding elements. Thus, if

$$\mathbf{M} = \begin{bmatrix} 1 & 0 & 2 \\ 0 & 1 & 0 \\ 2 & 0 & 1 \end{bmatrix} \text{ and } \mathbf{N} = \begin{bmatrix} 1 & 0 & -1 \\ 0 & 2 & 0 \\ 1 & 0 & 3 \end{bmatrix} \text{ then}$$

$$\mathbf{S} = \mathbf{M} + \mathbf{N} = \begin{bmatrix} 1+1 & 0+0 & 2-1 \\ 0+0 & 1+2 & 0+0 \\ 2-1 & 0+0 & 1+3 \end{bmatrix} = \begin{bmatrix} 2 & 0 & 1 \\ 0 & 3 & 0 \\ 1 & 0 & 4 \end{bmatrix}$$

Subtraction is performed in an analogous manner. In terms of the components, addition and subtraction are written as

$$S_{ij} = M_{ij} + N_{ij} \quad \text{and} \quad D_{ij} = M_{ij} - N_{ij}$$

Note that addition is commutative (i.e., $\mathbf{M} + \mathbf{N} = \mathbf{N} + \mathbf{M}$) and associative (i.e., $(\mathbf{A} + \mathbf{B}) + \mathbf{C} = \mathbf{A} + (\mathbf{B} + \mathbf{C})$). In *MatLab*, addition and subtraction are written as

```
S = M+N;
D = M-N;
```

(*MatLab* gda00_07)

Multiplication of two matrices is a more complicated operation and requires that the number of columns of the left-hand matrix equal the number of rows of the right-hand matrix. Thus, if the matrix $\mathbf{M}$ is $N \times K$ and the matrix $\mathbf{N}$ is $K \times M$, the product $\mathbf{P} = \mathbf{MN}$ is an $N \times M$ matrix defined according to the rule (Fig. I.2):

$$P_{ij} = \sum_{k=1}^{K} M_{ik} N_{kj}$$

The order of the indices is important. Matrix multiplication is in its standard form when all summations involve neighboring indices of adjacent quantities. Thus, for instance, the two instances of the summed variable k are *not* neighboring in the equation

$$Q_{ij} = \sum_{k=1}^{K} M_{ki} N_{kj}$$

and so the equation corresponds to $\mathbf{Q} = \mathbf{M}^T\mathbf{N}$ and not $\mathbf{Q} = \mathbf{MN}$. Matrix multiplication is not commutative (i.e., $\mathbf{MN} \neq \mathbf{NM}$) but is associative (i.e., $(\mathbf{AB})\mathbf{C} = \mathbf{A}(\mathbf{BC})$) and distributive (i.e., $\mathbf{A}(\mathbf{B} + \mathbf{C}) = \mathbf{AB} + \mathbf{AC}$). An important rule involving the matrix transpose is $(\mathbf{MN})^T = \mathbf{N}^T\mathbf{M}^T$ (note the reversal of the order).

Several special cases of multiplication involving vectors are noteworthy. Suppose that $\mathbf{a}$ and $\mathbf{b}$ are length-N column vectors. The combination $s = \mathbf{a}^T\mathbf{b}$ is a scalar number s and is called the *dot product* (or sometimes, *inner product*) of the vectors. It obeys the rule $\mathbf{a}^T\mathbf{b} = \mathbf{b}^T\mathbf{a}$. The dot product of a vector with itself is the square of its *Euclidian length*; that is, $\mathbf{a}^T\mathbf{a}$ is the sum of its squared elements of $\mathbf{a}$. The vector $\mathbf{a}$ is said to be a unit vector when $\mathbf{a}^T\mathbf{a} = 1$. The combination $\mathbf{T} = \mathbf{ab}^T$ is an $N \times N$ matrix; it is called the *outer product*. The product of a matrix and a vector is another vector, as in $\mathbf{c} = \mathbf{Ma}$. One interpretation of this relationship is that the matrix $\mathbf{M}$ "turns one vector into another." Note that the vectors

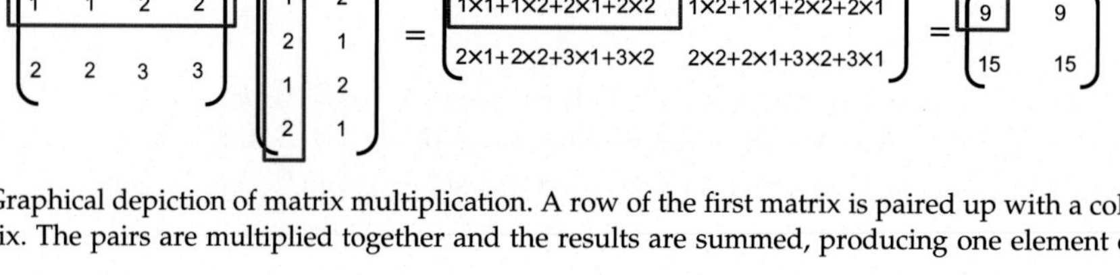

FIG. I.2 Graphical depiction of matrix multiplication. A row of the first matrix is paired up with a column of the second matrix. The pairs are multiplied together and the results are summed, producing one element of resultant matrix.

a and **c** can be of different length. An $M \times N$ matrix **M** turns the length-N vector **a** into a length-M vector **c**. The combination $s = \mathbf{a}^T \mathbf{M} \mathbf{a}$ is a scalar and is called a *quadratic form*, since it contains terms quadratic in the elements of **a**. Matrix multiplication, **P** = **MN**, has a useful interpretation in terms of dot products: P_{ij} is the dot product of the ith row of **M** with the jth column of **N**.

Any matrix is unchanged when multiplied by the *identity matrix*, conventionally denoted **I**. Thus, **a** = **Ia**, **M** = **IM** = **MI**, etc. This matrix has ones along its main diagonal, and zeroes elsewhere, as in

$$\mathbf{I} = \begin{bmatrix} 1 & 0 & 0 \\ 0 & 1 & 0 \\ 0 & 0 & 1 \end{bmatrix}$$

The elements of the identity matrix are usually written δ_{ij} and not I_{ij}, and the symbol δ_{ij} is usually called the *Kronecker delta* symbol, not the *elements of the identity matrix* (though that is exactly what it is). The equation **M** = **IM**, for an $N \times N$ matrix **M**, is written component-wise as

$$M_{ij} = \sum_{k=1}^{N} \delta_{ik} M_{kj}$$

This equation indicates that any summation containing a Kronecker delta symbol can be performed trivially. To obtain the result, one first identifies the variable that is being summed over (k in this case) and the variable that the summed variable is paired with in the Kronecker delta symbol (i in this case). The summation and the Kronecker delta symbol then are deleted from the equation, and all occurrences of the summed variable are replaced with the paired variable (all ks are replaced by is in this case). In *MatLab*, an $N \times N$ identity matrix can be created with the command

```
I = eye(N);
```

(*MatLab* gda00_07)

MatLab performs all multiplicative operations with ease. For example, suppose column vectors, **a** and **b**, and matrices, **M** and **N**, are defined as

$$\mathbf{a} = \begin{bmatrix} 1 \\ 3 \\ 5 \end{bmatrix} \text{ and } \mathbf{b} = \begin{bmatrix} 2 \\ 4 \\ 6 \end{bmatrix} \text{ and } \mathbf{M} = \begin{bmatrix} 1 & 0 & 2 \\ 0 & 1 & 0 \\ 2 & 0 & 1 \end{bmatrix} \text{ and } \mathbf{N} = \begin{bmatrix} 1 & 0 & -1 \\ 0 & 2 & 0 \\ -1 & 0 & 3 \end{bmatrix}$$

Then

$$s = \mathbf{a}^T \mathbf{b} = \begin{bmatrix} 1 \\ 3 \\ 5 \end{bmatrix}^T \begin{bmatrix} 2 \\ 4 \\ 6 \end{bmatrix} = \begin{bmatrix} 1 & 3 & 5 \end{bmatrix} \begin{bmatrix} 2 \\ 4 \\ 6 \end{bmatrix} = 2 \times 1 + 3 \times 4 + 5 \times 6 = 44$$

$$\mathbf{T} = \mathbf{a} \mathbf{b}^T = \begin{bmatrix} 1 \\ 3 \\ 5 \end{bmatrix} \begin{bmatrix} 2 \\ 4 \\ 6 \end{bmatrix}^T = \begin{bmatrix} 2 \times 1 & 4 \times 1 & 6 \times 1 \\ 2 \times 3 & 4 \times 3 & 6 \times 3 \\ 2 \times 5 & 4 \times 5 & 6 \times 5 \end{bmatrix} = \begin{bmatrix} 2 & 4 & 6 \\ 6 & 12 & 18 \\ 10 & 20 & 30 \end{bmatrix}$$

$$\mathbf{c} = \mathbf{Ma} = \begin{bmatrix} 1 & 0 & 2 \\ 0 & 1 & 0 \\ 2 & 0 & 1 \end{bmatrix} \begin{bmatrix} 1 \\ 3 \\ 5 \end{bmatrix} = \begin{bmatrix} 1\times1 & + & 0\times3 & + & 2\times5 \\ 0\times1 & + & 1\times3 & + & 0\times5 \\ 2\times1 & + & 0\times3 & + & 1\times5 \end{bmatrix} = \begin{bmatrix} 11 \\ 3 \\ 7 \end{bmatrix}$$

$$\mathbf{P} = \mathbf{MN} = \begin{bmatrix} 1 & 0 & 2 \\ 0 & 1 & 0 \\ 2 & 0 & 1 \end{bmatrix} \begin{bmatrix} 1 & 0 & -1 \\ 0 & 2 & 0 \\ -1 & 0 & 3 \end{bmatrix} = \begin{bmatrix} -1 & 0 & 5 \\ 0 & 2 & 0 \\ 1 & 0 & 1 \end{bmatrix}$$

corresponds to

```
s = a'*b;
T = a*b';
c = M*a;
P = M*N;
```

(*MatLab* gda00_07)

In *MatLab*, matrix multiplication is signified using the multiplications sign, * (the asterisk). There are cases, however, where one needs to violate the rules and multiply the quantities element-wise (e.g., create a vector, **d**, with elements $d_i = a_i b_i$). *MatLab* provides a special element-wise version of the multiplication sign, denoted .* (a period followed by an asterisk)

```
d = a.*b;
```

(*MatLab* gda00_07)

As described earlier, individual elements of vectors and matrices can be accessed by specifying the relevant row and column indices, in parentheses, e.g., a(2) is the second element of the column vector **a**, and M(2,3) is the second row, third column element of the matrix, **M**. Ranges of rows and columns can be specified using the : (colon) operator, e.g., M (:,2) is the second column of matrix, **M**; M(2,:) is the second row of matrix, **M**; and M (2:3,2:3) is the 2 × 2 submatrix in the lower right-hand corner of the 3×3 matrix, **M** (the expression, M(2:end,2:end), would work as well). These operations are further illustrated below:

$$\mathbf{a} = \begin{bmatrix} 1 \\ 2 \\ 3 \end{bmatrix} \quad \text{and} \quad \mathbf{M} = \begin{bmatrix} 1 & 2 & 3 \\ 4 & 5 & 6 \\ 7 & 8 & 9 \end{bmatrix}$$

$$s = a_2 = 2 \quad \text{and} \quad t = M_{23} = 6 \quad \text{and} \quad \mathbf{b} = \begin{bmatrix} M_{12} \\ M_{22} \\ M_{32} \end{bmatrix} = \begin{bmatrix} 2 \\ 5 \\ 8 \end{bmatrix}$$

$$\mathbf{c} = [M_{21} \ M_{22} \ M_{23}]^\mathrm{T} = \begin{bmatrix} 4 \\ 5 \\ 6 \end{bmatrix} \quad \text{and} \quad \mathbf{T} = \begin{bmatrix} M_{22} & M_{23} \\ M_{32} & M_{33} \end{bmatrix} = \begin{bmatrix} 5 & 6 \\ 8 & 9 \end{bmatrix}$$

correspond to

```
s = a(2);
t = M(2,3);
b = M(:,2);
c = M(2,:)';
T = M(2:3,2:3);
```

(*MatLab* gda00_08)

The colon notation can be used in other contexts in *MatLab* as well. For instance, [1:4] is the row vector [1, 2, 3, 4]. The syntax, 1:4, which omits the square brackets, works fine in *MatLab*. However, we will usually use square brackets, since they draw attention to the presence of a vector. Finally, we note that two colons can be used in sequence to indicate the spacing of elements in the resulting vector. For example, the expression [1:2:9] is the row vector [1, 3, 5, 7, 9] and that the expression [10:-1:1] is a row vector whose elements are in the reverse order from [1:10].

Matrix division is defined in analogy to reciprocals. If s is a scalar number, then multiplication by the reciprocal s^{-1} is equivalent to division by s. Here, the reciprocal obeys $s^{-1}s = ss^{-1} = 1$. The matrix analog to the reciprocal is called the *matrix inverse* and obeys

$$\mathbf{A}^{-1}\mathbf{A} = \mathbf{A}\mathbf{A}^{-1} = \mathbf{I}$$

It is defined only for square matrices. The calculation of the inverse of a matrix is complicated, and we will not describe it here, except to mention the 2×2 case

$$\begin{bmatrix} a & b \\ c & d \end{bmatrix}^{-1} = \frac{1}{ad - bc} \begin{bmatrix} d & -b \\ -c & a \end{bmatrix}$$

Just as the reciprocal s^{-1} is defined only when $s \neq 0$, the matrix inverse $\mathbf{A}^{-1}$ is defined only when a quantity called the *determinant* of $\mathbf{A}$, denoted $\det(\mathbf{A})$, is not equal to zero. The determinant of a square $N \times N$ matrix $\mathbf{M}$ is defined as

$$\det(\mathbf{M}) = \sum_{i=1}^{N}\sum_{j=1}^{N}\sum_{k=1}^{N}\cdots\sum_{q=1}^{N} \varepsilon^{ijk\cdots q} M_{1i}M_{2j}M_{3k}\ldots M_{Nq}$$

Here the quantity $\varepsilon^{ijk \cdots q}$ is +1 when $(i, j, k, \ldots, q)$ is an even permutation of $(1, 2, 3, \ldots, N)$, -1 when it is an odd permutation, and zero otherwise. Note that the determinant of an $N \times N$ is the sum of products of N elements of the matrix. In the case of a 2×2 matrix, the determinant contains products of two elements and is given by

$$\det\begin{bmatrix} a & b \\ c & d \end{bmatrix} = ad - bc$$

Note that the reciprocal of the 2×2 determinant appears in the formula for the 2×2 matrix inverse, implying that this matrix inverse does not exist when the determinant is zero. This is a general property of matrix inverses; they exist only when the matrix has nonzero determinant. In *MatLab*, the matrix inverse and determinant of a square matrix $\mathbf{A}$ are computed as

```
B = inv(A);
d = det(A);
```

(*MatLab* gda00_09)

In many of the formulas of inverse theory, the matrix inverse either premultiplies or postmultiplies other quantities, for instance:

$$\mathbf{c} = \mathbf{A}^{-1}\mathbf{b} \ \text{ and } \ \mathbf{D} = \mathbf{BA}^{-1}$$

These cases do not actually require the explicit calculation of $\mathbf{A}^{-1}$, just the combinations $\mathbf{A}^{-1}\mathbf{b}$ and $\mathbf{BA}^{-1}$, which are computationally simpler. *MatLab* provides generalizations of the division operator that implement these two cases:

```
c = A\b;
D = B/A;
```

(*MatLab* gda00_09)

A surprising amount of information on the structure of a matrix can be gained by studying how it affects a column vector that it multiplies. Suppose that $\mathbf{M}$ is an $N \times N$ square matrix and that it multiplies an *input* column vector, $\mathbf{v}$, producing an *output* column vector, $\mathbf{w} = \mathbf{Mv}$. We can examine how the output $\mathbf{w}$ compares to the input $\mathbf{v}$ as $\mathbf{v}$ is varied. One question of particular importance is

When is the output parallel to the input?

This question is called the *algebraic eigenvalue problem*. If $\mathbf{w}$ is parallel to $\mathbf{v}$, then $\mathbf{w} = \lambda\mathbf{v}$, where λ is a scalar proportionality factor. The parallel vectors satisfy the following equation:

$$\mathbf{Mv} = \lambda\mathbf{v} \ \text{ or } \ (\mathbf{M} - \lambda\mathbf{I})\mathbf{v} = 0$$

The trivial solution $\mathbf{v} = (\mathbf{M} - \lambda\mathbf{I})^{-1}0 = 0$ is not very interesting. A nontrivial solution is only possible when the matrix inverse $(\mathbf{M} - \lambda\mathbf{I})^{-1}$ does not exist. This is the case where the parameter λ is specifically chosen to make the determinant $\det(\mathbf{M} - \lambda\mathbf{I})$ exactly zero, since a matrix with zero determinant has no inverse. The determinant is calculated by adding together terms, each of which contains the product of N elements of the matrix. Since each element of the matrix contains, at most, one instance of λ, the product will contain powers of λ up to λ^N. Thus, the equation, $\det(\mathbf{M} - \lambda\mathbf{I}) = 0$, is an Nth order polynomial equation for λ. An Nth order polynomial equation has N roots, so we conclude that there must be N different proportionality factors, say λ_i, and N corresponding column vectors, say $\mathbf{v}^{(i)}$, that solve $\mathbf{Mv}^{(i)} = \lambda_i\mathbf{v}^{(i)}$. The column vectors, $\mathbf{v}^{(i)}$, are called the *characteristic vectors* (or *eigenvectors*) of the matrix, $\mathbf{M}$, and the proportionality factors, λ_i, are called the *characteristic values* (or *eigenvalues*). Eigenvectors are determined only up to an arbitrary multiplicative factor s, since if $\mathbf{v}^{(i)}$ is an eigenvector, so is $s\mathbf{v}^{(i)}$. Consequently, they are conventionally chosen to be unit vectors.

In the special case where $\mathbf{M}$ is symmetric, it can be shown that the eigenvalues, λ_i, are real and the eigenvectors are mutually perpendicular, $\mathbf{v}^{(i)T}\mathbf{v}^{(j)} = 0$ for $i \neq j$. The N eigenvalues can be arranged into a diagonal matrix, $\mathbf{\Lambda}$, whose elements are $[\mathbf{\Lambda}]_{ij} = \lambda_i\delta_{ij}$, where δ_{ij} is the Kronecker delta. The corresponding N eigenvectors $\mathbf{v}^{(i)}$ can be arranged as the columns of an $N \times N$ matrix $\mathbf{V}$, which satisfies, $\mathbf{V}^T\mathbf{V} = \mathbf{V}\mathbf{V}^T = \mathbf{I}$. The eigenvalue equation $\mathbf{Mv} = \lambda\mathbf{v}$ can then be succinctly written as

$$\mathbf{MV} = \mathbf{V\Lambda} \quad \text{or} \quad \mathbf{M} = \mathbf{V\Lambda V}^T$$

(The second equation is derived from the first by postmultiplying it by $\mathbf{V}^T$.) Thus, the matrix $\mathbf{M}$ can be reconstructed from its eigenvalues and eigenvectors. In *MatLab*, the matrix of eigenvalues $\mathbf{\Lambda}$ and matrix of eigenvectors $\mathbf{V}$ of a matrix $\mathbf{M}$ are computed as

```
[V,LAMBDA] = eig(M);
```

(*MatLab* gda00_09)

Here the eigenvalue matrix is called LAMBDA.

Many of the derivations of inverse theory require that a column vector $\mathbf{v}$ be considered a function of an independent variable, say x, and then differentiated with respect to that variable to yield the derivative $d\mathbf{v}/dx$. Such a derivative represents the fact that the vector changes from $\mathbf{v}$ to $\mathbf{v} + d\mathbf{v}$ as the independent variable changes from x to $x + dx$. Note that the resulting change $d\mathbf{v}$ is itself a vector. Derivatives are performed element-wise; that is,

$$\left[\frac{d\mathbf{v}}{dx}\right]_i = \frac{dv_i}{dx}$$

A somewhat more complicated situation is where the column vector $\mathbf{v}$ is a function of another column vector, say $\mathbf{y}$. The partial derivative

$$\frac{\partial v_i}{\partial y_j}$$

represents the change in the ith component of $\mathbf{v}$ caused by a change in the jth component of $\mathbf{y}$. Frequently, we will need to differentiate the linear function, $\mathbf{v} = \mathbf{My}$, where $\mathbf{M}$ is a matrix, with respect to $\mathbf{y}$:

$$\frac{\partial v_i}{\partial y_j} = \frac{\partial}{\partial y_j}\left[\sum_k M_{ik}y_k\right] = \sum_k M_{ik}\frac{\partial y_k}{\partial y_j}$$

Since the components of $\mathbf{y}$ are assumed to be independent, the derivative $\frac{\partial y_k}{\partial y_j}$ is zero except when $j = k$, in which case it is unity, which is to say $\frac{\partial y_k}{\partial y_j} = \delta_{kj}$. The expression for the derivative then simplifies to

$$\frac{\partial v_i}{\partial y_j} = \frac{\partial}{\partial y_j}\left[\sum_k M_{ik}y_k\right] = \sum_k M_{ik}\frac{\partial y_k}{\partial y_j} = \sum_k M_{ik}\delta_{kj} = M_{ij}$$

Thus the derivative of the linear function $\mathbf{v} = \mathbf{My}$ is the matrix $\mathbf{M}$. This relationship is the vector analog to the scalar case, where the linear function $v = my$ has the derivative $\frac{dv}{dy} = m$.

I.5 USEFUL MATLAB OPERATIONS

I.5.1 Loops

MatLab provides a looping mechanism, the `for` command, which can be useful when the need arises to sequentially access the elements of vectors and matrices. Thus, for example,

```
M = [ [1, 4, 7]', [2, 5, 8]', [3, 6, 9]' ];
for i = [1:3]
a(i) = M(i,i);
end
```

(*MatLab* gda00_10)

executes the a(i)=M(i,i) formula three times, each time with a different value of i (in this case, $i=1$, $i=2$, and $i=3$). The net effect is to copy the diagonal elements of the matrix **M** to the vector, **a**, that is, $a_i=M_{ii}$. Note that the end statement indicates the position of the bottom of the loop. Subsequent commands are not part of the loop and are executed only once.

Loops can be nested; that is, one loop can be inside another. Such an arrangement is necessary for accessing all the elements of a matrix in sequence. For example,

```
M = [ [1, 4, 7]', [2, 5, 8]', [3, 6, 9]' ];
for i = [1:3]
for j = [1:3]
N(i,4-j) = M(i,j);
end
end
```

(*MatLab* gda00_11)

copies the elements of the matrix, **M**, to the matrix, **N**, but reverses the order of the elements in each row, that is, $N_{i,4-j}=M_{i,j}$. Loops are especially useful in conjunction with *conditional* commands. For example,

```
a = [ 1, 2, 1, 4, 3, 2, 6, 4, 9, 2, 1, 4 ]';
for i = [1:12]
if ( a(i) >= 6 ) b(i) = 6;
else
b(i) = a(i);
end
end
```

(*MatLab* gda00_12)

sets $b_i=a_i$ if $a_i<6$ and sets $b_i=6$, otherwise (a process called *clipping* a vector, for it lops off parts of the vector that are larger than 6).

A purist might point out that *MatLab* syntax is so flexible that for loops are almost never really necessary. In fact, all three examples, above, can be computed with one-line formulas that omit for loops:

```
a = diag(M);
N = fliplr(M);
b=a; b(find(a>6))=6;
```

(*MatLab* gda00_13)

The first two formulas are quite simple, but rely upon the *MatLab* functions diag() (for "diagonal") and fliplr() (for "flip left-right"), whose existence we have not hitherto mentioned. The third formula, which used the find() function, requires further explanation.

The first part just copies the column vector **a** to **b**. In the second part, the (a>6) operation returns a vector of zeros and ones, depending upon whether the elements of the column vector **a** satisfy the inequality or not. The find() function uses this result and returns a list of the *indices* of the ones, that is, of the indices of the column vector **a** that match the condition. This list is then used to reset just those elements of **b** to 6, leaving the other elements unchanged.

One of the problems of a script-based environment is that learning the complete syntax of the scripting language can be pretty daunting. Writing a long script, such as one containing a for loop, will often be faster than searching through *MatLab* help files for a predefined function that implements the desired functionality in a single line of the script. When deciding between alternative ways of implementing a given functionality, you should always choose the one which *you* find clearest. Scripts that are terse or even computationally efficient are not necessarily a virtue, especially if they are difficult to debug. You should avoid creating formulas that are so inscrutable that you are not sure whether they will function correctly. Of course, the degree of inscrutability of any given formula will depend upon your level of familiarity with *MatLab*. Your repertoire of techniques will grow as you become more practiced.

I.5.2 Loading Data From a File

MatLab can read and write files with a variety for formats, but we start here with the simplest and most common, the text file. As an example, we load a global temperature dataset compiled by the National Aeronautics and Space Administration. The author's recommendation is that you always keep a file of notes about any data set that you work with, and that these notes include information on where you obtained the data set and any modifications that you subsequently made to it:

> The text file global_temp.txt contains global temperature change data from NASA's web site *http://data.giss.nasa.gov/gistemp*. It has two columns of data, time (in calendar years) and temperature anomaly (in degrees C) and is 46 lines long. Information about the data is in the file global_temp_notes.txt. The citation for this data is Hansen et al. (2010).

We reproduce the first few lines of global_temp.txt, here:

```
1965 -0.11
1966 -0.03
1967 -0.01
... ...
```

The data are read into to *MatLab* as follows:

```
D = load('../data/global_temp.txt');
t = D(:,1);
d = D(:,2);
```

(*MatLab* gda00_14)

The load() function reads the data into a 46 × 2 matrix, **D**. Note that the filename is given as ../data/global_temp.txt, as contrasted to just global_temp.txt, since the script is run from the ch00 folder while the data are in the data folder. The filename is surrounded by single quotes to indicate that it is a *character string* and not a variable name. The subsequent two lines break out **D** into two separate column vectors, **t** of time and **d** of temperature data. This step is not strictly speaking necessary, but fewer mistakes will be made if the different variables in the dataset have each their own name.

I.5.3 Plotting Data

MatLab's plotting commands are very powerful, but they are also very complicated. We present here a set of commands for making a simple *x*–*y* plot that is intermediate between a very crude, unlabeled plot, and an extremely artistic one. The reader may wish to adopt either a simpler or a more complicated version of this set, depending upon need and personal preference. The plot of the global temperature data shown in Fig. I.3 was created with the commands:

```
figure(1);
set(gcf, 'pos',[10, 10, 600, 300]);
clf;
set(gca,'LineWidth',3);
set(gca,'FontSize',14);
hold on;
axis( [1965, 2020, -0.5, 1.0] );
plot(t,d,'r-','LineWidth',3);
plot(t,d,'ko','LineWidth',3);
xlabel('calendar year');
ylabel('temperature anomaly, deg C');
ylabel('temperature anomaly, deg C');
title('global temperature data');
```

(*MatLab* gda00_14)

FIG. I.3 Global temperature data for the time period 1965–2016. See text for further discussion. MatLab *script gda00_14.*

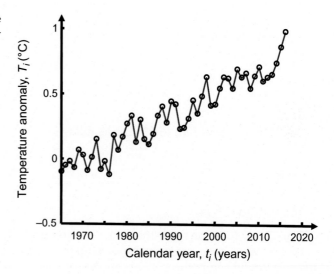

The figure(1)command directs *MatLab* to create a figure window and to label it Figure 1. The set(gcf,...) command changes the width and height of the figure to 600 and 300, respectively. The clf (for clear figure) command erases any previous plots in the figure window, ensuring that it is blank. We prefer axes with thicker lines and a bigger font than *MatLab's* default, and so reset them to 3 and 14, respectively, with two calls to the set(gca,...) command. The hold on command instructs *MatLab* to overlay multiple plots in the same window, overriding the default that a second plot erases the first. The axis([1965, 2020, -0.5, 1.0]) command manually sets the axes to an appropriate range. If this command is omitted, *MatLab* chooses the scaling of the axes based on the range of the data. The two plot(...) commands plot the time t and temperature d data in two different ways: first as a red line (indicated by the 'r-') and the second with black circles (the 'ko'). We also increase the width of the lines with the 'LineWidth', 3 directive. Finally, the horizontal and vertical axes are labeled using the xlabel() and ylabel() commands, and the plot is titled with the title() command. Note that the text is surrounded with single quotes, to indicate that it is a character string.

Creating Character Strings Containing the Values of Variables

The results of computations are most understandable if described in a combination of words and numbers. The sprintf()function (for "string print formatted") creates a character string that includes both text and the value of a variable. The string can then be used as a plot label, a filename, as text displayed in the Command Window, or for a wide variety of similar purposes. Thus, for instance,

```
title(sprintf('temperature data from %d to %d',t(1),t(N)));
```

(*MatLab* gda00_15)

creates a plot title 'temperature data from 1965 to 2010.' The sprintf() command can be used in any function that expects a character string, including title(), xlabel(), ylabel(), load(), and the function disp(), not mentioned until now, which writes a character string to the Command Window. The sprintf()function is fairly inscrutable, and we refer readers to the *MatLab* help pages for a detailed description. Briefly, the function uses *placeholders* that start with the character % to indicate where in the character string the value of the variable should be placed. Thus,

```
disp(sprintf('Number of data: %d', N));
```

(*MatLab* gda00_15)

writes the character string 'Number of data: 46' (since the value of N was 46) to the Command Window. The %d is the *placeholder for an integer*. It is replaced with '46,' the value of N. If the variable is fractional, as contrasted to integer, the *floating-point placeholder*, %f, is used instead. For example,

```
% display first few lines
disp('first few lines of data');
for i=[1:4] disp(sprintf('%f %f',t(i),d(i)));
end
```

(*MatLab* gda00_15)

writes the first four lines of data to the Command Window. Note that two placeholders have been used in the same *format string*. When displaying text to the Command Window, an alternative function is available that combines the functionality of `sprintf()` and `disp()`

```
fprintf('%f %f\n',t(i),d(i));
```

which is equivalent to

```
disp(sprintf('%f %f',t(i),d(i)));
```

The `\n` (for *newline*) at the end of the format string `'a=%f\n'` indicates that subsequent characters should be written to a new line in the command window, rather than being appended onto the end of the current line.

MatLab Live Scripts

Staring with Release 2016a, *MatLab* provides a file format called a *Live Script*, with file suffix `.mlx`, which provides a method for storing a script, its output and figures and ancillary text and graphics in a single file. The Live Script can be viewed without having to be re-run, and it, or portions of it, can be re-run as needed to update the output and figures. It provides a simple method of organizing your work into a single linear document—a format that is easier to understand, present, and archive than an m-script that generates Command Window output and/or a large number of figure windows. Live Scripts are no harder to create, edit, and run than m-scripts; we recommend them for homework problems and research projects. However, they currently have some limitations compared to m-scripts. The one that is most important in the context of this book is that a Live Script with nested `for`-loops executes significantly more slowly than an m-script and can sometimes have an unacceptably long execution time. Performance will probably improve in future releases, but until then, you will probably need to use a mix of m-scripts and Live Scripts in your work. Both are provided as a companion to this book.

References

Hansen, J., Ruedy, R., Sato, M., Lo, K., 2010. Global surface temperature change. Rev. Geophys. 48RG4004.

Menke, W., Menke, J., 2011. Environmental Data Analysis With MatLab. Academic Press, Elsevier Inc., Oxford. 263 pp.

Part-Enander, E., Sjoberg, A., Melin, B., Isaksson, P., 1996. The Matlab Handbook. Addison-Wesley, New York. 436 pp.

Pratap, R., 2009. Getting Started with MATLAB: A Quick Introduction for Scientists and Engineers. Oxford University Press, Oxford. 256 pp.

CHAPTER

1

Describing Inverse Problems

1.1 FORMULATING INVERSE PROBLEMS

The starting place in most inverse problems is a description of the data. Since in most inverse problems the data are simply a list of numerical values, a vector provides a convenient means of their representation. If N measurements are performed in a particular experiment, for instance, one might consider these numbers as the elements of a vector $\mathbf{d}$ of length N.

The purpose of the data analysis is to gain *knowledge* through systematic examination of data. While knowledge can take many forms, we assume here that it is primarily numerical in nature. We analyze data so to infer, as best we can, the values of numerical quantities— *model parameters*. Model parameters are chosen to be *meaningful*; that is, they are chosen to

© 2018 Elsevier Inc. All rights reserved.

capture the essential character of the processes that are being studied. The model parameters can be represented as the elements of a vector $\mathbf{m}$, which is of length M

$$\text{date}: \quad \mathbf{d} = [d_1, d_2, d_3, d_4, \dots, d_N]^T$$
$$\text{model parameters}: \quad \mathbf{m} = [m_1, m_2, m_3, m_4, \dots, m_M]^T \tag{1.1}$$

Here, T signifies transpose.

The basic statement of an inverse problem is that the model parameters and the data are in some way related. This relationship is called the *quantitative model* (or *model*, or *theory*, for short). Usually, the model takes the form of one or more formulas that the data and model parameters are expected to follow.

If, for instance, one were attempting to determine the density of an object, such as a rock, by measuring its mass and volume, there would be $N = 2$ data—mass and volume (say, d_1 and d_2, respectively)—and $M = 1$ unknown model parameter, density (say, m_1). The model would be the statement that density times volume equals mass, which can be written compactly by the vector equation $d_2 m_1 = d_1$. Note that the model parameter, density, is more meaningful than either mass or volume, in that it represents an intrinsic property of a substance that is related to its chemistry. The data—mass and volume—are easy to measure, but they are less fundamental because they depend on the size of the object, which is usually incidental.

In more realistic situations, the data and model parameters are related in more complicated ways. Most generally, the data and model parameters might be related by one or more implicit equations such as

$$\begin{aligned} f_1(\mathbf{d}, \mathbf{m}) &= 0 \\ f_2(\mathbf{d}, \mathbf{m}) &= 0 \\ &\vdots \\ f_L(\mathbf{d}, \mathbf{m}) &= 0 \end{aligned} \quad \text{or} \quad \mathbf{f}(\mathbf{d}, \mathbf{m}) = 0 \tag{1.2}$$

where L is the number of equations. In this example concerning the measuring of density, $L = 1$ and $d_2 m_1 - d_1 = 0$ would constitute the one equation of the form $f_1(\mathbf{d}, \mathbf{m}) = 0$. These implicit equations, which can be compactly written as the vector equation $\mathbf{f}(\mathbf{d}, \mathbf{m}) = 0$, summarize what is known about how the measured data and the unknown model parameters are related. The purpose of inverse theory is to solve, or "invert," these equations for the model parameters, or whatever kinds of answers might be possible or desirable in any given situation.

No claims are made either that the equations $\mathbf{f}(\mathbf{d}, \mathbf{m}) = 0$ contain enough information to specify the model parameters uniquely or that they are even consistent. One of the purposes of inverse theory is to answer these kinds of questions and provide means of dealing with the problems that they imply. In general, $\mathbf{f}(\mathbf{d}, \mathbf{m}) = 0$ can consist of arbitrarily complicated (nonlinear) functions of the data and model parameters. In many problems, however, the equation takes on one of several simple forms. It is convenient to give names to some of these special cases, since they commonly arise in practical problems; we shall give them special consideration in later chapters.

1.1.1 Implicit Linear Form

The function $\mathbf{f}$ is linear in both data and model parameters and can therefore be written as the matrix equation

$$f(\mathbf{d}, \mathbf{m}) = 0 = \mathbf{F}\begin{bmatrix} \mathbf{d} \\ \mathbf{m} \end{bmatrix} = \mathbf{Fx} \tag{1.3}$$

where $\mathbf{F}$ is an $L \times (M+N)$ matrix and the vector $\mathbf{x} = [\mathbf{d}^T, \mathbf{m}^T]^T$ is a concatenation of $\mathbf{d}$ and $\mathbf{m}$, that is, $\mathbf{x} = [d_1, d_2, \ldots, d_N, m_1, m_2, \ldots, m_M]^T$.

1.1.2 Explicit Form

In many instances, it is possible to separate the data from the model parameters and thus to form $L = N$ equations that are linear in the data (but still nonlinear in the model parameters through a vector function $\mathbf{g}$).

$$f(\mathbf{d}, \mathbf{m}) = 0 = \mathbf{d} - \mathbf{g}(\mathbf{m}) \tag{1.4}$$

1.1.3 Explicit Linear Form

In the explicit linear form, the function $\mathbf{g}$ is also linear, leading to the $N \times M$ matrix equation (where $L = N$)

$$f(\mathbf{d}, \mathbf{m}) = 0 = \mathbf{d} - \mathbf{Gm} \tag{1.5}$$

This form is equivalent to a special case of the matrix $\mathbf{F}$ in Section 1.1.1:

$$\mathbf{F} = [\mathbf{I}, -\mathbf{G}] \tag{1.6}$$

1.2 THE LINEAR INVERSE PROBLEM

The simplest and best-understood inverse problems are those that can be represented with the explicit linear equation $\mathbf{Gm} = \mathbf{d}$. This equation, therefore, forms the foundation of the study of discrete inverse theory. As will be shown later, many important inverse problems that arise in the physical sciences involve precisely this equation. Others, while involving more complicated equations, can often be solved through linear approximations.

The matrix $\mathbf{G}$ is called the data kernel, in analogy to the theory of integral equations, in which the analogs of the data and model parameters are two continuous functions $d(x)$ and $m(x)$, where x is some independent variable. Continuous inverse theory lies between these two extremes, with discrete data but a continuous model function.

Discrete inverse theory:

$$d_i = \sum_{j=1}^{M} G_{ij} m_j \tag{1.7a}$$

Continuous inverse theory:

$$d_i = \int G_i(x) m(x) \, dx \tag{1.7b}$$

Integral equation theory:

$$d(y) = \int G(y, x) m(x) \, dx \tag{1.7c}$$

The main difference among discrete inverse theory, continuous inverse theory, and integral equation theory is whether the model m and data d are treated as continuous functions or discrete parameters. The data d_i in inverse theory are necessarily discrete, since inverse theory is concerned with deducing knowledge from observational data, which always has a discrete nature. Both continuous inverse problems and integral equations can be converted to discrete inverse problems by approximating the model $m(x)$ as a vector of its values at a set of M closely spaced points

$$\mathbf{m} = [m(x_1), m(x_2), m(x_3), \ldots, m(x_M)]^{\mathrm{T}} \tag{1.8}$$

and the integral as a Riemann summation (or by some other quadrature formula).

1.3 EXAMPLES OF FORMULATING INVERSE PROBLEMS

1.3.1 Example 1: Fitting a Straight Line

Suppose that N temperature measurements T_i are made at times t_i in the atmosphere (Fig. 1.1). The data are then a vector $\mathbf{d}$ of N measurements of temperature, where $\mathbf{d} = [T_1, T_2, T_3, \ldots, T_N]^{\mathrm{T}}$. The times t_i are not, strictly speaking, data. Instead, they provide some auxiliary information that describes the geometry of the experiment. This distinction will be further clarified later.

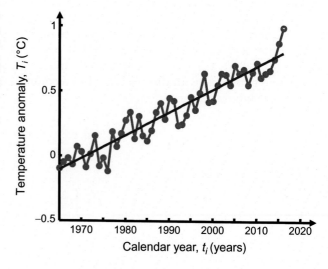

FIG. 1.1 *(Red)* Average global temperature for the time period, 1965–2010. The inverse problem is to determine the rate of increase of temperature and its confidence interval. *(Blue)* Straight line fit to data. The slope of the line is 0.015±0.002 (95%)°C/year. *Data from Hansen, J., Ruedy, R., Sato, M., Lo, K., 2010. Global surface temperature change. Rev. Geophys. 48, RG4004. doi:10.1029/2010RG000 345. MatLab script gda01_01.*

Suppose that we assume a model in which temperature is a linear function of time: $T = a + bt$. The intercept a and slope b then form the two model parameters of the problem, $\mathbf{m} = [a, b]^{\mathrm{T}}$. According to the model, each temperature observation must satisfy $T_i = a + bt_i$:

$$
\begin{aligned}
T_1 &= a + bt_1 \\
T_2 &= a + bt_2 \\
&\vdots \\
T_N &= a + bt_N
\end{aligned}
\tag{1.9}
$$

These equations can be arranged as the matrix equation $\mathbf{d} = \mathbf{Gm}$

$$
\begin{bmatrix} T_1 \\ T_2 \\ \vdots \\ T_N \end{bmatrix}
=
\begin{bmatrix} 1 & t_1 \\ 1 & t_2 \\ \vdots & \vdots \\ 1 & t_N \end{bmatrix}
\begin{bmatrix} a \\ b \end{bmatrix}
\tag{1.10}
$$

In *MatLab*, the matrix $\mathbf{G}$ is computed as:

```
G=[ones(N,1), t];
```
(*MatLab* script gda01_01)

1.3.2 Example 2: Fitting a Parabola

If the model in Example 1 is changed to assume a quadratic variation of temperature with time of the form $T = a + bt + ct^2$, then a new model parameter c is added to the problem, and $\mathbf{m} = [a,b,c]^{\mathrm{T}}$. The number of model parameters is now $M = 3$. The data and model parameters are supposed to satisfy

$$
\begin{aligned}
T_1 &= a + bt_1 + ct_1^2 \\
T_2 &= a + bt_2 + ct_2^2 \\
&\vdots \\
T_N &= a + bt_N + ct_N^2
\end{aligned}
\tag{1.11}
$$

These equations can be arranged into the matrix equation

$$
\begin{bmatrix} T_1 \\ T_2 \\ \vdots \\ T_N \end{bmatrix}
=
\begin{bmatrix} 1 & t_1 & t_1^2 \\ 1 & t_2 & t_2^2 \\ \vdots & \vdots & \vdots \\ 1 & t_N & t_N^2 \end{bmatrix}
\begin{bmatrix} a \\ b \\ c \end{bmatrix}
\tag{1.12}
$$

This matrix equation has the explicit linear form $\mathbf{d} = \mathbf{Gm}$. Note that, although the equation is linear in the data and model parameters, it is not linear in the auxiliary variable t.

The equation has a very similar form to the equation of the previous example, which brings out one of the underlying reasons for employing matrix notation: it can often emphasize similarities between superficially different problems. In *MatLab*, the matrix $\mathbf{G}$ is computed as:

```
G=[ones(N,1), t, t.^2];
```
(*MatLab* script gda01_02)

Note the use of the element-wise power, signified ".^," to compute t_i^2.

1.3.3 Example 3: Acoustic Tomography

Suppose that a wall is assembled from a rectangular array of bricks (Fig. 1.2) and that each brick is composed of a different type of clay. If the acoustic velocities of the different clays differ, one might attempt to distinguish the different kinds of bricks by measuring the travel time of sound across the various rows and columns of bricks in the wall. The data in this problem are $N=8$ measurements of travel times, $d=[T_1,T_2,T_3,\ldots,T_8]^T$. The model assumes that each brick is composed of a uniform material and that the travel time of sound across each brick is proportional to the width and height of the brick. The proportionality factor is the brick's *slowness* s_i, thus giving $M=16$ model parameters $\mathbf{m}=[s_1,s_2,s_3,\ldots,s_{16}]^T$, where the ordering is according to the numbering scheme of the figure. The data and model parameters are related by

$$\begin{aligned}
\text{row } 1: \quad & T_1 = hs_1 + hs_2 + hs_3 + hs_4 \\
\text{row } 2: \quad & T_2 = hs_5 + hs_6 + hs_7 + hs_8 \\
& \vdots \\
\text{column } 4: \quad & T_8 = hs_4 + hs_8 + hs_{12} + hs_{16}
\end{aligned} \tag{1.13}$$

and the matrix equation is

$$\begin{bmatrix} T_1 \\ T_2 \\ \vdots \\ T_8 \end{bmatrix} = h \begin{bmatrix} 1 & 1 & 1 & 1 & 0 & 0 & 0 & 0 & 0 & 0 & 0 & 0 & 0 & 0 & 0 & 0 \\ 0 & 0 & 0 & 0 & 1 & 1 & 1 & 1 & 0 & 0 & 0 & 0 & 0 & 0 & 0 & 0 \\ \vdots & \vdots & \vdots & \vdots & \vdots & \vdots & \vdots & \vdots & \vdots & \vdots & \vdots & \vdots & \vdots & \vdots & \vdots & \vdots \\ 0 & 0 & 0 & 1 & 0 & 0 & 0 & 1 & 0 & 0 & 0 & 1 & 0 & 0 & 0 & 1 \end{bmatrix} \begin{bmatrix} s_1 \\ s_2 \\ \vdots \\ s_{16} \end{bmatrix} \tag{1.14}$$

Here, the bricks are assumed to be of width *and* height h. The *MatLab* code for constructing $\mathbf{G}$ is:

```
G=zeros(N,M);
for i = [1:4]
for j = [1:4]
% measurements over rows
k = (i-1)*4 + j;
G(i,k)=h;
% measurements over columns
k = (j-1)*4 + i;
G(i+4,k)=h;
end
end
```
(*MatLab* script gda01_03)

FIG. 1.2 The travel time of acoustic waves *(blue line)* through the rows and columns of a square array of bricks is measured with acoustic source S and receiver R placed on the edges of the square. The inverse problem is to infer the acoustic properties of the bricks, here depicted by the *colors*. Although the overall pattern is spatially variable, individual bricks are assumed to be homogeneous.

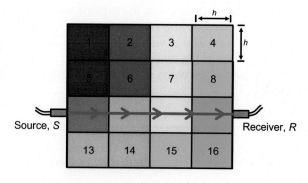

1.3.4 Example 4: X-ray Imaging

Tomography is the process of forming images of the interior of an object from measurements made along rays passed through that object ("tomo" comes from the Greek word for "slice"). The computerized tomography scanner is an X-ray imaging device that has revolutionized the diagnosis of brain tumors and many other medical conditions. The scanner solves an inverse problem for the X-ray opacity of body tissues using measurements of the amount of radiation absorbed from many crisscrossing beams of X-rays (Fig. 1.3).

The basic physical model underlying this device is the idea that the intensity of X-rays diminishes with the distance traveled, at a rate proportional to the intensity of the X-ray beam, and an absorption coefficient that depends on the type of tissue:

$$\frac{\mathrm{d}I}{\mathrm{d}s} = -c(x, y)I \tag{1.15}$$

Here, I is the intensity of the beam, s the distance along the beam, and $c(x,y)$ the absorption coefficient, which varies with position. If the X-ray source has intensity I_0, then the intensity at the ith detector is

$$I_i = I_0 \exp\left\{-\int_{\mathrm{beam}i} c(x, y)\,\mathrm{d}s\right\} \approx I_0\left\{1 - \int_{\mathrm{beam}i} c(x, y)\,\mathrm{d}s\right\} \tag{1.16a}$$

$$I_0 - I_i = I_0 \int_{\mathrm{beam}i} c(x, y)\,\mathrm{d}s \tag{1.16b}$$

Note that Eq. (1.16a) is a nonlinear function of the unknown absorption coefficient $c(x,y)$ and that the absorption coefficient varies continuously along the beam. This is a nonlinear problem in continuous inverse theory. However, it can be linearized, for small net absorption,

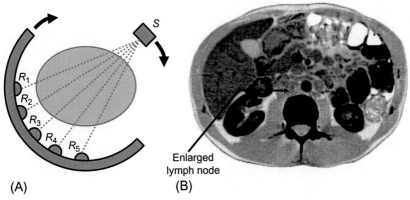

FIG. 1.3 (A) An idealized computed tomography (CT) medical scanner measures the X-ray absorption along lines *(blue)* passing through the body of the patient *(orange)*. After a set of measurements are made, the source S and receivers R_i are rotated, and the measurements are repeated so that data along many *crisscrossing lines* are collected. The inverse problem is to determine the X-ray opacity as a function of position in the body. (B) Actual CT image of a patient infected with *Mycobacterium genavense*. From de Lastours, V., Guillemain, R., Mainardi, J.-L., Aubert, A., Chevalier, P., Lefort, A., Podglajen, I., 2008. *Early diagnosis of disseminated* Mycobacterium genavense *infection. Emerg. Infect. Dis. 14(2), 346.*

by approximating the exponential with the first two terms in its Taylor series expansion, that is, $\exp(-x) \approx 1 - x$.

We now convert this problem to a discrete inverse problem of the form $\mathbf{Gm} = \mathbf{d}$. We assume that the continuously varying absorption coefficient can be adequately represented by a grid of many small square boxes (or *pixels*), each of which has a constant absorption coefficient. With these pixels numbered 1 through M, the model parameters are then the vector $\mathbf{m} = [c_1, c_2, c_3, \dots, c_M]^T$. The integral can then be written as the sum

$$\Delta I_i = \frac{I_0 - I_i}{I_0} = \sum_{j=1}^{M} \Delta s_{ij} c_j \tag{1.17}$$

Here, the data $d_i = \Delta I_i$ represent the differences between the X-ray intensities at the source and at the detector, and $G_{ij} = \Delta s_{ij}$ is the distance the ith beam travels in the jth pixel.

The inverse problem can then be summarized by the matrix equation

$$\begin{bmatrix} \Delta I_1 \\ \Delta I_2 \\ \vdots \\ \Delta I_N \end{bmatrix} = \begin{bmatrix} \Delta s_{11} & \Delta s_{12} & \dots & \Delta s_{1M} \\ \Delta s_{21} & \Delta s_{22} & \dots & \Delta s_{2M} \\ \dots & \dots & \dots & \dots \\ \Delta s_{N1} & \Delta s_{N2} & \dots & \Delta s_{NM} \end{bmatrix} \begin{bmatrix} c_1 \\ c_2 \\ \dots \\ c_M \end{bmatrix} \tag{1.18}$$

Since each beam passes through only a few of the many boxes, many of the Δs_{ij} are zero. Such a matrix is said to be *sparse*.

Computations with sparse matrices can be made extremely efficient by storing only the nonzero elements and by never explicitly multiplying or adding the zero elements (since the result is a foregone conclusion). However, special software support is necessary to gain this efficiency, since the computer must keep track of the zero elements. In *MatLab*, matrices need to be declared as sparse:

```
G = spalloc( N, M, MAXNONZEROELEMENTS);
```
(*MatLab* script gda01_03)

Once so defined, many normal matrix operations, including addition and multiplication, are efficiently computed without further user intervention. We will discuss this technique further in subsequent chapters, for its use makes practical the solving of very large inverse problems (say, with millions of model parameters). Further examples are given in Menke and Menke (2011).

1.3.5 Example 5: Spectral Curve Fitting

Not every inverse problem can be adequately represented by the discrete linear equation $\mathbf{Gm} = \mathbf{d}$. Consider, for example, a spectrogram containing a set of emission or absorption peaks, that vary with some auxiliary variable z (Fig. 1.4). The positions f, area A, and width c of the peaks are of interest because they reflect the chemical composition of the sample. Denoting the shape of peak j as $p(z, f_j, A_j, c_j)$, the model is that the spectrum consists of a sum of q such peaks

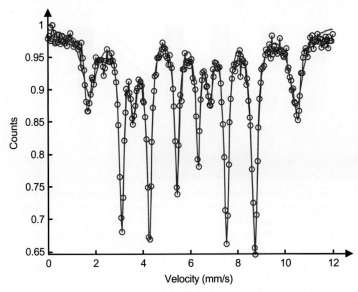

FIG. 1.4 Example of a Mossbauer spectroscopy experiment performed by the *Spirit* rover on Martian soil. *(Red)* Absorption peaks reflect the concentration of different iron-bearing minerals in the soil. The inverse problem is to determine the position and area of each peak, which can be used to determine the concentration of the minerals. *(Blue)* The sum of 10 Lorentzian curves fit to the data. *Data courtesy of NASA and the University of Mainz. MatLab script gda01_04.*

$$d_i = \sum_{j=1}^{q} p\left(z_i, f_j, A_j, c_j\right) = \sum_{j=1}^{q} \frac{A_j c_j^2}{\left(z_i - f_i\right)^2 + c_j^2} \tag{1.19}$$

Here, the peak shape $p(z_i, f_j, A_j, c_j)$ is taken to be a *Lorentzian*. The data and model are therefore related by the *nonlinear* explicit equation $\mathbf{d} = \mathbf{g(m)}$, where $\mathbf{m}$ is a vector of the position, area, and width of each peak. This equation is inherently nonlinear.

1.3.6 Example 6: Factor Analysis

Another example of a nonlinear inverse problem is that of determining the composition of chemical end members on the basis of the chemistry of a suite of mixtures of the end members. Consider a simplified "ocean" (Fig. 1.5) in which sediments are composed of mixtures of several chemically distinct rocks eroded from the continents. One expects the fraction of chemical j in the ith sediment sample S_{ij} to be related to the amount of end-member rock in sediment sample $i(C_{ik})$ and to the amount of the jth chemical in the end-member rock (F_{kj}) as

$$\begin{bmatrix} \text{sample} \\ \text{composition} \end{bmatrix} = \sum_{\text{end members}} \begin{bmatrix} \text{amount of} \\ \text{end member} \end{bmatrix} \begin{bmatrix} \text{end member} \\ \text{composition} \end{bmatrix}$$

$$S_{ij} = \sum_{k=1}^{p} C_{ik} F_{kj} \quad \text{or} \quad \mathbf{S} = \mathbf{CF} \tag{1.20}$$

In a typical experiment, the number of end members p, the end-member composition $\mathbf{F}$, and the amount of end members in the samples $\mathbf{C}$ are all unknown model parameters. Since the data $\mathbf{S}$ are on one side of the equations, this problem is also of the explicit nonlinear type. Note that basically the problem is to factor a matrix $\mathbf{S}$ into two other matrices $\mathbf{C}$ and $\mathbf{F}$.

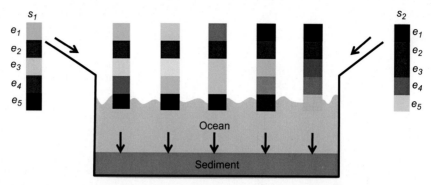

FIG. 1.5   Sediment on the floor of this idealized ocean is a mixture of rocks eroded from several sources s_i. The sources are characterized by chemical elements, e_1 through e_5, depicted here with color bars. The chemical composition of the sediments is a simple mixture of the composition of the sources. The inverse problem is to determine the number and composition of sources from observations of the composition of the sediments. *MatLab* script gda01_05.

This factoring problem is a well-studied part of the theory of matrices, and methods are available to solve it. As will be discussed in Chapter 10, this problem (which is often called *factor analysis*) is very closely related to the algebraic eigenvalue problem.

1.3.7 Example 7: Correcting for an Instrument Response

Because of physical limitations, many types of geophysical sensors output a time-varying signal $d(t)$ that differs from the true input value of the parameter $m(t)$ that is being measured. Consequently, the inverse problem of estimating $m(t)$ from observations of $d(t)$ is a common one; it must be solved before the output of the sensor can be fully understood and interpreted.

A seismometer is a good example of a sensor with an output that does not exactly match its input. Most seismometer models are much more sensitive to short period ground vibrations than to long period ones. Consequently, a simple spike in ground motion is output from the sensor as a pulse of more complicated shape (Fig. 1.6). When the input spike has unit area, the output pulse is called the instrument's *response* $g(t)$. Seismometers (and many other geophysical sensors) behave linearly. The amplitude of the output pulse is proportional to the amplitude of the input spike and, when several spikes occur in rapid succession, the sensor outputs several overlapping response functions.

When represented as a time series **m**, smoothly varying ground motion can be thought of as superposition of a sequence of spikes of amplitude m_i. The seismometer output is a sequence of overlapping response functions of corresponding amplitude (Fig. 1.7):

$$d_1 = g_1 m_1$$
$$d_2 = g_1 m_2 + g_2 m_1$$
$$d_3 = g_1 m_3 + g_2 m_2 + g_3 m_1 \qquad (1.21)$$
$$d_4 = g_1 m_4 + g_2 m_3 + g_3 m_2 + g_4 m_1$$
$$\vdots$$

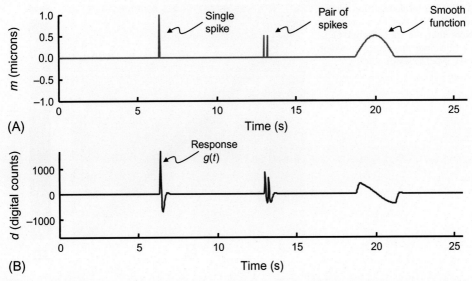

FIG. 1.6 (A) Hypothetical ground displacement $m(t)$ consisting of a single spike, followed by a pair of closely spaced spikes, followed by a smooth function. (B) Output $d(t)$ of a typical seismometer to the ground displacement. While the output has some correspondence to the input, the shapes of features are different. The output associated with a single spike is called the seismometer's response $g(t)$. *MatLab* script gda01_06.

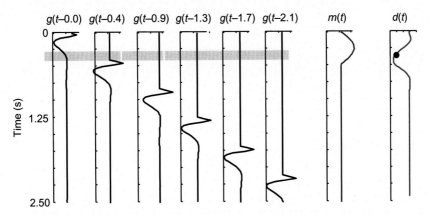

FIG. 1.7 Graphical depiction of the convolution relationship $\mathbf{d} = \mathbf{g} * \mathbf{m}$. A horizontal band *(gray bar)* though time-shifted versions of the response $g(t)$ *(black curves)* are "dotted into" the input $m(t)$ *(blue curve)* to yield one point *(black circle)* of the output $m(t)$ *(red curve)*.

Note that the *current* datum d_i always includes the term $g_1 m_i$; that is, the first element of the response scaled by the current input m_i. The other terms represent the contribution of the past inputs $m_{i-1}, m_{i-2}, m_{i-3} \cdots$. Eq. (1.21) can be succinctly written as:

$$d_i = \sum_{j=1}^{\infty} g_{i-j+1} m_j \qquad (1.22)$$

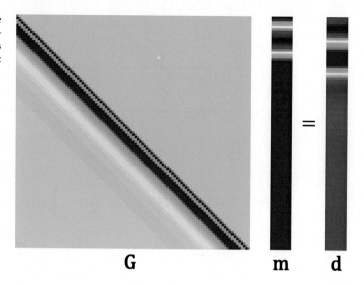

G   m   d

and is called the *convolution* of **g** and **m**. It is written as **d** = **g** * **m**, where the asterisk * denotes the convolution operation (*not* multiplication). The convolution can be put into the standard linear form **d** = **Gm** by defining a data kernel with elements $G_{ij} = g_{i-j+1}$:

$$
\begin{bmatrix} d_1 \\ d_2 \\ d_3 \\ \cdots \\ d_N \end{bmatrix} = \begin{bmatrix} g_1 & 0 & 0 & \cdots & 0 \\ g_2 & g_1 & 0 & \cdots & 0 \\ g_3 & g_2 & g_1 & \cdots & 0 \\ \cdots & \cdots & \cdots & \cdots & 0 \\ g_N & g_{N-1} & g_{N-2} & \cdots & g_1 \end{bmatrix} \begin{bmatrix} m_1 \\ m_2 \\ m_3 \\ \cdots \\ m_N \end{bmatrix}
$$

(1.22)

Matrices with constant diagonals, like **G** in Eq. (1.22), are called *Toeplitz* matrices (Fig. 1.8). The process of estimating **m** from observations of *d* is called *deconvolution*, and is an important type of inverse problem.

1.4 SOLUTIONS TO INVERSE PROBLEMS

We shall use the terms *solution* and *answer* to indicate broadly whatever information we are able to determine about the problem under consideration. As we shall see, there are many different points of view regarding what constitutes a solution to an inverse problem. Of course, one generally wants to know the numerical values of the model parameters (we call this kind of answer an *estimate* of the model parameters). Unfortunately, this issue is made complicated by the ubiquitous presence of measurement error and also by the possibility that some model parameters are not constrained by any observation. The solution of inverse problems rarely leads to exact information about the values of the model parameters. More typically, the practitioner of inverse theory is forced to make various compromises between the kind of information he or she actually wants and the kind of information that can in fact be obtained from any given data set. These compromises lead to other kinds of "answers" that are more abstract than simple estimates of the model parameters. Part of the practice of inverse theory is identifying what features of a solution are most valuable and making the compromises that emphasize these features. Some of the possible forms an "answer" to an inverse problem might take are described later.

1.4.1 Estimates of Model Parameters

The simplest kind of solution to an inverse problem is an estimate $\mathbf{m}^{est}$ of the model parameters. An estimate is simply a set of numerical values for the model parameters, $\mathbf{m}^{est} = [1.4, 2.9, \ldots, 1.0]^T$, for example. Estimates are generally the most useful kind of solution to an inverse problem. Nevertheless, in many situations, they can be very misleading. For instance, estimates in themselves give no insight into the quality of the solution. Depending on the structure of the particular problem, measurement errors might be averaged out (in which case the estimates might be meaningful) or amplified (in which case the estimates might be nonsense). In other problems, many solutions might exist. To single out arbitrarily only one of these solutions and call it $\mathbf{m}^{est}$ gives the false impression that a unique solution has been obtained.

1.4.2 Bounding Values

One remedy to the problem of defining the quality of an estimate is to state additionally some bounds that define its certainty. These bounds can be either absolute or probabilistic. Absolute bounds imply that the true value of the model parameter lies between two stated values, for example, $1.3 \leq m_1 \leq 1.5$. Probabilistic bounds imply that the estimate is likely to be between the bounds, with some given degree of certainty. For instance, $m_1^{est} = 1.4 \pm 0.1$ (95%) might mean that there is a 95% probability that the true value of the model parameter m_1^{true} lies between 1.3 and 1.5.

When they exist, bounding values can often provide the supplementary information needed to interpret properly the solution to an inverse problem. There are, however, many instances in which bounding values do not exist.

1.4.3 Probability Density Functions

A generalization of the stating of bounding values is the stating of the complete probability density function $p(\mathbf{m})$ for model parameters, either as an analytic function or as values on an M-dimensional grid. The usefulness of this technique depends, in part, on the complexity of $p(\mathbf{m})$. If the probability density functions $p(m_i)$ for an individual model parameter m_i has only one peak (Fig. 1.9A), then it provides little more information than an estimate based on the position of the peak's center with error bounds based on the peak's shape. On the other hand, if the probability density function is very complicated (Fig. 1.9C), it is basically uninterpretable (except in the sense that it implies that the model parameter cannot be well

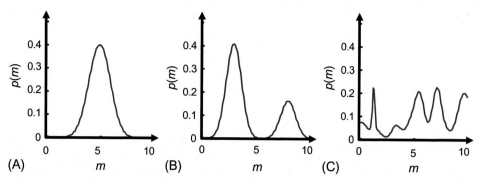

FIG. 1.9   Three hypothetical probability density functions for a model parameter, m. (A) The first is so simple that its properties can be summarized by its central position, at $m=5$, and the width of its peak. (B) The second implies that the model parameter has two probable ranges of values, one near $m=3$ and the other near $m=8$. (C) The third is so complicated that it provides no easily interpretable information about the model parameter. *MatLab* script gda01_08.

estimated). Only in those exceptional instances in which it has some intermediate complexity (Fig. 1.9B) does it really provide information toward the solution of an inverse problem.

1.4.4 Sets of Realizations of Model Parameters

Except in the well-understood Gaussian (Normal) case, which we will discuss later in the book, most probability density functions are exceedingly difficult to compute. A large set of realizations $\mathbf{m}^{(i)}$ of model parameter vectors drawn from $p(\mathbf{m})$ are somewhat easier to compute and can serve as an alternative. The set, itself, might be considered the solution to the inverse problem, since many of the properties of the probability density function can be inferred from it. However, this set might need to be extremely large to capture the properties of $p(\mathbf{m})$, especially when the number M of model parameters is large. Deriving useful knowledge from, say, a billion examples of possible $\mathbf{m}$s is a challenging task.

1.4.5 Weighted Averages of Model Parameters

In many instances, it is possible to identify combinations or averages of the model parameters that are in some sense better determined than the model parameters themselves. For instance, given $\mathbf{m}=[m_1, m_2]^T$, it may turn out that $\langle m \rangle = 0.2m_1 + 0.8m_2$ is better determined than either m_1 or m_2. Unfortunately, one might not have the slightest interest in such an average, be it well determined or not, because it may not have physical significance. It may not contribute useful knowledge.

Averages *can* be of considerable interest when the model parameters represent a discretized version of some continuous function. If the weights are large only for a few physically adjacent parameters, then the average is said to be *localized*. The meaning of the average in such a case is that, although the data cannot resolve the model parameters at a particular point, they can resolve the average of the model parameters in the *neighborhood* of that point.

In the following chapters, we shall derive methods for determining each of these different kinds of solutions to inverse problems. We note here, however, that there is a great deal of underlying similarity between these types of "answers." In fact, it will turn out that the same numerical "answer" will be interpretable as any of several classes of solutions.

1.5 PROBLEMS

1.1 Suppose that you determine the masses of 100 objects by weighing the first, then weighing the first and second together, and then weighing the rest in triplets: the first, second, and third; the second, third, and fourth; and so forth. (A) Identify the data and model parameters in this problem. How many of each are there? (B) Write down the matrix G in the equation $d = Gm$ that relates the data to the model parameters. (C) How sparse is G? What percent of it is zero?

1.2 Suppose that you determine the height of 50 objects by measuring the first, and then stacking the second on top of the first and measuring their combined height, stacking the third on top of the first two and measuring their combined height, and so forth. (A) Identify the data and model parameters in this problem. How many of each are there? (B) Write down the matrix G in the equation $d = Gm$ that relates the data to the model parameters. (C) How sparse is G? What percent of it is zero?

1.3 Write a *MatLab* script to compute G in the case of the cubic equation, $T = a + bz + cz^2 + dz^3$. Assume that *11* zs are equally spaced from 0 to 10.

1.4 Let the data d be the running average of the model parameters, m, computed by averaging groups of three neighboring points; that is, $d_i = (m_{i-1} + m_i + m_{i+1})/3$. (A) What is the matrix G in the equation $d = Gm$ in this case? (B) What problems arise at the top and bottom rows of the matrix and how can you deal with them? (C) How sparse is G? What percent of it is zero?

1.5 Simplify Eq. (1.20) by assuming that there is only one sample S_{1j} whose composition is measured. Consider the case where the composition of the p factors is known, but their proportions in the sample are unknown, and rewrite Eq. (10.2) in the form $d = Gm$. *Hint*: You might start by taking the transpose of Eq. (1.20).

Reference

Menke, W., Menke, J., 2011. Environmental Data Analysis with MatLab. Academic Press, Elsevier Inc., Oxford, UK. 263 pp.

Further Reading

de Lastours, V., Guillemain, R., Mainardi, J.-L., Aubert, A., Chevalier, P., Lefort, A., Podglajen, I., 2008. Early diagnosis of disseminated *Mycobacterium genavense* infection. Emerg. Infect. Dis. 14 (2), 346.

Hansen, J., Ruedy, R., Sato, M., Lo, K., 2010. Global surface temperature change. Rev. Geophys. 48RG4004. https://doi.org/10.1029/2010RG000345.

Some Comments on Probability Theory

2.1 NOISE AND RANDOM VARIABLES

In the preceding chapter, we represented the results of an experiment as a vector $\mathbf{d}$ whose elements were individual measurements. Usually, however, a single number is insufficient to represent a single observation. Measurements contain noise, and if an observation were to be performed several times, each measurement would be different (Fig. 2.1). Information about the range and shape of this scatter must also be provided to characterize the data completely.

The concept of a *random variable* is used to describe this property. Each random variable has definite and precise properties, governing the range and shape of the scatter of values one observes. These properties cannot be measured directly; however, one can only make individual measurements, or *realizations*, of the random variable and try to estimate its true properties from these data.

© 2018 Elsevier Inc. All rights reserved.

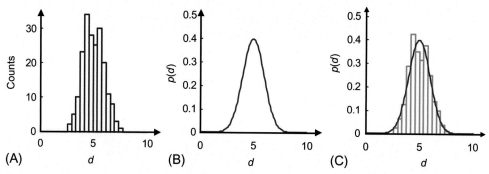

FIG. 2.1 (A) Histogram showing data from 200 repetitions of an experiment in which datum d is measured. Noise causes observations to scatter about their mean value, $\langle d \rangle = 5$. (B) Probability density function (p.d.f.), $p(d)$, of the data. (C) Histogram *(blue)* and p.d.f. *(red)* superimposed. Note that the histogram has a shape similar to the p.d.f. *MatLab* script gda02_01.

FIG. 2.2 The shaded area $p(d)\Delta d$ of the probability density function $p(d)$ gives the probability, P, that the observation will fall between d and $d + \Delta d$. *MatLab* script gda02_02.

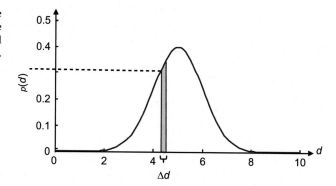

The true properties of the random variable d are specified by a *probability density function* $p(d)$ (abbreviated *p.d.f.*). This function gives the probability that a particular realization of the random variable will have a value in the neighborhood of d. The probability that the measurement is between d and $d + dd$ is $p(d) \, dd$ (Fig. 2.2). (Our choice of the variable name "d" for "data" makes the differential "dd" look a bit funny, but we will just have to live with it.)

Since each measurement must have some value, the probability that d lies somewhere between $-\infty$ and $+\infty$ is complete certainty (usually given the value of 100% or unity), which is written as

$$\int_{-\infty}^{+\infty} p(d)\mathrm{d}d = 1 \tag{2.1}$$

The probability P that d lies in some specific range, say between d_1 and d_2, is the integral of $p(d)$ over that range:

$$P(d_1, d_2) = \int_{d_1}^{d_2} p(d)\mathrm{d}d \tag{2.2}$$

The special case of $d_1 = -\infty$, $d_2 = d$, which represents the probability that the value of the random variable is less than or equal to a given value d, is called the *cumulative distribution function* $P(d)$ (abbreviated *c.d.f.*). Note that the numerical value of P represents an actual probability, while the numerical value of p does not.

In *MatLab*, we use a vector d evenly spaced values with sampling Dd to represent the random variable and we use a vector p to represent the probability density function at corresponding values of d. The total probability Ptotal (which should be unity) and the cumulative probability distribution P are calculated as

```
Ptotal = Dd * sum(p);
P = Dd * cumsum(p);
```

(*MatLab* script gda02_03)

Here, we are employing the Riemann approximation for an integral, $\int p(d)\mathrm{d}d \approx \Delta d \sum_i p(d_i)$. Note that the sum() function returns a scalar, the sum of the elements of p, whereas the cumsum() function returns a vector, the running sum of the elements of p.

The probability density function $p(d)$ completely describes the random variable, d. Unfortunately, it is a continuous function that may be quite complicated. A few numbers that summarize the major properties of the probability density function can be very helpful. One such kind of number indicates the typical numerical value of a measurement. The most likely measurement is the one with the highest probability, that is, the value of d at which $p(d)$ is peaked (Fig. 2.3). However, if the distribution is skewed, this *maximum likelihood point* may not be a good indication of the typical measurement, since a wide range of other values also have high probability. In such instances, the *mean*, or *expected* measurement, $\langle d \rangle$, is a better characterization of a typical measurement. This number is the "balancing point" of the distribution and is given by

$$\langle d \rangle = E(d) = \int_{-\infty}^{+\infty} d\,p(d)\mathrm{d}d \tag{2.3}$$

Another property of a distribution is its overall width. Wide distributions imply very noisy data, and narrow ones imply relatively noise-free data. One way of measuring the width of a distribution is to multiply it by a function that is zero near the center (peak) of the distribution and that grows on either side of the peak (Fig. 2.4). If the distribution is narrow, then the resulting function will be everywhere small; if the distribution is wide, then the result will be large.

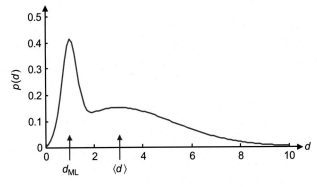

FIG. 2.3 The maximum likelihood point d_{ML} of the probability density function $p(d)$ gives the most probable value of the datum d. In general, this value can be different than the mean datum $\langle d \rangle$ which is at the "balancing point" of the distribution. *MatLab* script gda02_04.

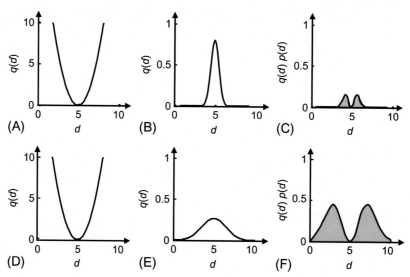

FIG. 2.4   (A and D) Parabola of the form $q(d) = (d - \langle d \rangle)^2$ is used to measure the width of two probability density functions $p(d)$ (B and E), which have the same mean $\langle d \rangle$ but different widths. The product qp is everywhere small for the narrow function (C) but had two large peaks for the wider distribution (F). The area (*shaded orange*) under qp is a measure of the width of the function $p(d)$ and is called the variance. The variances of (A) and (F) are $(0.5)^2$ and $(1.5)^2$, respectively. *MatLab* script gda02_05.

A quantitative measure of the width of the peak is the area under the resulting function. If one chooses the parabola $(d - \langle d \rangle)^2$ as the function, where $\langle d \rangle = E(d)$ is the expected value of the random variable, then this measure is called the *variance* σ^2 of the distribution and is written as

$$\sigma^2 = \int_{-\infty}^{+\infty} (d - \langle d \rangle)^2 p(d) \mathrm{d}d \tag{2.4}$$

The square root of the variance, σ, is a measure of the width of the distribution. In *MatLab*, the expected value and variance are computed as

```
Ed = Dd * sum(d.*p);
sigma2 = Dd * sum(((d-Ed).^2).*p);
```

(*MatLab* script gda02_05)

Here d is a vector of equally spaced values of the random variable d, with spacing Dd, and p is the corresponding value of the probability density function.

As we will discuss further in Chapter 5, the mean and variance can be estimated from a set of N realizations of data d_i as

$$\langle d \rangle^{\mathrm{est}} = \frac{1}{N} \sum_{i=1}^{N} d_i \quad \text{and} \quad (\sigma^2)^{\mathrm{est}} = \frac{1}{N-1} \sum_{i=1}^{N} (d_i - \langle d \rangle^{\mathrm{est}})^2 \tag{2.5}$$

The quantity $\langle d \rangle^{\mathrm{est}}$ is called the *sample mean* and the quantity σ^{est} is called the *sample standard deviation*. In *MatLab*, these estimates can be computed using the mean(dr) and std(dr) functions, where dr is a vector of N realizations of the random variable d.

2.2 CORRELATED DATA

Experiments usually involve the collection of more than one datum. We therefore need to quantify the probability that a set of random variables will take on a given value. The joint probability density function $p(\mathbf{d})$ is the probability that the first datum will be in the neighborhood of d_1, that the second will be in the neighborhood of d_2, etc. If the data are independent—that is, if there are no patterns in the occurrence of the values between pairs of random variables—then this joint distribution is just the product of the individual distributions (Fig. 2.5)

$$p(\mathbf{d}) = p(d_1)p(d_2)p(d_3)\cdots p(d_N) \tag{2.6}$$

The probability density function for a single random variable, say d_i, irrespective of all the others, is computed by integrating $p(\mathbf{d})$ over all the other variables:

$$p(d_i) = \int_{-\infty}^{+\infty} \cdots \int_{-\infty}^{+\infty} p(\mathbf{d})\,\mathrm{d}d_j\mathrm{d}d_k\ldots\mathrm{d}d_l \tag{2.7}$$
$$(N-1 \text{ intergrals})$$

In some experiments, measurements *are* correlated. High values of one datum tend to occur consistently with either high or low values of another datum (Fig. 2.6). The joint

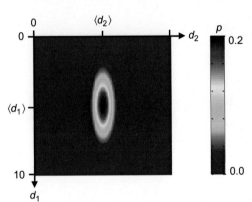

FIG. 2.5 The probability density function $p(d_1, d_2)$ is displayed as an image, with values given by the accompanying color bar. These data are uncorrelated, since especially large values of d_2 are no more or less likely if d_1 is large or small. In this example, the variance of d_1 and d_2 are 2.25 and 0.25, respectively. *MatLab* script gda02_06.

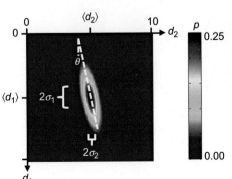

FIG. 2.6 The probability density function $p(d_1, d_2)$ is displayed as an image, with values given by the accompanying color bar. These data are positively correlated, since large values of d_2 are especially probable if d_1 is large. The function has means $\langle d_1 \rangle = 5$ and $\langle d_2 \rangle = 5$ and widths in the coordinate directions $\sigma_1 = 1.5$ and $\sigma_2 = 0.5$. The angle θ is a measure of the degree of correlation and is related to the covariance $\mathrm{cov}(d_1, d_2) = 0.4$. *MatLab* script gda02_07.

distribution for such data must be constructed to take this correlation into account. Given a joint distribution $p(d_1, d_2)$ for two random variables d_1 and d_2, one can test for correlation by selecting a function that divides the (d_1, d_2) plane into four quadrants of alternating sign, centered on the mean of the distribution (Fig. 2.7). If one multiplies the distribution by this function, and then sums up the area, the result will be zero for uncorrelated distributions, since they tend to lie equally in all four quadrants. Correlated distributions will have either positive or negative area, since they tend to be concentrated in two opposite quadrants (Fig. 2.8). If $[d_1 - \langle d_1 \rangle][d_2 - \langle d_2 \rangle]$ is used as the function, the resulting measure of correlation is called the covariance:

$$\text{cov}(d_1, d_2) = \int_{-\infty}^{+\infty} \int_{-\infty}^{+\infty} [d_1 - \langle d_1 \rangle][d_2 - \langle d_2 \rangle] p(d_1, d_2) \mathrm{d}d_1 \mathrm{d}d_2 \tag{2.8}$$

Note that the covariance of a datum with itself is just the variance. The covariance, therefore, characterizes the basic shape of the joint distribution.

When there are many data given by the vector $\mathbf{d}$, it is convenient to define a vector of expected values and a matrix of covariances as

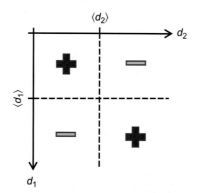

FIG. 2.7   The function $q(d_1, d_2) = (d_1 - \langle d_2 \rangle)(d_2 - \langle d_2 \rangle)$ divides the (d_1, d_2) plane into four quadrants of alternating sign.

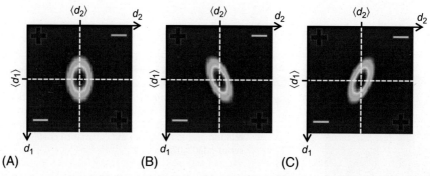

FIG. 2.8   The probability density function $p(d_1, d_2)$ displayed as an image, when the data are (A) uncorrelated, (B) positively correlated, and (C) negatively correlated. The *dashed lines* indicated the four quadrants of alternating sign used to determine the correlation (see Fig. 2.7). *MatLab* script gda02_08.

$$\langle d \rangle_i = \int_{-\infty}^{+\infty} \cdots \int_{-\infty}^{+\infty} d_i p(\mathbf{d}) \mathrm{d}d_1 \dots \mathrm{d}d_N$$

$$[\mathrm{cov}\ \mathbf{d}]_{ij} = \int_{-\infty}^{+\infty} \cdots \int_{-\infty}^{+\infty} [d_i - \langle d_i \rangle][d_j - \langle d_j \rangle] p(\mathbf{d}) \mathrm{d}d_1 \cdots \mathrm{d}d_N \tag{2.9}$$

Henceforth, we will abbreviate these multidimensional integrals as $\int d^N d$. The diagonal elements of the covariance matrix are variances. They are measures of the scatter in the data. The off-diagonal elements are covariances. They indicate the degree to which pairs of data are correlated. Notice that the integral for the mean can be written in terms of the univariate probability density function $p(d_i)$ and the integral for the variance can be written in terms of the bivariate probability density function $p(d_i, d_j)$, since the other dimension of $p(\mathbf{d})$ is just "integrated away" to unity:

$$\langle d \rangle_i = \int_{-\infty}^{+\infty} d_i p(d_i) \mathrm{d}d_i$$

$$[\mathrm{cov}\ \mathbf{d}]_{ij} = \int_{-\infty}^{+\infty} \int_{-\infty}^{+\infty} [d_i - \langle d_i \rangle][d_j - \langle d_j \rangle] p(d_i, d_j) \mathrm{d}d_i \mathrm{d}d_j \tag{2.10}$$

The covariance matrix can be estimated from a set of N realizations of data. Suppose that there are N different types of data and that K realizations of them have been observed. The data can be organized into a matrix $\mathbf{D}$, with the N columns referring to the different data types and the K rows to the different realizations. The *sample covariance* is then

$$[\mathrm{cov}\ \mathbf{d}]_{ij}^{\mathrm{est}} = \frac{1}{K} \sum_{k=1}^{K} \left(D_{ki} - \langle D_i \rangle^{\mathrm{est}} \right) \left(D_{kj} - \langle D_j \rangle^{\mathrm{est}} \right) \tag{2.11}$$

Here $\langle D_i \rangle^{\mathrm{est}}$ is the sample mean of the ith data type. The *MatLab* function cov(D) implements this formula.

2.3 FUNCTIONS OF RANDOM VARIABLES

The basic premise of inverse theory is that the data and model parameters are related. Any method that solves the inverse problem—that estimates a model parameter on the basis of data—will map errors from the data to the estimated model parameters. Thus the *estimates* of the model parameters are themselves random variables, which are described by a distribution $p(\mathbf{m}^{\mathrm{est}})$. Whether or not the *true* model parameters are random variables depends on the problem. It is appropriate to consider them deterministic quantities in some problems and random variables in others. *Estimates* of the model parameters, however, are always random variables.

We need the tools to transform probability density functions from $p(\mathbf{d})$ to $p(\mathbf{m})$ when the relationship $\mathbf{m}(\mathbf{d})$ is known. We start simply and consider just one datum and one model parameter, related by the simple function $m(d) = 2d$. Now suppose that $p(d)$ is *uniform* on the interval $(0,1)$; that is, d has equal probability of being anywhere in this range. The probability density function is constant and must have amplitude $p(d) = 1$, since the total probability must be unity (width $\times$ height $= 1 \times 1 = 1$). The probability density function $p(m)$ is also uniform,

FIG. 2.9 (A) The uniform probability density function $p(d)=1$ on the interval $0<d<1$. (B) The transformed probability density function $p(m)$, given the relationship $m=2d$. Note that a patch (*shaded rectangle*) of probability in m is wider and lower than the equivalent patch in d.

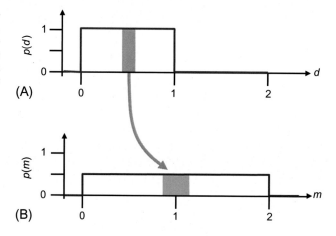

but on the interval $(0, 2)$, since m is twice d. Thus, $p(m)=\frac{1}{2}$, since its total probability must also be unity (width $\times$ height $=2\times\frac{1}{2}=1$) (Fig. 2.9). This result shows that $p(m)$ is not merely $p[d(m)]$, but rather must include a factor that accounts for the stretching (or shrinking) of the m-axis with respect to the d-axis.

This stretching factor can be derived by transforming the integral for total probability:

$$1=\int_{d_{min}}^{d_{max}}p(d)\mathrm{d}d=\int_{d(m_{mn})}^{d(m_{max})}p[d(m)]\frac{\mathrm{d}d}{\mathrm{d}m}\mathrm{d}m=\int_{m_{min}}^{m_{max}}p(m)\mathrm{d}m \qquad (2.12)$$

By inspection, $p(m)=p[d(m)]\mathrm{d}d/\mathrm{d}m$, so the stretching factor is $\mathrm{d}d/\mathrm{d}m$. The limits (d_{min}, d_{max}) transform to (m_{min}, m_{max}). However, depending upon the function $m(d)$, we may find that $m_{min}>m_{max}$; that is, the direction of integration might be reversed ($m(d)=1/d$ would be one such case). We handle this problem by adding an absolute value sign

$$p(m)=p[d(m)]\left|\frac{\mathrm{d}d}{\mathrm{d}m}\right| \qquad (2.13)$$

together with the understanding that the integration is always performed in the direction of positive m. Note that in the case above, with $p(d)=1$ and $m(d)=2d$, we find $\mathrm{d}d/\mathrm{d}m=\frac{1}{2}$ and (as expected) $p(m)=1\times\frac{1}{2}=\frac{1}{2}$.

In general, probability density functions change shape when transformed from d to m. Consider, for example, the uniform probability density function $p(d)=1$ on the interval $(0, 1)$ together with the function $m(d)=d^2$ (Fig. 2.10).—We find $d=m^{1/2}$, $\mathrm{d}d/\mathrm{d}m=\frac{1}{2}m^{-1/2}$, and $p(m)=\frac{1}{2}m^{-1/2}$, with m defined on the interval $(0, 1)$. Thus, while $p(d)$ is uniform, $p(m)$ has a peak (actually an integrable singularity) at $m=0$ (Fig. 2.10B).

The general case of transforming $p(\mathbf{d})$ to $p(\mathbf{m})$, given the functional relationship $\mathbf{d}(\mathbf{m})$, is more complicated but is derived using the rule for transforming multidimensional integrals that is analogous to Eq. (2.13). This rule states that the volume element transforms as $\mathrm{d}^N d = J(\mathbf{m})\mathrm{d}^N m$ where $J(\mathbf{m})=|\det(\partial\mathbf{d}/\partial\mathbf{m})|$ is the *Jacobian determinant*, that is, the absolute value of the determinant of the matrix whose elements are $[\partial\mathbf{d}/\partial\mathbf{m}]_{ij}=\partial d_i/\partial m_j$:

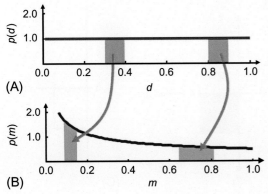

(A)

(B)

FIG. 2.10   (A) The uniform probability density function $p(d)=1$ on the interval $0<d<1$. (B) The transformed probability density function $p(m)$, given the relationship $m=d^2$. Note areas of equal probability, which are of equal height and width in the variable d are transformed into areas of unequal height and width in the variable, m. *MatLab* script gda02_09.

$$1 = \int p(\mathbf{d})\mathrm{d}^N d = \int p[\mathbf{d}(\mathbf{m})]\left|\det\left[\frac{\partial \mathbf{d}}{\partial \mathbf{m}}\right]\right|\mathrm{d}^N m = \int p[\mathbf{d}(\mathbf{m})]J(\mathbf{m})\mathrm{d}^N m$$
$$= \int p(\mathbf{m})\mathrm{d}^N m \tag{2.14}$$

Hence, by inspection, we find that the probability density function transforms as

$$p(\mathbf{m}) = p[\mathbf{d}(\mathbf{m})]\left|\det\left[\frac{\partial \mathbf{d}}{\partial \mathbf{m}}\right]\right| = p[\mathbf{d}(\mathbf{m})]J(\mathbf{m}) \tag{2.15}$$

Note that for the linear transformation $\mathbf{m}=\mathbf{Md}$, the Jacobian is constant, with the value $J=|\det(\mathbf{M}^{-1})|=|\det(\mathbf{M})|^{-1}$. As an example, consider a two-dimensional probability density function that is uniform on the intervals (0, 1) for d_1 and (0, 1) for d_2, together with the transformation $m_1=d_1+d_2$, $m_2=d_1-d_2$. As is shown in Fig. 2.11, $p(\mathbf{d})$ corresponds to a square of unit area in the (d_1,d_2) plane and $p(\mathbf{m})$ corresponds to a square of area 2 in the (m_1,m_2) plane. In order that the total area be unity in both cases, we must have $p(\mathbf{d})=1$ and $p(\mathbf{m})=\frac{1}{2}$. The transformation matrix is

$$M=\begin{bmatrix} 1 & 1 \\ 1 & -1 \end{bmatrix} \quad \text{so } |\det(M)|=2 \text{ and } J=\frac{1}{2} \tag{2.16}$$

Thus, by Eq. (2.15), we find that $p(\mathbf{m})=p(\mathbf{d})J=1\times\frac{1}{2}=\frac{1}{2}$, which agrees with our expectations.

Note that we can convert $p(\mathbf{d})$ to a univariate distribution $p(d_1)$ by integrating over d_2. Since the sides of the square are parallel to the coordinate axes, the integration yields the uniform probability density function, $p(d_1)=1$ (Fig. 2.11C). Similarly, we can convert $p(\mathbf{m})$ to a univariate distribution $p(m_1)$ by integrating over m_2. However, because the sides of the square are oblique to the coordinate axes, $p(m_1)$ is a triangular—not a uniform—probability density function (Fig. 2.11D).

Transforming a probability density function $p(\mathbf{d})$ is straightforward, but tedious. Fortunately, in the case of the linear function $\mathbf{m}=\mathbf{Md}+\mathbf{v}$, where $\mathbf{M}$ and $\mathbf{v}$ are an arbitrary matrix and vector, respectively, it is possible to make some statements about the properties of the results without explicitly calculating the transformed probability density function $p(\mathbf{m})$. In particular, the mean and covariance can be shown, respectively, to be

$$\langle \mathbf{m} \rangle = \mathbf{M}\langle \mathbf{d} \rangle + \mathbf{v} \tag{2.17a}$$

FIG. 2.11 (A) The uniform probability density function $p(d_1, d_2) = 1$ on the interval $0 < d_1 < 1$, $0 < d_2 < 1$. Also shown are (m_1, m_2) axes, where $m_1 = d_1 + d_2$ and $m_2 = d_1 - d_2$. (B) The transformed probability density function $p(m_1, m_2) = 0.25$. (C) The univariate distribution $p(d_1)$ is formed by integrating $p(d_1, d_2)$ over d_2. It is a uniform distribution of amplitude 1. (D) The univariate distribution $p(m_1)$ is formed by integrating $p(m_1, m_2)$ over m_2. It is a triangular distribution of peak amplitude 0.5.

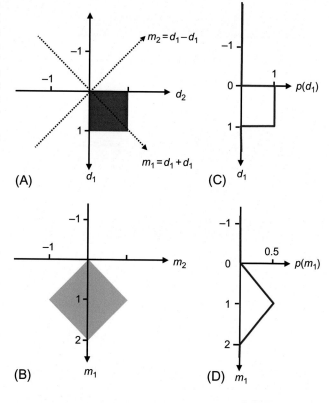

and

$$[\text{cov } \mathbf{m}] = \mathbf{M}[\text{cov } \mathbf{d}]\mathbf{M}^{\mathrm{T}} \qquad (2.17\text{b})$$

These rules are derived by transforming the definition of the mean and variance:

$$\langle m \rangle_i = \int m_i p(\mathbf{m}) \mathrm{d}^N m = \int \sum_j M_{ij} d_j p[\mathbf{d}(\mathbf{m})] \left| \det\left[\frac{\partial \mathbf{d}}{\partial \mathbf{m}}\right] \right| \left| \det\left[\frac{\partial \mathbf{m}}{\partial \mathbf{d}}\right] \right| \mathrm{d}^N d$$

$$= \sum_j M_{ij} \int d_j p(\mathbf{d}) \mathrm{d}^N d = \sum_j M_{ij} \langle d_j \rangle \qquad (2.18)$$

$$[\text{cov} \, \mathbf{m}]_{ij} = \int \left(m_i - \langle m \rangle_i \right) \left(m_j - \langle m_j \rangle \right) p(\mathbf{m}) \mathrm{d}^N m$$

$$= \int \sum_p \left(M_{ip} d_p - M_{ip} \langle d_p \rangle \right) \sum_q \left(M_{iq} d_q - M_{iq} \langle d_q \rangle \right)$$

$$p[\mathbf{d}(\mathbf{m})] \left| \det\left[\frac{\partial \mathbf{d}}{\partial \mathbf{m}}\right] \right| \left| \det\left[\frac{\partial \mathbf{m}}{\partial \mathbf{d}}\right] \right| \mathrm{d}^N d \qquad (2.19)$$

$$= \sum_p \sum_q M_{ip} M_{jq} \int \left(d_q - \langle d_q \rangle \right) \left(d_q - \langle d_q \rangle \right) p(\mathbf{d}) \mathrm{d}^N d$$

$$= \sum_p \sum_q M_{ip} [\text{cov} \, \mathbf{d}]_{pq} M_{jq}$$

Eq. (2.17b) is very important, because the covariance of the data is a measure of the amount of measurement error. The equation functions as a rule for *error propagation*; that is, given [cov **d**] representing measurement error, it provides a way to compute [cov **m**] representing the corresponding error in the model parameters. While the rule requires that the data and the model parameters be linearly related, it is independent of the functional form of the probability density function $p(\mathbf{d})$. Furthermore, it can be shown to be correct even when the matrix **M** is not square.

As an example, consider a model parameter m_1, which is linearly related to the data by

$$m_1 = \frac{1}{N}\sum_{i=1}^{N} d_i = \frac{1}{N}[1, 1, 1, ..., 1]\mathbf{d} \tag{2.20}$$

Note that this formula is the sample mean, as defined in Eq. (2.5). This formula implies that matrix $\mathbf{M} = [1, 1, 1, ..., 1]/N$ and vector $\mathbf{v} = 0$. Suppose that the data are uncorrelated and all have the same mean $\langle d \rangle$ and variance σ_d^2. Then we see that $\langle m_1 \rangle = \mathbf{M}\langle \mathbf{d} \rangle + \mathbf{v} = \langle d \rangle$ and $\mathbf{var}(m_1) = \mathbf{M}[\text{cov } \mathbf{d}]\mathbf{M}^T = \sigma_d^2/N$. The model parameter m_1 has a probability density function $p(m_1)$ with the same mean as **d**; that is, $\langle m_1 \rangle = \langle d \rangle$. Hence it is an estimate of the mean of the data. Its variance $\sigma_m^2 = \sigma_d^2/N$ is less than the variance of **d**. The square root of the variance, which is a measure of the width of the $p(m_1)$, is proportional to $N^{-1/2}$. Thus, accuracy of determining the mean of a group of data increases as the number of observations increases, albeit slowly (because of the square root).

In the case of uncorrelated data with uniform variance (i.e., [cov **d**] $= \sigma_d^2\mathbf{I}$), the covariance of the model parameters is [cov **m**] $= \mathbf{M}[\text{cov } \mathbf{d}]\mathbf{M}^T = \sigma_d^2\mathbf{M}\mathbf{M}^T$. In general, $\mathbf{M}\mathbf{M}^T$, while symmetric, is not diagonal. Not only do the model parameters have unequal variance, but they are also correlated. Strongly correlated model parameters are usually undesirable, but (as we will discuss later) good experimental design can sometimes eliminate them.

2.4 GAUSSIAN PROBABILITY DENSITY FUNCTIONS

The probability density function for a particular random variable can be arbitrarily complicated, but in many instances, data possess the rather simple *Gaussian* (or *Normal*) probability density function

$$p(d) = \frac{1}{(2\pi)^{1/2}\sigma}\exp\left[-\frac{(d - \langle d \rangle)^2}{2\sigma^2}\right] \tag{2.21}$$

This probability density function has mean $\langle d \rangle$ and variance σ^2 (Fig. 2.12). The Gaussian probability density function is so common because it is the limiting probability density function for the sum of random variables. The *central limit theorem* shows (with certain limitations) that regardless of the probability density function of a set of independent random variables, the probability density function of their sum tends to a Gaussian distribution as the number of summed variables increases. As long as the noise in the data comes from several sources of comparable size, it will tend to follow a Gaussian probability density function. This behavior is exemplified by the sum of the two uniform probability density functions in Section 2.3.

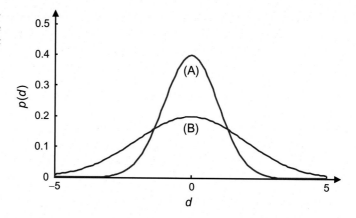

FIG. 2.12 Gaussian (Normal) distribution with zero mean and $\sigma = 1$ for curve (A) and $\sigma = 2$ for curve (B). *MatLab* script gda02_10.

The probability density function of their sum is more nearly Gaussian than the individual probability density functions (it being triangular instead of rectangular).

The joint probability density function for two independent Gaussian variables is just the product of two univariate probability density functions. When the data are correlated (say, with mean $\langle \mathbf{d} \rangle$ and covariance $[\text{cov}\,\mathbf{d}]$), the joint probability density function is more complicated, since it must express the degree of correlation. The appropriate generalization can be shown to be

$$p(\mathbf{d}) = \frac{1}{(2\pi)^{N/2}(\det[\text{cov}\,\mathbf{d}])^{1/2}} \exp\left(-\frac{1}{2}[\mathbf{d} - \langle\mathbf{d}\rangle]^{\mathrm{T}}[\text{cov}\,\mathbf{d}]^{-1}[\mathbf{d} - \langle\mathbf{d}\rangle]\right) \qquad (2.22)$$

Note that this probability density function reduces to Eq. (2.21) in the special case of $N=1$ (where $[\text{cov}\,\mathbf{d}]$ becomes σ_d^2). It is perhaps not apparent that the general case has an area of unity, a mean of $\langle\mathbf{d}\rangle$, and a covariance matrix of $[\text{cov}\,\mathbf{d}]$. However, these properties can be derived by inserting Eq. (2.22) into the relevant integral and by transforming to the new variable $\mathbf{y} = [\text{cov}\,\mathbf{d}]^{-\frac{1}{2}}[\mathbf{d} - \langle\mathbf{d}\rangle]$ (whence the integral becomes substantially simplified).

When $p(\mathbf{d})$ (Eq. 2.22) is transformed using the linear rule $\mathbf{m} = \mathbf{M}\mathbf{d}$, the resulting $p(\mathbf{m})$ is also Gaussian in form with mean $\langle\mathbf{m}\rangle = \mathbf{M}\langle\mathbf{d}\rangle$ and covariance matrix $[\text{cov}\,\mathbf{m}] = \mathbf{M}[\text{cov}\,\mathbf{d}]\mathbf{M}^{\mathrm{T}}$. Thus, all linear functions of Gaussian random variables are themselves Gaussian.

In Chapter 5, we will show that the information contained in each of two probability density functions can be combined by multiplying the two distributions. Interestingly, the product of two Gaussian probability density functions is itself Gaussian (Fig. 2.13). Given Gaussian $p_A(\mathbf{d})$ with mean $\langle\mathbf{d}_A\rangle$ and covariance $[\text{cov}\,\mathbf{d}]_A$ and Gaussian $p_B(\mathbf{d})$ with mean $\langle\mathbf{d}_B\rangle$ and covariance $[\text{cov}\,\mathbf{d}]_B$, the product $p_C(\mathbf{d}) = p_A(\mathbf{d})\,p_B(\mathbf{d})$ is Gaussian with mean and variance (e.g., Menke and Menke, 2011, their Section 5.4)

$$\langle\mathbf{d_C}\rangle = \left([\text{cov}\,\mathbf{d}]_A^{-1} + [\text{cov}\,\mathbf{d}]_B^{-1}\right)^{-1}\left([\text{cov}\,\mathbf{d}]_A^{-1}\langle\mathbf{d_A}\rangle + [\text{cov}\,\mathbf{d}]_B^{-1}\langle\mathbf{d_B}\rangle\right)$$
$$[\text{cov}\,\mathbf{d}]_C^{-1} = [\text{cov}\,\mathbf{d}]_A^{-1} + [\text{cov}\,\mathbf{d}]_B^{-1} \qquad (2.23)$$

The idea that the model and data are related by an explicit relationship $\mathbf{g}(\mathbf{m}) = \mathbf{d}$ can be reinterpreted in light of this probabilistic description of the data. We can no longer assert that

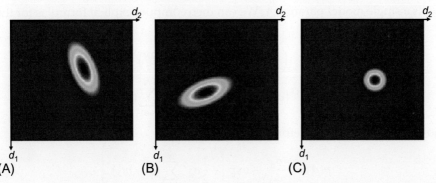

FIG. 2.13 (A) A Normal probability density function $p_A(d_1,d_2)$. (B) Another Normal probability density function $p_B(d_1,d_2)$. (C) The product of these two functions $p_C(d_1,d_2) = p_A(d_1,d_2)p_B(d_1,d_2)$ is Normal. *MatLab* script gda02_11.

this relationship can hold for the data themselves, since they are random variables. Instead, we assert that this relationship holds for the mean data: $g(m) = \langle d \rangle$. The distribution for the data can then be written as

$$p(\mathbf{d}) = \frac{1}{(2\pi)^{N/2}(\det[\operatorname{cov}\mathbf{d}])^{1/2}} \exp\left(-\frac{1}{2}[\mathbf{d} - \mathbf{g}(\mathbf{m})]^{\mathrm{T}}[\operatorname{cov}\mathbf{d}]^{-1}[\mathbf{d} - \mathbf{g}(\mathbf{m})]\right) \qquad (2.24)$$

The model parameters now have the interpretation of a set of unknown quantities that define the shape of the distribution for the data. One approach to inverse theory (which will be pursued in Chapter 5) is to use the data to determine the distribution and thus the values of the model parameters.

For the Gaussian distribution (Eq. 2.24) to be sensible, $\mathbf{g}(\mathbf{m})$ must not be a function of any random variables. This is why we differentiated between data and auxiliary variables in Chapter 1; the latter must be known exactly. If the auxiliary variables are themselves uncertain, then they must be treated as data and the inverse problem becomes an implicit one with a much more complicated distribution than the above problem exhibits.

As an example of constructing the distribution for a set of data, consider an experiment in which the temperature d_i in some small volume of space is measured N times. If the temperature is assumed not to be a function of time and space, the experiment can be viewed as the measurement of N realizations of the same random variable or as the measurement of one realization of N distinct random variables that all have the same distribution. We adopt the second viewpoint.

If the data are independent Gaussian random variables with mean $\langle d \rangle$ and variance σ_d^2 so that $[\operatorname{cov}\mathbf{d}] = \sigma_d^2\mathbf{I}$, then we can represent the assumption that all the data have the same mean by an equation of the form $\mathbf{Gm} = \mathbf{d}$:

$$\begin{bmatrix} 1 \\ 1 \\ \cdots \\ 1 \end{bmatrix} [m_1] = \begin{bmatrix} d_1 \\ d_2 \\ \cdots \\ d_N \end{bmatrix} \qquad (2.25)$$

where m_1 is a single model parameter. We can then compute explicit formulas for the expressions in $p(\mathbf{d})$ as

$$\left(\det[\operatorname{cov}\mathbf{d}]^{-1}\right)^{1/2} = \left(\sigma_{\mathrm{d}}^{-2N}\right)^{1/2} = \sigma_{\mathrm{d}}^{-N}$$

$$[\mathbf{d} - \mathbf{Gm}]^{\mathrm{T}}[\operatorname{cov}\mathbf{d}]^{-1}[\mathbf{d} - \mathbf{Gm}] = \sigma_{\mathrm{d}}^{-2}\sum_{i=1}^{N}(d_i - m_1)^2$$

(2.26)

The joint distribution is therefore

$$p(\mathbf{d}) = \frac{\sigma_{\mathrm{d}}^{-N}}{(2\pi)^{N/2}}\exp\left[-\frac{1}{2}\sigma_{\mathrm{d}}^{-2}\sum_{i=1}^{N}(d_i - m_1)^2\right]$$

(2.27)

2.5 TESTING THE ASSUMPTION OF GAUSSIAN STATISTICS

In the following chapters, we shall derive methods of solving inverse problems that are applicable whenever the data exhibit Gaussian statistics. In many instances, the assumption that the data follow this distribution is a reasonable one; nevertheless, having some means to test it is important.

First, consider a set of N Gaussian random variables x_i each with zero mean and unit variance. Suppose we construct a new random variable

$$\chi_K^2 = \sum_{i=1}^{K} x_i^2$$

(2.28)

by summing squares of x_i. The function relating the x_i to $\chi^2{}_K$ is nonlinear, so $\chi^2{}_K$ does not have a Gaussian probability density function, but rather a different one (which we will not derive here) with the functional form

$$p(\chi_K^2) = \frac{1}{2^{K/2}\left(\dfrac{K}{2} - 1\right)!}\left[\chi_K^2\right]^{(K/2)-1}\exp\left(-\frac{1}{2}\chi_K^2\right)$$

(2.29)

It is called the *chi-squared probability density function*. It can be shown to be unimodal with mean K and variance $2K$. We shall make use of it in the discussion to follow.

We begin by supposing that we have some method of solving the inverse problem for the estimated model parameters. Assuming further that the model is explicit, we can compute the variation of the data about its estimated mean—a quantity we refer to as the error $\mathbf{e} = \mathbf{d} - \mathbf{g}(\mathbf{m}^{\mathrm{est}})$. Does this error follow an uncorrelated Gaussian distribution with uniform variance?

To test the hypothesis that it does, we first make a histogram of the N errors e_i, in which the histogram intervals have been chosen so that there are about the same number of errors e_i in each interval. This histogram is then normalized to unit area, and the area A_i^{est} of each of the, say, p intervals is noted. We then compare these areas (which are all in the range from zero to unity) with the areas A_i predicted by a Gaussian distribution with the same mean and variance as the e_i. The overall difference between these areas can be quantified by using

$$\left(\chi_K^2\right)^{est} = N \sum_{i=1}^{p} \frac{\left(A_i^{est} - A_i\right)^2}{A_i} \tag{2.30}$$

If the data followed a Gaussian distribution exactly, then $\left(\chi_K^2\right)^{est}$ should be close to zero (it will not *be* zero since there are always random fluctuations). We therefore need to inquire whether the $\left(\chi_K^2\right)^{est}$ measured for any particular data set is sufficiently far from zero that it is improbable that the data follow the Gaussian distribution. This is done by computing the theoretical distribution of $\left(\chi_K^2\right)^{est}$ and testing whether $\left(\chi_K^2\right)^{est}$ is probable. The usual rule for deciding that the data do not follow the assumed distribution is that values greater than or equal to $\left(\chi_K^2\right)^{est}$ occur less than 5% of the time (if many realizations of the entire experiment were performed).

The quantity $\left(\chi_K^2\right)^{est}$ can be shown to follow approximately a χ_K^2 distribution with $K = p - 3$ degrees of freedom, regardless of the type of distribution involved. The reason that the degrees of freedom are $p - 3$ rather than p is that three constraints have been introduced into the problem: that the area of the histogram is unity and that the mean and variance of the Gaussian distribution match those of the data. This test is known as *Pearson's chi-squared test* (Fig. 2.14). In *MatLab*, the probability P that χ_K^2 is greater than or equal to $\left(\chi_K^2\right)^{est}$ is computed as

```
P = 1−chi2cdf( x2est, K );
```

(*MatLab* script gda02_12)

Here `chi2cdf()` is the cumulative chi-squared distribution, that is, the probability that χ_K^2 is less than or equal to $\left(\chi_K^2\right)^{est}$.

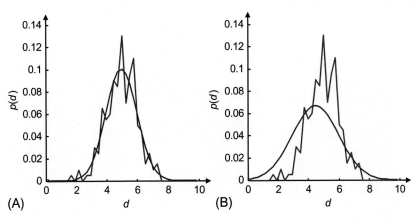

FIG. 2.14 Example of Pierson's chi-squared test. The *red curve* is a probability density function (p.d.f.) estimated by binning 200 realizations of a random variable d drawn from a Gaussian population with a mean of 5 and a variance of 1^2. (A) Gaussian p.d.f. with the same mean and variance as the empirical one. (B) Gaussian p.d.f. with a mean of 4.5 and a variance of 1.5^2. According to the test, χ^2 values exceeding the observed value occur extremely frequently (75% of the time) for (A) but extremely infrequently (0.003%) for (B). *MatLab* script gda02.12.

2.6 CONDITIONAL PROBABILITY DENSITY FUNCTIONS

Consider a scenario in which we are measuring the diameter d_1 and weight d_2 of sand grains drawn randomly from a pile of sand. We can consider d_1 and d_2 random variables described by a joint probability density function $p(d_1,d_2)$. The variables d_1 and d_2 will be correlated, since large grains will also tend to be heavy. Now, suppose that $p(d_1,d_2)$ is known. Once we draw a sand grain from the pile and weigh it, we already know something about its diameter, since diameter is correlated with weight. The quantity that embodies this information is called the *conditional* probability density function of d_1, given d_2, and is written $p(d_1|d_2)$.

The *conditional* probability density function of $p(d_1|d_2)$ is not the same as $p(d_1,d_2)$, although it is related to it. The key difference is that $p(d_1|d_2)$ is really only a probability density function in the variable d_1, with the variable d_2 just providing auxiliary information. Thus, the integral of $p(d_1|d_2)$ with respect to d_1 needs to be unity, regardless of the value of d_2. Thus, we must normalize $p(d_1,d_2)$ by dividing it by the total probability that d_1 occurs, given a specific value for d_2

$$p(d_1|d_2) = \frac{p(d_1,d_2)}{\int p(d_1,d_2)\mathrm{d}d_1} = \frac{p(d_1,d_2)}{p(d_2)} \tag{2.31}$$

Here, we have used the fact that $p(d_2) = \int p(d_1,d_2)\mathrm{d}d_1$. The same logic allows us to calculate the conditional probability density function for d_2, given d_1

$$p(d_2|d_1) = \frac{p(d_1,d_2)}{\int p(d_1,d_2)\mathrm{d}d_2} = \frac{p(d_1,d_2)}{p(d_1)} \tag{2.32}$$

See Fig. 2.15 for an example. Combining these two equations yields

$$p(d_1,d_2) = p(d_1|d_2)p(d_2) = p(d_2|d_1)p(d_1) \tag{2.33}$$

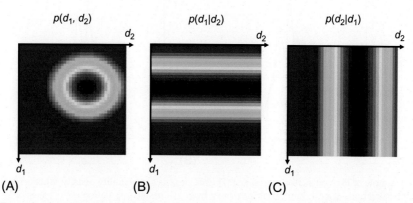

FIG. 2.15 Example of conditional probability density functions. (A) A Gaussian joint probability density function $p(d_1,d_2)$. (B) The corresponding conditional probability density function $p(d_1|d_2)$. (C) The corresponding conditional probability density function $p(d_2|d_1)$. *MatLab* script gda02_13.

This result shows that the two conditional probability density functions are related but that they are *not* equal: $p(d_1|d_2) \neq p(d_2|d_1)$. The two equations can be further rearranged into a result called *Bayes theorem*

$$p(d_1|d_2) = \frac{p(d_2|d_1)p(d_1)}{p(d_2)} = \frac{p(d_2|d_1)p(d_1)}{\int p(d_1,d_2)dd_1} = \frac{p(d_2|d_1)p(d_1)}{\int p(d_2|d_1)p(d_1)dd_1}$$

$$p(d_2|d_1) = \frac{p(d_1|d_2)p(d_2)}{p(d_1)} = \frac{p(d_1|d_2)p(d_2)}{\int p(d_1,d_2)dd_2} = \frac{p(d_1|d_2)p(d_2)}{\int p(d_1|d_2)p(d_2)dd_2}$$

(2.34)

Note that only the denominators of the three fractions in each equation are different. They correspond to three different but equivalent ways of writing $p(d_1)$ and $p(d_2)$.

As an example, consider the case where diameter can take on only two discrete values, small (S) and big (B), and when weight can take on only two values, light (L) and heavy (H). A hypothetical joint probability function is

$$P(d_1,d_2) = \begin{bmatrix} d_1|d_2 & L & H \\ S & 0.8000 & 0.0010 \\ B & 0.1000 & 0.0990 \end{bmatrix}$$

(2.35)

In this scenario, about 90% of the small sand grains are light, about 99% of the large grains are heavy, and small/light grains are much more common than big/heavy ones. Univariate distributions are computed by summing over rows or columns, and in Eq. (2.7):

$$P(d_1) = \begin{bmatrix} d_1 \\ S & 0.8010 \\ B & 0.1990 \end{bmatrix} \quad \text{and} \quad P(d_2) = \begin{bmatrix} d_2 & L & H \\ & 0.9000 & 0.1000 \end{bmatrix}$$

(2.36)

According to Eq. (2.34), the conditional distributions are

$$P(d_1|d_2) = \begin{bmatrix} d_1|d_2 & L & H \\ S & 0.8888 & 0.0100 \\ B & 0.1111 & 0.0990 \end{bmatrix} \quad \text{and} \quad P(d_2|d_1) = \begin{bmatrix} d_1|d_2 & L & H \\ S & 0.9986 & 0.0012 \\ B & 0.5025 & 0.4974 \end{bmatrix}$$

(2.37)

Now suppose that we pick one sand grain from the pile, measure its diameter, and determine that it is big. What is the probability that it is heavy? We may be tempted to think that the probability is very high, since weight is highly correlated to size. But this reasoning is incorrect because heavy grains are about equally divided between the big and small size categories. The correct probability is given by $P(H|B)$, which is 49.74%.

Bayes theorem offers some insight into what is happening. Eq. (2.34), adapted for discrete values by interpreting the integral as a sum, becomes

$$P(H|B) = \frac{P(B|H)P(H)}{P(B|L)P(L) + P(B|H)P(H)} = \frac{0.9900 \times 0.1000}{0.1111 \times 0.9000 + 0.9900 \times 0.1000}$$

$$= \frac{0.0990}{0.1000 + 0.0990} = \frac{0.0990}{0.1990} = 0.4974$$

(2.38)

The numerator of Eq. (2.38) represents the big, heavy grains and the denominator represents *all* the ways that one can get big grains, that is, the sum of big, heavy grains and big, light grains. In the scenario, light grains are extremely common, and although only a small fraction of them are heavy, their number affects the probability very significantly.

This analysis, called *Bayesian Inference*, allows us to assess the importance of any given measurement. Before having measured the size of the sand grain, our best estimate of whether it is heavy is 10%, because heavy grains make up 10% of the total population (i.e., $P(H) = 0.10$). After the measurement, the probability rises to 49.74%, which is about a factor of five more certain. As we will see in Chapter 5, Bayesian Inference plays an important role in the solution of inverse problems.

2.7 CONFIDENCE INTERVALS

The confidence of a particular observation is the probability that one realization of the random variable falls within a specified distance of the true mean. Confidence is therefore related to the distribution of area in $p(d)$. If most of the area is concentrated near the mean, then the interval for, say, 95% confidence will be very small; otherwise, the confidence interval will be large. The width of the confidence interval is related to the variance. Distributions with large variances will also tend to have large confidence intervals. Nevertheless, the relationship is not direct, since variance is a measure of width, not area. The relationship is easy to quantify for the simplest univariate distributions. For instance, Gaussian probability density functions have 68% confidence intervals 1σ wide and 95% confidence intervals 2σ wide. Other types of simple distributions have similar relationships. If one knows that a particular Gaussian random variable has $\sigma = 1$, then if a realization of that variable has the value 50, one can state that there is a 95% chance that the mean of the random variable lies between 48 and 52. One might symbolize this by $\langle d \rangle = 50 \pm 2$ (95%).

The concept of confidence intervals is more difficult to work with when one is dealing with joint probability density functions of several correlated random variables. One must define some volume in the space of data and compute the probability that the true means of the data are within the volume. One must also specify the shape of that volume. The more complicated the distribution, the more difficult it is to choose an appropriate shape and calculate the probability within it.

Even in the case of the Gaussian multivariate probability density functions, statements about confidence levels need to be made carefully, as is illustrated by the following scenario. Suppose that the Gaussian probability density function $p(d_1, d_2)$ represents two measurements, say the length and diameter of a cylinder, and suppose that these measurements are uncorrelated with equal variance, σ_d^2. As we might expect, the univariate probability density function $p(d_1) = \int p(d_1, d_2) dd_2$ has variance, σ_d^2, and so the probability, P_1, that d_1 falls between $d_1 - \sigma_d$ and $d_1 + \sigma_d$, is 0.68 or 68%. Similarly, the probability, P_2, that d_2 falls between $d_2 - \sigma_d$ and $d_2 + \sigma_d$, is also 68%. But P_1 represents the probability of d_1, irrespective of the value of d_2, and P_2 represents the probability of d_2, irrespective of the value of d_1. The probability, P, that *both* d_1 and d_2 simultaneously fall within their respective one-sigma confidence intervals is $P = P_1 P_2 = (0.68)^2 = 0.46$ or 46%, which is significantly smaller than 68%.

One occasionally encounters a journal article containing a table of many (say 100) estimated parameters, each one with a stated 2σ error bound. The probability that *all one hundred* measurements fall within their respective bounds is $(0.95)^{100}$ or 0.6%—which is pretty close to zero!

2.8 COMPUTING REALIZATIONS OF RANDOM VARIABLES

The ability to create a vector of realizations of a random variable is very important. For instance, it can be used to simulate noise when testing a data analysis method on synthetic data (i.e., artificially prepared data with well-controlled properties). And it can be used to generate a suite of possible models, to test against data.

MatLab provides a function `random()` that can generate realizations drawn from many different probability density functions. For instance,

```
m = random('Normal',mbar,sigma,N,1);
```

(*MatLab* script gda02_14)

creates a vector `m` of `N` Gaussian-distributed (Normally distributed) data with mean `mbar` and variance `sigma^2`.

In cases where no predefined function is available, it is possible to transform an available distribution, say $p(d)$, into the desired distribution, say $q(m)$, using the transformation rule

$$p[d(m)]\frac{\mathrm{d}d}{\mathrm{d}m} = q(m) \tag{2.39}$$

Most software environments provide a predefined function for realizations of a uniform distribution on the interval (0,1). Then, since $p(d)=1$, Eq. (2.39) is a differential equation for $d(m)$

$$\frac{\mathrm{d}d}{\mathrm{d}m} = q(m) \quad \text{or} \quad d = \int q(m)\mathrm{d}m = Q(m) \tag{2.40}$$

Here, $Q(m)$ is the cumulative probability distribution corresponding to $q(m)$. The transformation is then $m = Q^{-1}(d)$; that is, one must invert the cumulative probability distribution to give the value of m for which the probability d occurs. Thus, the transformation requires that the inverse cumulative probability distribution be known.

MatLab provides a `norminv()` function that calculates the inverse cumulative probability distribution in the Gaussian case, as well as a `random('unif',...)` function that returns realizations of the uniform probability density function. Thus,

```
d = random('unif',0,1,N,1);
m = norminv(d,mbar,sigma);
```

(*MatLab* script gda02_14)

creates `N` realizations of a Gaussian probability density function (Fig. 2.16). Such an approach offers no advantage in the Gaussian case, since the `random('Normal',...)` function

FIG. 2.16 Gaussian probability density function $p(m)$ with mean 5 and variance 1^2. *(Red curve)* Computed by binning 1000 realizations of a random variable generated using *MatLab*'s `random ("Normal",...)` function. *(Blue)* Computed by binning 1000 realizations of a random variable generated by transforming a uniform distribution. *(Black)* Exact formula. *MatLab* script gda02_14.

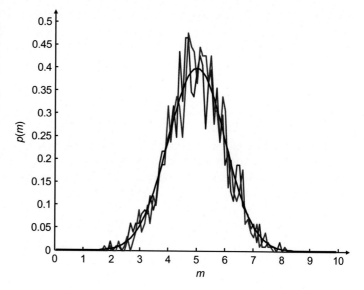

is available. It is of practical use in cases not supported by *MatLab*, as long as an appropriate `Qinv()` function can be provided.

Another method of producing a vector of realizations of a random variable is the *Metropolis-Hastings algorithm*. It is a useful alternative to the transformation method described earlier, especially since it requires evaluating only the probability density function $p(d)$ and not its cumulative inverse. It is an iterative algorithm that builds the vector **d** element by element. The first element, d_1, is set to an arbitrary number, such as zero. Subsequent elements are generated in sequence, with an element d_i, generating a successor d_{i+1} according to this algorithm: First, randomly draw a *proposed* successor d' from a conditional probability density function $q(d'|d_i)$. The exact form of $q(d'|d_i)$ is arbitrary; however, it must be chosen so that d' is typically in the neighborhood of d_i. One possible choice is the Gaussian function

$$q(d'|d_i) = \frac{1}{(2\pi)^{1/2}\sigma} \exp\left\{ -\frac{(d'-d_i)^2}{2\sigma^2} \right\} \tag{2.41}$$

Here, σ represents the size of the neighborhood, that is, the typical deviation of d' away from d_i. Second, generate a random number α drawn from a uniform distribution on the interval $(0,1)$. Third, accept the proposed successor and set $d_{i+1}=d'$ if

$$\alpha < \frac{p(d')q(d_i|d')}{p(d_i)q(d'|d_i)} \tag{2.42}$$

Otherwise, set $d_{i+1}=d_i$. When repeated many times, this algorithm leads to a vector **d** that has approximately the probability density function $p(d)$ (Fig. 2.17). Note that the conditional probability density functions cancel from Eq. (2.42) when $q(d'|d_i)=q(d_i|d')$, as is the case for the Gaussian in Eq. (2.41).

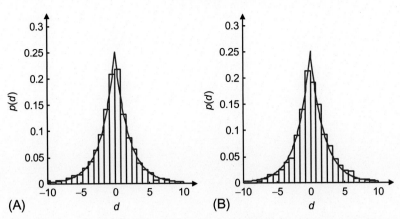

FIG. 2.17 Histograms *(blue curves)* of 5000 realizations of a random variable d for the probability density function *(red curves)* $p(d) = \frac{1}{2} c \exp(-|d|/c)$ with $c = 2$. (A) Realizations computed by transforming data drawn from a uniform distribution and (B) realizations computed using the Metropolis-Hastings algorithm. *MatLab* script gda02_15.

2.9 PROBLEMS

2.1 What is the mean and variance of the uniform distribution $p(d) = 1$ on the interval (0,1)?

2.2 Suppose d is a Gaussian random variable with zero mean and unit variance. What is the probability density function of $E = d^2$? Hint: Since the sign of d gets lost when it is squared, you can assume that $p(d)$ is one-sided, that is, defined for only $d \geq 0$ and with twice the amplitude of the usual Gaussian.

2.3 Write a *MatLab* script that uses the `random()` function to create a vector **d** of $N = 1000$ realizations of a Gaussian-distributed random variable with mean $\langle d \rangle = 4$ and variance $\sigma_d^2 = 2^2$. Count up the number of instances where $d_i > (\langle d \rangle + 2\sigma_d)$. Is this about the number you expected?

2.4 Suppose that the data are uncorrelated with uniform variance, [cov **d**] $= \sigma_d^2 \mathbf{I}$, and that the model parameters are linear functions of the data, $\mathbf{m} = \mathbf{Md}$. (A) What property must **M** have for the model parameters to be uncorrelated with uniform variance σ_m^2? (B) Express this property in terms of the rows of the **M**.

2.5 Use the transformation method to compute realizations of the probability density function $p(m) = 3m^2$ on the interval (0,1), starting from realizations of the uniform distribution $p(d) = 1$. Check your results by plotting a histogram.

Reference

Menke, W., Menke, J., 2011. Environmental Data Analysis with MatLab. Academic Press, Elsevier Inc., Oxford, UK, 263 pp.

3

Solution of the Linear, Gaussian Inverse Problem, Viewpoint 1: The Length Method

Geophysical Data Analysis
https://doi.org/10.1016/B978-0-12-813555-6.00003-4

© 2018 Elsevier Inc. All rights reserved.

3.1 THE LENGTHS OF ESTIMATES

The simplest of methods for solving the linear inverse problem $\mathbf{Gm} = \mathbf{d}$ is based on measures of the size, or length, of the estimated model parameters $\mathbf{m}^{\text{est}}$ and of the predicted data $\mathbf{d}^{\text{pre}} = \mathbf{Gm}^{\text{est}}$.

To see that measures of length can be relevant to the solution of inverse problems, consider the simple problem of fitting a straight line to data (Fig. 3.1). This problem is often solved by the so-called method of least squares. In this method, one tries to pick the model parameters (intercept and slope) so that the predicted data are as close as possible to the observed data. For each observation, one defines a prediction error, or misfit, $e_i = d_i^{\text{obs}} - d_i^{\text{pre}}$. The best-fit line is then the one with model parameters that lead to the smallest overall error E, defined as

$$E = \sum_{i=1}^{N} e_i^2 = \mathbf{e}^{\mathrm{T}}\mathbf{e} \tag{3.1}$$

The total error E (the sum of the squares of the individual errors) is exactly the squared Euclidean length of the vector $\mathbf{e}$, or $E = \mathbf{e}^{\mathrm{T}}\mathbf{e}$.

The method of least squares estimates the solution of an inverse problem by finding the model parameters that minimize a particular measure of the length of the prediction error, $\mathbf{e} = \mathbf{d}^{\text{obs}} - \mathbf{d}^{\text{pre}}$, namely, its Euclidean length. As will be detailed later, it is the simplest of the methods that use measures of length as the guiding principle in solving an inverse problem.

3.2 MEASURES OF LENGTH

Note that although the Euclidean length is one way of quantifying the size or length of a vector, it is by no means the only possible measure. For instance, one could equally well quantify length by summing the absolute values of the elements of the vector.

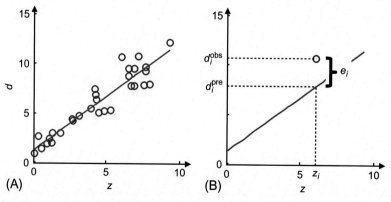

FIG. 3.1   (A) Least squares fitting of a straight line to (z, d) data. (B) The error e_i for each observation is the difference between the observed and predicted datum: $e_i = d_i^{\text{obs}} - d_i^{\text{pre}}$. *MatLab* script gda03_01.

The term *norm* is used to refer to some measure of length or size and is indicated by a set of double vertical bars; that is, $\|\mathbf{e}\|$ is the norm of the vector $\mathbf{e}$. The most commonly employed norms are those based on the sum of some power of the elements of a vector and are given the name L_n, where n is the power:

$$L_1 \text{ norm}: \quad \|\mathbf{e}\|_1 = \left[\sum_i |e_i|^1 \right] \tag{3.2a}$$

$$L_2 \text{ norm}: \quad \|\mathbf{e}\|_2 = \left[\sum_i |e_i|^2 \right]^{1/2} \tag{3.2b}$$

$$L_n \text{ norm}: \quad \|\mathbf{e}\|_n = \left[\sum_i |e_i|^n \right]^{1/n} \tag{3.2c}$$

Successively higher norms give the largest element of $\mathbf{e}$ successively larger weight. The limiting case of $n \to \infty$ gives nonzero weight to only the largest element (Fig. 3.2); therefore, it is equivalent to the selection of the vector element with largest absolute value as the measure of length and is written as

$$L_\infty \text{ norm}: \quad \|\mathbf{e}\|_\infty = \max_i |e_i| \tag{3.2d}$$

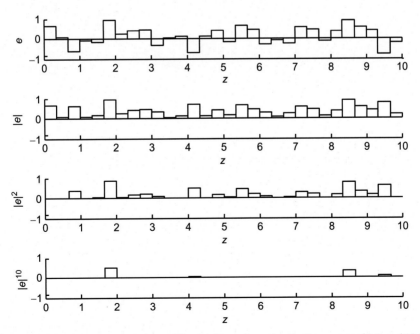

FIG. 3.2 Hypothetical prediction error, $e_i(z_i)$, and its absolute value, raised to the powers of 1, 2, and 10. While most elements of $|e_i|$ are numerically significant, only a few elements of $|e_i|^{10}$ are. *MatLab* script gda03_02.

The method of least squares uses the L_2 norm to quantify length. It is appropriate to inquire why this, and not some other choice of norm, is used. The answer involves the way in which one chooses to weight data *outliers* that fall far from the average trend (Fig. 3.3). If the data are very accurate, then the fact that one prediction falls far from its observed value is important. A high-order norm is used, since it weights the larger errors preferentially. On the other hand, if the data are expected to scatter widely about the trend, then no significance can be placed upon a few large prediction errors. A low-order norm is used, since it gives more equal weight to errors of different sizes.

As will be discussed in more detail later, the L_2 norm implies that the data obey Gaussian statistics. Gaussians are rather short-tailed functions, so it is appropriate to place considerable weight on any data that have a large prediction error.

The likelihood of an observed datum falling far from the trend depends on the shape of the distribution for that datum. Long-tailed distributions imply many scattered (improbable) points. Short-tailed distributions imply very few scattered points (Fig. 3.4). The choice of a norm, therefore, implies an assertion that the data obey a particular type of statistics.

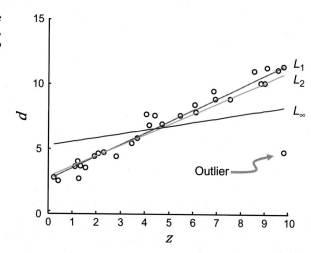

FIG. 3.3   Straight line fits to (z, d) pairs where the error is measured under the L_1, L_2, and L_∞ norms. The L_1 norm gives the least weight to the one outlier. *MatLab* script gda03_03.

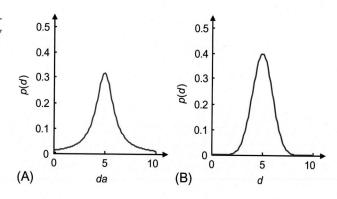

FIG. 3.4   (A) Long-tailed probability density function. (B) Short-tailed probability density function. *MatLab* script gda03_04.

Even though many measurements have approximately Gaussian statistics, most data sets generally have a few outliers that are wildly improbable. The occurrence of these points demonstrates that the assumption of Gaussian statistics is in error, especially in the tails of the distribution. If one applies least squares to this kind of problem, the estimates of the model parameters can be completely erroneous. Least squares weights large errors so heavily that even one "bad" data point can completely throw off the result. In these situations, methods based on the L_1 norm give more reliable estimates. (Methods that can tolerate a few bad data are said to be *robust*.)

Matrix norms can be defined in a manner similar to vector norms (see Eq. 3.3d). Vector and matrix norms obey the following relationships:

Vector norms:

$$\|\mathbf{x}\| > 0 \quad \text{as long as} \quad \mathbf{x} \neq 0 \tag{3.3a}$$

$$\|a\mathbf{x}\| = [a]\,\|\mathbf{x}\| \tag{3.3b}$$

$$\|\mathbf{x} + \mathbf{y}\| \leq \|\mathbf{x}\| + \|\mathbf{y}\| \tag{3.3c}$$

Matrix norms:

$$\|\mathbf{A}\|_2 = \left(\sum_{i=1}^{N} \sum_{j=1}^{N} A_{ij}^2 \right)^{1/2} \tag{3.3d}$$

$$\|c\mathbf{A}\| = |c|\,\|\mathbf{A}\| \tag{3.3e}$$

$$\|\mathbf{A}\mathbf{x}\| \leq \|\mathbf{A}\|\,\|\mathbf{x}\| \tag{3.3f}$$

$$\|\mathbf{A}\mathbf{B}\| \leq \|\mathbf{A}\|\,\|\mathbf{B}\| \tag{3.3g}$$

$$\|\mathbf{A} + \mathbf{B}\| \leq \|\mathbf{A}\| + \|\mathbf{B}\| \tag{3.3h}$$

Eqs. (3.3c), (3.3h) are called *triangle inequalities* because of their similarity to Pythagoras's law for right triangles.

The number of nonzero elements in a vector is a measure of its simplicity. It is sometimes denoted $\|\mathbf{e}\|_0^0$, since:

$$\|\mathbf{e}\|_0^0 \equiv \lim_{n \to 0} \|\mathbf{e}\|_n^n = \sum_{i=1} |e_i|^0 = \text{number of nonzero elements in } \mathbf{e} \tag{3.4}$$

(at least if one defines 0^0 as zero). However, this usage must be understood as an informal shorthand, for $\|\mathbf{e}\|_0^0$ does not obey all the rules of a norm (such as Eq. 3.3b).

3.3 LEAST SQUARES FOR A STRAIGHT LINE

The elementary problem of fitting a straight line to data illustrates the basic procedures applied in this technique. The model is the assertion that the data can be described by the linear equation $d_i = m_1 + m_2 z_i$. Note that there are two model parameters, $M = 2$, and that typically there are many more than two data, $N > M$. Since a line is defined by precisely two points, it is clearly impossible to choose a straight line that passes through every one of the data, except in the instance that they all lie precisely on the same straight line. Collinearity rarely occurs when measurements are influenced by noise.

As we shall discuss in more detail later, the fact that the equation $d_i = m_1 + m_2 z_i$ cannot be satisfied for every i means that the inverse problem is *overdetermined*; that is, it has no solution for which $\mathbf{e} = 0$. One therefore seeks values of the model parameters that solve $d_i = m_1 + m_2 z_i$ approximately, where the goodness of the approximation is defined by the error

$$E = \mathbf{e}^T \mathbf{e} = \sum_{i=1}^{N} (d_i - m_1 - m_2 z_i)^2 \tag{3.5}$$

This problem is then the elementary calculus problem of locating the minimum of the function $E(m_1, m_2)$ and is solved by setting the derivatives of E to zero and solving the resulting equations.

$$\frac{\partial E}{\partial m_1} = \frac{\partial}{\partial m_1} \sum_{i=1}^{N} [d_i - m_1 - m_2 z_i]^2 = 2Nm_1 + 2m_2 \sum_{i=1}^{N} z_i - 2 \sum_{i=1}^{N} d_i = 0$$

$$\frac{\partial E}{\partial m_2} = \frac{\partial}{\partial m_2} \sum_{i=1}^{N} [d_i - m_1 - m_2 z_i]^2 = 2m_1 \sum_{i=1}^{N} z_i + 2m_2 \sum_{i=1}^{N} z_i^2 - 2 \sum_{i=1}^{N} z_i d_i = 0 \tag{3.6}$$

These two equations are then solved simultaneously for m_1 and m_2, yielding the classic formulas for the least squares fitting of a line.

3.4 THE LEAST SQUARES SOLUTION OF THE LINEAR INVERSE PROBLEM

Least squares can be extended to the general linear inverse problem in a very straightforward manner. Again, one computes the derivative of the error E with respect to one of the model parameters, say, m_q, and sets the result to zero. The error E is

$$E = \mathbf{e}^T \mathbf{e} = (\mathbf{d} - \mathbf{Gm})^T (\mathbf{d} - \mathbf{Gm}) = \sum_{i=1}^{N} \left[d_i - \sum_{j=1}^{M} G_{ij} m_j \right] \left[d_i - \sum_{k=1}^{M} G_{ik} m_k \right] \tag{3.7}$$

Note that the indices on the sums within the parentheses are different dummy variables, to prevent confusion. Multiplying out the terms and reversing the order of the summations lead to

$$E = \sum_{j=1}^{M} \sum_{k=1}^{M} m_j m_k \sum_{i=1}^{N} G_{ij} G_{ik} - 2 \sum_{j=1}^{M} m_j \sum_{i=1}^{N} G_{ij} d_i + \sum_{i=1}^{N} d_i d_i \tag{3.8}$$

The derivatives $\partial E / \partial m_q$ are now computed. Performing this differentiation term by term gives

$$\frac{\partial}{\partial m_q}\left[\sum_{j=1}^{M}\sum_{k=1}^{M}m_j m_k \sum_{i=1}^{N}G_{ij}G_{ik}\right] = \sum_{j=1}^{M}\sum_{k=1}^{M}\left[\delta_{jq}m_k + m_j\delta_{kq}\right]\sum_{i=1}^{N}G_{ij}G_{ik}$$

$$= 2\sum_{k=1}^{M}m_k \sum_{i=1}^{N}G_{iq}G_{ik} \tag{3.9}$$

for the first term. Since the model parameters are independent variables, derivatives of the form $\partial m_i/\partial m_j$ are either unity, when $i=j$, or zero, when $i\neq j$. Thus, $\partial m_i/\partial m_j$ is just the Kronecker delta δ_{ij} (see Section I.4) and the formula containing it can be simplified trivially. The second term gives

$$-2\frac{\partial}{\partial m_q}\left[\sum_{j=1}^{M}m_j\sum_{i=1}^{N}G_{ij}d_i\right] = -2\sum_{j=1}^{M}\delta_{jq}\sum_{i=1}^{N}G_{ij}d_i = -2\sum_{i=1}^{N}G_{iq}d_i \tag{3.10}$$

Since the third term does not contain any ms, it is zero as

$$\frac{\partial}{\partial m_q}\left[\sum_{i=1}^{N}d_i d_i\right] = 0 \tag{3.11}$$

Combining the three terms gives

$$\frac{\partial E}{\partial m_q} = 0 = 2\sum_{k=1}^{M}m_k\sum_{i=1}^{N}G_{iq}G_{ik} - 2\sum_{i=1}^{N}G_{iq}d_i \tag{3.12}$$

Writing this equation in matrix notation yields

$$\mathbf{G}^T\mathbf{G}\mathbf{m} - \mathbf{G}^T\mathbf{d} = 0 \tag{3.13}$$

Note that the quantity $\mathbf{G}^T\mathbf{G}$ is a square $M \times M$ matrix and that it multiplies a vector $\mathbf{m}$ of length M. The quantity $\mathbf{G}^T\mathbf{d}$ is also a vector of length M. This equation is therefore a square matrix equation for the unknown model parameters. Presuming that $[\mathbf{G}^T\mathbf{G}]^{-1}$ exists (an important question that we shall return to later), we have the following estimate for the model parameters:

$$\mathbf{m}^{est} = [\mathbf{G}^T\mathbf{G}]^{-1}\mathbf{G}^T\mathbf{d} \tag{3.14}$$

which is the least squares solution to the inverse problem $\mathbf{Gm}=\mathbf{d}$.

When the dimensions of $\mathbf{G}$ are small (say, N and M less than a few hundred), the least squares solution is computed as

```
mest = (G'*G)\(G'*d);
```

(*MatLab* script gda03_05)

However, for larger problems, the computational cost of computing $\mathbf{G}^T\mathbf{G}$ can be prohibitive. Furthermore, $\mathbf{G}^T\mathbf{G}$ is rarely as sparse as $\mathbf{G}$ itself. In this case, an iterative matrix solver, such as the *biconjugate gradient algorithm*, is preferred. This algorithm only requires products of the form $\mathbf{G}^T\mathbf{G}\mathbf{v}$, where $\mathbf{v}$ is a vector constructed by the algorithm, and this product can be performed as $\mathbf{G}^T(\mathbf{Gv})$ so that $\mathbf{G}^T\mathbf{G}$ is never explicitly calculated (e.g., Menke and Menke, 2011, their Section 5.8). *MatLab* provides a `bicg()` function, which is as follows:

```
tol = 1e-6;
maxit = 3*N;
mest2 = bicg(@leastsquaresfcn, G'*dobs, tol, maxit);
```

(*MatLab* script gda03_05)

The algorithm involves iteratively improving a solution, with the iterations terminating when the error is less than a tolerance `tol` or when `maxit` iterations have been performed (whichever comes first). The first argument is a *handle* to a user-provided `leastsquaresfcn ()` function that calculates $\mathbf{G}^T(\mathbf{Gv})$:

```
function y = leastsquaresfcn(v,transp_flag)
global G;
temp = G*v;
y = G'*temp;
return
```

(*MatLab* script gda03_05)

This function is stored in the file `leastsquaresfcn.m`. In order for this function to perform correctly, *MatLab* must be instructed that the **G** being used within it is the same as is defined in the main script. This is accomplished by placing the commands

```
clear G;
global G;
```

(*MatLab* script gda03_05)

at the beginning of the main script. Although this algorithm will work for any **G**, it is particularly useful when **G** has been defined as sparse using the `spalloc()` function. An example is given in *MatLab* script gda03_06.

3.5 SOME EXAMPLES

3.5.1 The Straight Line Problem

In the straight line problem, the model is $d_i = m_1 + m_2 z_i$, so the equation $\mathbf{Gm} = \mathbf{d}$ has the form

$$
\begin{bmatrix} 1 & z_1 \\ 1 & z_2 \\ \vdots & \vdots \\ 1 & z_N \end{bmatrix} \begin{bmatrix} m_1 \\ m_2 \end{bmatrix} = \begin{bmatrix} d_1 \\ d_2 \\ \vdots \\ d_N \end{bmatrix} \tag{3.15}
$$

In *MatLab*, the matrix **G** can be created with the command

```
G = [ones(N,1), z];
```

(*MatLab* script gda03_05)

The matrix products required by the least squares solution are

$$\mathbf{G}^{\mathrm{T}}\mathbf{G} = \begin{bmatrix} 1 & 1 & \cdots & 1 \\ z_1 & z_2 & \cdots & z_N \end{bmatrix} \begin{bmatrix} 1 & z_1 \\ 1 & z_2 \\ \vdots & \vdots \\ 1 & z_N \end{bmatrix} = \begin{bmatrix} N & \sum_{i=1}^{N} z_i \\ \sum_{i=1}^{N} z_i & \sum_{i=1}^{N} z_i^2 \end{bmatrix} \quad (3.16)$$

and

$$\mathbf{G}^{\mathrm{T}}\mathbf{d} = \begin{bmatrix} 1 & 1 & \cdots & 1 \\ z_1 & z_2 & \cdots & z_N \end{bmatrix} \begin{bmatrix} d_1 \\ d_2 \\ \vdots \\ d_N \end{bmatrix} = \begin{bmatrix} \sum_{i=1}^{N} d_i \\ \sum_{i=1}^{N} d_i z_i \end{bmatrix} \quad (3.17)$$

This gives the least squares solution

$$\mathbf{m}^{\mathrm{est}} = \left[\mathbf{G}^{\mathrm{T}}\mathbf{G}\right]^{-1}\mathbf{G}^{\mathrm{T}}\mathbf{d} = \begin{bmatrix} N & \sum_{i=1}^{N} z_i \\ \sum_{i=1}^{N} z_i & \sum_{i=1}^{N} z_i^2 \end{bmatrix}^{-1} \begin{bmatrix} \sum_{i=1}^{N} d_i \\ \sum_{i=1}^{N} d_i z_i \end{bmatrix} \quad (3.18)$$

3.5.2 Fitting a Parabola

The problem of fitting a parabola is a trivial generalization of fitting a straight line (Fig. 3.4). Now the model is $d_i = m_1 + m_2 z_i + m_3 z_i^2$, so the equation $\mathbf{Gm} = \mathbf{d}$ has the form

$$\begin{bmatrix} 1 & z_1 & z_1^2 \\ 1 & z_2 & z_2^2 \\ \vdots & \vdots & \vdots \\ 1 & z_N & z_N^2 \end{bmatrix} \begin{bmatrix} m_1 \\ m_2 \\ m_3 \end{bmatrix} = \begin{bmatrix} d_1 \\ d_2 \\ \vdots \\ d_N \end{bmatrix} \quad (3.19)$$

In *MatLab*, the matrix $\mathbf{G}$ can be created with the command

```
G = [ones(N,1), z, z.^2];
```

(*MatLab* script gda03_07)

The matrix products required by the least squares solution are as follows:

$$\mathbf{G}^{\mathrm{T}}\mathbf{G} = \begin{bmatrix} 1 & 1 & \cdots & 1 \\ z_1 & z_2 & \cdots & z_N \\ z_1^2 & z_2^2 & \cdots & z_N^2 \end{bmatrix} \begin{bmatrix} 1 & z_1 & z_1^2 \\ 1 & z_2 & z_2^2 \\ \vdots & \vdots & \vdots \\ 1 & z_N & z_N^2 \end{bmatrix}$$

$$= \begin{bmatrix} N & \sum_{i=1}^{N} z_i & \sum_{i=1}^{N} z_i^2 \\ \sum_{i=1}^{N} z_i & \sum_{i=1}^{N} z_i^2 & \sum_{i=1}^{N} z_i^3 \\ \sum_{i=1}^{N} z_i^2 & \sum_{i=1}^{N} z_i^3 & \sum_{i=1}^{N} z_i^4 \end{bmatrix} \quad (3.20)$$

and

$$\mathbf{G}^{\mathrm{T}}\mathbf{d}=\begin{bmatrix} 1 & 1 & \cdots & 1 \\ z_1 & z_2 & \cdots & z_N \\ z_1^2 & z_2^2 & \cdots & z_N^2 \end{bmatrix}\begin{bmatrix} d_1 \\ d_2 \\ \vdots \\ d_N \end{bmatrix}=\begin{bmatrix} \sum_{i=1}^{N} d_i \\ \sum_{i=1}^{N} z_i d_i \\ \sum_{i=1}^{N} z_i^2 d_i \end{bmatrix} \tag{3.21}$$

giving the least squares solution

$$\mathbf{m}^{\mathrm{est}}=\left[\mathbf{G}^{\mathrm{T}}\mathbf{G}\right]^{-1}\mathbf{G}^{\mathrm{T}}\mathbf{d}=\begin{bmatrix} N & \sum_{i=1}^{N} z_i & \sum_{i=1}^{N} z_i^2 \\ \sum_{i=1}^{N} z_i & \sum_{i=1}^{N} z_i^2 & \sum_{i=1}^{N} z_i^3 \\ \sum_{i=1}^{N} z_i^2 & \sum_{i=1}^{N} z_i^3 & \sum_{i=1}^{N} z_i^4 \end{bmatrix}^{-1}\begin{bmatrix} \sum_{i=1}^{N} d_i \\ \sum_{i=1}^{N} z_i d_i \\ \sum_{i=1}^{N} z_i^2 d_i \end{bmatrix} \tag{3.22}$$

An example of using a quadratic fit to examine Kepler's third law is shown in Fig. 3.5.

3.5.3 Fitting a Plane Surface

To fit a plane surface, two auxiliary variables, say, x and y, are needed. The model is

$$d_i = m_1 + m_2 x_i + m_3 y_i \tag{3.23}$$

so the equation $\mathbf{Gm}=\mathbf{d}$ has the form

$$\begin{bmatrix} 1 & x_1 & y_1 \\ 1 & x_2 & y_2 \\ \vdots & \vdots & \vdots \\ 1 & x_N & y_N \end{bmatrix}\begin{bmatrix} m_1 \\ m_2 \\ m_3 \end{bmatrix}=\begin{bmatrix} d_1 \\ d_2 \\ \vdots \\ d_N \end{bmatrix} \tag{3.24}$$

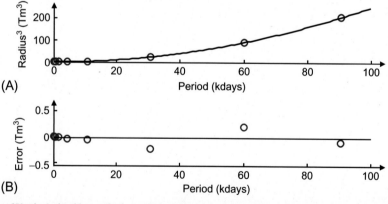

FIG. 3.5   Test of Kepler's third law, which states that the cube of the orbital radius of a planet equals the square of its orbital period. (A) Data *(red circles)* for our solar system are least squares fit with a quadratic formula $d_i = m_1 + m_2 z_i + m_3 z_i^2$, where d_i is radius cubed and z_i is period. (B) Error of fit. A separate graph is used so that the error can be plotted at a meaningful scale. *Data courtesy of Wikipedia. MatLab* script gda03_07.

In *MatLab*, the matrix **G** can be created with the command

```
G = [ones(N,1), x, y];
```

(*MatLab* script gda03_08)

forming the matrix products $\mathbf{G}^T\mathbf{G}$

$$\mathbf{G}^T\mathbf{G} = \begin{bmatrix} 1 & 1 & \cdots & 1 \\ x_1 & x_2 & \cdots & x_N \\ y_1 & y_2 & \cdots & y_N \end{bmatrix} \begin{bmatrix} 1 & x_1 & y_1 \\ 1 & x_2 & y_2 \\ \vdots & \vdots & \vdots \\ 1 & x_N & y_N \end{bmatrix}$$

$$= \begin{bmatrix} N & \sum_{i=1}^N x_i & \sum_{i=1}^N y_i \\ \sum_{i=1}^N x_i & \sum_{i=1}^N x_i^2 & \sum_{i=1}^N x_i y_i \\ \sum_{i=1}^N y_i & \sum_{i=1}^N x_i y_i & \sum_{i=1}^N y_i^2 \end{bmatrix} \tag{3.25}$$

and

$$\mathbf{G}^T\mathbf{d} = \begin{bmatrix} 1 & 1 & \cdots & 1 \\ x_1 & x_2 & \cdots & x_N \\ y_1 & y_2 & \cdots & y_N \end{bmatrix} \begin{bmatrix} d_1 \\ d_2 \\ \vdots \\ d_N \end{bmatrix} = \begin{bmatrix} \sum_{i=1}^N d_i \\ \sum_{i=1}^N x_i d_i \\ \sum_{i=1}^N y_i d_i \end{bmatrix} \tag{3.26}$$

giving the least squares solution

$$\mathbf{m}^{\mathrm{est}} = \left[\mathbf{G}^T\mathbf{G}\right]^{-1}\mathbf{G}^T\mathbf{d} = \begin{bmatrix} N & \sum_{i=1}^N x_i & \sum_{i=1}^N y_i \\ \sum_{i=1}^N x_i & \sum_{i=1}^N x_i^2 & \sum_{i=1}^N x_i y_i \\ \sum_{i=1}^N y_i & \sum_{i=1}^N x_i y_i & \sum_{i=1}^N y_i^2 \end{bmatrix}^{-1} \begin{bmatrix} \sum_{i=1}^N d_i \\ \sum_{i=1}^N x_i d_i \\ \sum_{i=1}^N y_i d_i \end{bmatrix} \tag{3.27}$$

An example of using a planar fit to examine earthquakes on a geologic fault is shown in Fig. 3.6.

3.5.4 Inverting Reflection Coefficients for Interface Properties

The seismological properties of an elastic material, such as the Earth, can be described by three material parameters: density ρ; compressional velocity α; and shear velocity β. When two homogenous layers are in contact, their contrast can be characterized by the change $\Delta\rho$, $\Delta\alpha$, and $\Delta\beta$ in these material properties, with respect to their average values of ρ, α, and β. The interface can be studied by directing a compressional wave ("*P*" wave) or a shear wave ("*S*" wave) at it, and observing the *P* and *S* waves that are reflected from it. When a unit-amplitude *P* wave is incident on the interface, the coefficients R_{PP} and R_{PS} give the amplitudes of the reflected *P* and *S* waves, respectively (Fig. 3.7A). When a unit-amplitude *S* wave is incident on the interface, the coefficients R_{SP} and R_{SS} give the amplitudes of the reflected *P* and

FIG. 3.6 *(Circles)* Earthquakes in the Kurile subduction zone, northwest Pacific Ocean. The *x*-axis points north and the *y*-axis east. The earthquakes scatter about a dipping planar surface *(colored grid)*, determined using least squares. *Data courtesy of the U.S. Geological Survey. MatLab script gda03_08.*

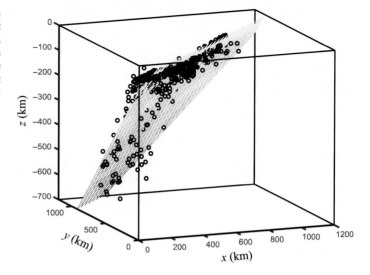

FIG. 3.7 (A) Schematic diagrams of *P* and *S* wave reflected from a layer interface. In this scenario, measurements of reflection coefficients are used to infer interface properties. (B) True jumps in density ρ, compressional velocity α and shear velocity β across the interface *(black bars)*, together with corresponding estimates *(red bars)* and 95% confidence intervals *(orange bars)*. (C) True reflections coefficients $R_{PP}(\theta)$ *(bold black curve)*, $R_{PS}(\theta)$, *(bold red curve)*, $R_{SP}(\theta)$ *(bold green curve)*, and $R_{SS}(\theta)$ *(bold blue curve)* as a function of angle of incidence θ, together with corresponding noisy versions *(thin curves)* and the predictions of the model *(circles)*.

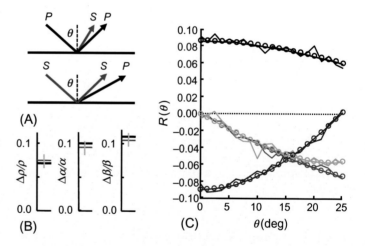

S waves, respectively. These coefficients depend upon the angle of incidence of the incident waves, say θ_P and θ_S, as well as the changes in the material parameters.

A common problem in seismology is estimating the properties of the interface using measurements of the reflection coefficients made at a suite of angles of incidence. This inverse problem has $M=3$ unknowns that represent the fractional change in material properties across the interface:

$$\mathbf{m} = \begin{bmatrix} m_1 \\ m_2 \\ m_3 \end{bmatrix} = \begin{bmatrix} \Delta\rho/\rho \\ \Delta\alpha/\alpha \\ \Delta\beta/\beta \end{bmatrix} \tag{3.28}$$

For a weak interface, the reflection coefficients can be shown to be the linear functions of these model parameters (Aki and Richards, 2009, their Eq. 5.46):

$$R_{PP} = R_{PP}^{(1)} m_1 + R_{PP}^{(2)} m_2 + R_{PP}^{(3)} m_3$$
$$R_{PS} = R_{PS}^{(1)} m_1 + R_{PS}^{(2)} m_2 + R_{PS}^{(3)} m_3$$
$$R_{SP} = R_{SP}^{(1)} m_1 + R_{SP}^{(2)} m_2 + R_{SP}^{(3)} m_3$$
$$R_{SS} = R_{SS}^{(1)} m_1 + R_{SS}^{(2)} m_2 + R_{SS}^{(3)} m_3$$

$$(3.29)$$

Each of the 12 coefficients $R_{PP}^{(1)}, R_{PP}^{(2)}, \ldots, R_{SS}^{(3)}$ quantify the contribution of one model parameter to one reflection coefficients. They are complicated functions of angle of incidence and the layer properties:

with $R_{PP}^{(1)} = \frac{1}{2}(1 - 4\beta^2 p^2)$ and $R_{PP}^{(2)} = \dfrac{1}{2\cos^2\theta_P}$ and $R_{PP}^{(3)} = -4\beta^2 p^2$ and

$$R_{PS}^{(1)} = C\left(1 - 2\beta^2 p^2 - 2\beta^2 \frac{\cos\theta_P}{\alpha} \frac{\cos\theta_S}{\beta}\right) \text{ and } R_{PS}^{(2)} = 0$$

$$\text{and } R_{PS}^{(3)} = 4C\beta^2\left(p^2 - \frac{\cos\theta_P}{\alpha} \frac{\cos\theta_S}{\beta}\right)$$

$$R_{SP}^{(1)} = DR_{PS}^{(1)} \text{ and } R_{SP}^{(2)} = DR_{PS}^{(2)} \text{ and } R_{SP}^{(3)} = DR_{PS}^{(3)}$$

$$R_{SS}^{(1)} = -\frac{1}{2}(1 - 4\beta^2 p^2) \text{ and } R_{SS}^{(2)} = 0 \text{ and } R_{SS}^{(3)} = -\left(\frac{1}{2\cos^2\theta_S} - 4\beta^2 p^2\right)$$

and $p = \dfrac{\sin\theta_P}{\alpha} = \dfrac{\sin\theta_S}{\beta}$ and $\cos\theta_P = \sqrt{1 - \alpha^2 p^2}$ and $\cos\theta_S = \sqrt{1 - \beta^2 p^2}$

and $C = \dfrac{-p\alpha}{2\cos\theta_S}$ and $D = \dfrac{\cos\theta_S}{\alpha}\dfrac{\beta}{\cos\theta_P}$

$$(3.30)$$

An experiment in which each reflection coefficient is measured at a suite of K angles (Fig. 3.7C) has $N = 4K$ data and $M = 3$ unknowns. The data vector is formed by concatenating the measurements of each of the reflection coefficients, and the data kernel is built from Eq. (3.29), ordering its rows in a corresponding fashion:

$$\mathbf{d}^{obs} = \begin{bmatrix} R_{PP}(\theta_1) \\ \vdots \\ R_{PP}(\theta_K) \\ R_{PS}(\theta_1) \\ \vdots \\ R_{PS}(\theta_K) \\ R_{SP}(\theta_1) \\ \vdots \\ R_{SP}(\theta_K) \\ R_{SS}(\theta_1) \\ \vdots \\ R_{SS}(\theta_K) \end{bmatrix} \text{ and } \mathbf{G} = \begin{bmatrix} R_{PP}^{(1)}(\theta_1) & R_{PP}^{(2)}(\theta_1) & R_{PP}^{(3)}(\theta_1) \\ \vdots & \vdots & \vdots \\ R_{PP}^{(1)}(\theta_K) & R_{PP}^{(2)}(\theta_K) & R_{PP}^{(3)}(\theta_K) \\ R_{PS}^{(1)}(\theta_1) & R_{PS}^{(2)}(\theta_1) & R_{PS}^{(3)}(\theta_1) \\ \vdots & \vdots & \vdots \\ R_{PS}^{(1)}(\theta_K) & R_{PS}^{(2)}(\theta_K) & R_{PS}^{(3)}(\theta_K) \\ R_{SP}^{(1)}(\theta_1) & R_{SP}^{(2)}(\theta_1) & R_{SP}^{(3)}(\theta_1) \\ \vdots & \vdots & \vdots \\ R_{SP}^{(1)}(\theta_K) & R_{SP}^{(2)}(\theta_K) & R_{SP}^{(3)}(\theta_K) \\ R_{SS}^{(1)}(\theta_1) & R_{SS}^{(2)}(\theta_1) & R_{SS}^{(3)}(\theta_1) \\ \vdots & \vdots & \vdots \\ R_{SS}^{(1)}(\theta_K) & R_{SS}^{(2)}(\theta_K) & R_{SS}^{(3)}(\theta_K) \end{bmatrix}$$

$$(3.31)$$

The least squares estimate of the model parameters is then $\mathbf{m}^{est} = [\mathbf{G}^T\mathbf{G}]^{-1}\mathbf{G}^T\mathbf{d}^{obs}$ (Fig. 3.7B)

3.6 THE EXISTENCE OF THE LEAST SQUARES SOLUTION

The least squares solution arose from consideration of an inverse problem that had no exact solution. Since there was no exact solution, we chose to do the next best thing: to estimate the solution by those values of the model parameters that gave the best approximate solution (where "best" meant minimizing the L_2 prediction error). By writing a single formula $\mathbf{m}^{est} = [\mathbf{G}^T\mathbf{G}]^{-1}\mathbf{G}^T\mathbf{d}$, we implicitly assumed that there was only one such "best" solution. As we shall prove later, least squares fails if the number of solutions that give the same minimum prediction error is greater than one.

To see that least squares fails for problems with nonunique solutions, consider the straight line problem with only one data point (Fig. 3.8). It is clear that this problem is nonunique; many possible lines can pass through the point, and each has zero prediction error. The solution then contains the following expression:

$$[\mathbf{G}^T\mathbf{G}]^{-1} = \begin{bmatrix} N & \sum_{i=1}^{N} z_i \\ \sum_{i=1}^{N} z_i & \sum_{i=1}^{N} z_i^2 \end{bmatrix}^{-1} \rightarrow \begin{bmatrix} 1 & z_1 \\ z_1 & z_1^2 \end{bmatrix}^{-1} \tag{3.32}$$

The inverse of a matrix is proportional to the reciprocal of the determinant of the matrix so that

$$[\mathbf{G}^T\mathbf{G}]^{-1} \propto \left(\det \begin{bmatrix} 1 & z_1 \\ z_1 & z_1^2 \end{bmatrix} \right)^{-1} = \frac{1}{z_1^2 - z_1^2} \tag{3.33}$$

This expression clearly is singular. The formula for the least squares solution fails.

The question of whether the equation $\mathbf{Gm} = \mathbf{d}$ provides enough information to specify uniquely the model parameters serves as a basis for classifying inverse problems. A classification system based on this criterion is discussed in Sections 3.6.1–3.6.3.

3.6.1 Underdetermined Problems

When the equation $\mathbf{Gm} = \mathbf{d}$ does not provide enough information to determine uniquely all the model parameters, the problem is said to be *underdetermined*. As we saw in the earlier

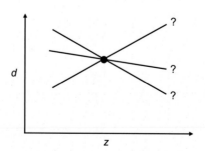

FIG. 3.8   An infinity of different lines can pass through a single point. The prediction error for each is $E = 0$.

example, this can happen if there are several solutions that have zero prediction error. From elementary linear algebra, we know that underdetermined problems occur when there are more unknowns than data, that is, when $M > N$. We must note, however, that there is no special reason why the prediction error must be zero for an underdetermined problem. Frequently, the data uniquely determine some of the model parameters but not others. For example, consider the acoustic experiment in Fig. 3.9. Since no measurements are made of the acoustic slowness in the second brick, it is clear that this model parameter is completely unconstrained by the data. In contrast, the acoustic slowness of the first brick is *overdetermined*, since in the presence of measurement noise, no choice of s_1 can satisfy the data exactly. The equation describing this experiment is

$$
h \begin{bmatrix} 1 & 0 \\ 1 & 0 \\ \vdots & \vdots \\ 1 & 0 \end{bmatrix} \begin{bmatrix} s_1 \\ s_2 \end{bmatrix} = \begin{bmatrix} d_1 \\ d_2 \\ \vdots \\ d_N \end{bmatrix}
\tag{3.34}
$$

where s_i is the slowness in the ith brick, h the brick width, and the d_i the measurements of travel time. If one were to attempt to solve this problem with least squares, one would find that the term $[\mathbf{G}^T\mathbf{G}]^{-1}$ is singular. Even though $M < N$, the problem is still underdetermined since the data kernel has a very poor structure. Although this is a rather trivial case in which only some of the model parameters are underdetermined, in realistic experiments, the problem arises in more subtle forms.

We shall refer to underdetermined problems that have nonzero prediction error as *mixed-determined problems*, to distinguish them from *purely underdetermined problems* that have zero prediction error.

3.6.2 Even-Determined Problems

In even-determined problems, there is exactly enough information to determine the model parameters. There is only one solution, and it has zero prediction error.

3.6.3 Overdetermined Problems

When there is too much information contained in the equation $\mathbf{Gm} = \mathbf{d}$ for it to possess an exact solution, we speak of it as being *overdetermined*. This is the case in which we can employ

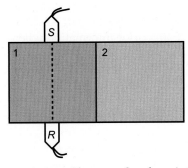

FIG. 3.9 An acoustic travel time experiment with source S and receiver R in a medium consisting of two discrete bricks. Because of poor experimental geometry, the acoustic waves *(dashed line)* pass only through brick 1. The slowness of brick 2 is completely underdetermined.

least squares to select a "best" approximate solution. Overdetermined problems typically have more data than unknowns, that is, $N > M$, although for the reasons discussed earlier it is possible to have problems that are to some degree overdetermined even when $N < M$ and to have problems that are to some degree underdetermined even when $N > M$.

To deal successfully with the full range of inverse problems, we shall need to be able to characterize whether an inverse problem is under- or overdetermined (or some combination of the two). We shall develop quantitative methods for making this characterization in Chapter 7. For the moment, we assume that it is possible to characterize the problem intuitively on the basis of the kind of experiment the problem represents.

3.7 THE PURELY UNDERDETERMINED PROBLEM

Suppose that an inverse problem $\mathbf{Gm} = \mathbf{d}$ has been identified as one that is purely underdetermined. For simplicity, assume that there are fewer equations than unknown model parameters, that is, $N < M$, and that there are no inconsistencies in these equations. It is therefore possible to find more than one solution for which the prediction error E is zero. (In fact, we shall show that underdetermined linear inverse problems have an infinite number of such solutions.) Although the data provide information about the model parameters, they do not provide enough to determine them uniquely.

We must have some means of singling out precisely one of the infinite number of solutions with zero prediction error E to obtain a unique solution $\mathbf{m}^{\text{est}}$ to the inverse problem. To do this, we must add to the problem some information not contained in the equation $\mathbf{Gm} = \mathbf{d}$. This extra information is called prior (or *a priori*) information (Jackson, 1979). Prior information can take many forms, but in each case, it quantifies expectations about the character of the solution that are not based on the actual data.

For instance, in the case of fitting a straight line through a single data point, one might have the expectation that the line also passes through the origin. This prior information now provides enough information to solve the inverse problem uniquely, since two points (one datum, one prior) determine a line.

Another example of prior information concerns expectations that the model parameters possess a given sign or lie in a given range. For instance, suppose the model parameters represent density at different points in the earth. Even without making any measurements, one can state with certainty that the density is everywhere positive, since density is an inherently positive quantity. Furthermore, since the interior of the earth can reasonably be assumed to be rock, its density must have values in some range known to characterize rock, say, between 1000 and 10,000 kg/m³. If one can use this prior information when solving the inverse problem, it may greatly reduce the range of possible solutions—or even cause the solution to be unique.

There is something unsatisfying about having to add prior information to an inverse problem to single out a solution. Where does this information come from, and how certain is it? There are no firm answers to these questions. In certain instances, one might be able to identify reasonable prior assumptions; in other instances, one might not. Clearly, the importance of the prior information depends greatly on the *use* one plans for the estimated model parameters. If one simply wants one example of a solution to the problem, the choice of prior information is unimportant. However, if one wants to develop arguments that depend on the uniqueness of the estimates, the validity of the prior assumptions is of paramount importance. These problems are the price one must pay for estimating the model parameters of a nonunique inverse problem. As will be shown in Chapter 6, there are other kinds of "answers" to inverse problems that do not depend on prior information (localized averages, for example). However, these "answers" invariably are not as easily interpretable as estimates of model parameters.

The first kind of prior assumption we shall consider is the expectation that the solution to the inverse problem is *simple*, where the notion of simplicity is quantified by some measure of the length of the solution. One such measure is simply the Euclidean length of the solution, $L = \mathbf{m}^T\mathbf{m} = \sum m_i^2$. A solution is therefore defined to be simple if it is small when measured under the L_2 norm. Admittedly, this measure is perhaps not a particularly realistic measure of simplicity. It can be useful occasionally, and we shall describe shortly how it can be generalized to more realistic measures. One instance in which solution length may be realistic is when the model parameters describe the velocity of various points in a moving fluid. The length L is then a measure of the kinetic energy of the fluid. In certain instances, it may be appropriate to find that velocity field in the fluid that has the smallest possible kinetic energy of those solutions satisfying the data.

We pose the following problem: Find the $\mathbf{m}^{\text{est}}$ that minimizes $L = \mathbf{m}^T\mathbf{m} = \sum m_i^2$ subject to the constraint that $\mathbf{e} = \mathbf{d} - \mathbf{Gm} = 0$. This problem can easily be solved by the method of Lagrange multipliers (see Section 14.1). We minimize the function as

$$\Phi(\mathbf{m}) = L + \sum_{i=1}^{N} \lambda_i e_i = \sum_{i=1}^{M} m_i^2 + \sum_{i=1}^{N} \lambda_i \left[d_i - \sum_{j=1}^{M} G_{ij} m_j \right] \tag{3.35}$$

with respect to m_q, where λ_i are the Lagrange multipliers. Taking the derivatives yields

$$\frac{\partial \Phi}{\partial m_q} = \sum_{i=1}^{M} 2 \frac{\partial m_i}{\partial m_q} m_i - \sum_{i=1}^{N} \lambda_i \sum_{j=1}^{M} G_{ij} \frac{\partial m_j}{\partial m_q} = 2 m_q - \sum_{i=1}^{N} \lambda_i G_{iq} \tag{3.36}$$

Setting this result to zero and rewriting it in matrix notation yield the equation $2\mathbf{m} = \mathbf{G}^T\boldsymbol{\lambda}$, which must be solved along with the constraint equation $\mathbf{Gm} = \mathbf{d}$. Plugging the first equation into the second gives $\mathbf{d} = \mathbf{Gm} = \mathbf{G}[\mathbf{G}^T\boldsymbol{\lambda}/2]$. We note that the matrix $\mathbf{GG}^T$ is a square $N \times N$ matrix. If its inverse exists, we can then solve this equation for the Lagrange multipliers, $\boldsymbol{\lambda} = 2[\mathbf{GG}^T]^{-1}\mathbf{d}$. Then inserting this expression into the first equation yields the solution

$$\mathbf{m}^{\text{est}} = \mathbf{G}^T \left[\mathbf{GG}^T \right]^{-1} \mathbf{d} \tag{3.37}$$

We shall discuss the conditions under which this solution exists later. As we shall see, one condition is that the equation $\mathbf{Gm} = \mathbf{d}$ be purely underdetermined—that it contain no inconsistencies.

3.8 MIXED-DETERMINED PROBLEMS

Most inverse problems that arise in practice are neither completely overdetermined nor completely underdetermined. For instance, in the X-ray tomography problem, there may

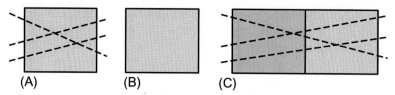

FIG. 3.10   (A) The X-ray opacity of the brick is overdetermined, since measurements of X-ray intensity are made along three different paths *(dashed lines)*. (B) The opacity is underdetermined, since no measurements have been made. (C) The average opacity of these two bricks is overdetermined, but since each path has an equal length in either brick, the individual opacities are underdetermined.

be one box through which several rays pass (Fig. 3.10A). The X-ray opacity of this box is clearly overdetermined. On the other hand, there may be boxes that have been missed entirely (Fig. 3.10B). These boxes are completely undetermined. There may also be boxes that cannot be individually resolved because every ray that passes through one also passes through an equal distance of the other (Fig. 3.10C). These boxes are also underdetermined, since only their mean opacity is determined.

Ideally, we would like to sort the unknown model parameters into two groups: those that are overdetermined and those that are underdetermined. Actually, to do this, we need to form a new set of model parameters that are linear combinations of the old. For example, in the two-box problem above, the average opacity $m'_1 = \frac{1}{2}(m_1 + m_2)$ is completely overdetermined, whereas the difference in opacity $m'_2 = \frac{1}{2}(m_1 + m_2)$ is completely underdetermined. We want to perform this partitioning from an arbitrary equation $\mathbf{Gm} = \mathbf{d} \rightarrow \mathbf{G'm'} = \mathbf{d'}$, where $\mathbf{m'}$ is partitioned into an upper part $\mathbf{m'^o}$ that is overdetermined and a lower part $\mathbf{m'^u}$ that is underdetermined:

$$\begin{bmatrix} \mathbf{G'^o} & 0 \\ 0 & \mathbf{G'^u} \end{bmatrix} \begin{bmatrix} \mathbf{m'^o} \\ \mathbf{m'^u} \end{bmatrix} = \begin{bmatrix} \mathbf{d'^o} \\ \mathbf{d'^u} \end{bmatrix} \tag{3.38}$$

If this can be achieved, we could determine the overdetermined model parameters by solving the upper equations in the least squares sense and determine the underdetermined model parameters by finding those that have minimum L_2 solution length. In addition, we would have found a solution that added as little prior information to the inverse problem as possible.

This partitioning process can be accomplished through singular-value decomposition of the data kernel, a process that we shall discuss in Chapter 7. Since it is a relatively time-consuming process, we first examine an approximate process that works if the inverse problem is not too underdetermined.

Instead of partitioning $\mathbf{m}$, suppose that we determine a solution that minimizes some combination Φ of the prediction error and the solution length for the unpartitioned model parameters:

$$\Phi(\mathbf{m}) = E + \varepsilon^2 L = \mathbf{e}^T \mathbf{e} + \varepsilon^2 \mathbf{m}^T \mathbf{m} \tag{3.39}$$

where the weighting factor ε^2 determines the relative importance given to the prediction error and solution length. If ε is made large enough, this procedure will clearly minimize the underdetermined part of the solution. Unfortunately, it also tends to minimize the overdetermined part of the solution. As a result, the solution will not minimize the prediction error E

and will not be a very good estimate of the true model parameters. If ε is set to zero, the prediction error will be minimized, but no prior information will be provided to single out the underdetermined model parameters. It may be possible, however, to find some compromise value for ε that will approximately minimize E while approximately minimizing the length of the underdetermined part of the solution. There is no simple method of determining what this compromise ε should be (without solving the partitioned problem); it must be determined by trial and error. By minimizing $\Phi(\mathbf{m})$ with respect to the model parameters in a manner exactly analogous to the least squares derivation, we obtain

$$\left[\mathbf{G}^{\mathrm{T}}\mathbf{G} + \varepsilon^2\mathbf{I}\right]\mathbf{m}^{\mathrm{est}} = \mathbf{G}^{\mathrm{T}}\mathbf{d} \quad \text{or} \quad \mathbf{m}^{\mathrm{est}} = \left[\mathbf{G}^{\mathrm{T}}\mathbf{G} + \varepsilon^2\mathbf{I}\right]^{-1}\mathbf{G}^{\mathrm{T}}\mathbf{d} \tag{3.40}$$

This estimate of the model parameters is called the *damped least squares* solution. The concept of error has been generalized to include not only prediction error but also error in fitting the prior information (that the solution length is zero, in this case). The underdeterminacy of the inverse problem is said to have been damped.

3.9 WEIGHTED MEASURES OF LENGTH AS A TYPE OF PRIOR INFORMATION

There are many instances in which $L = \mathbf{m}^{\mathrm{T}}\mathbf{m}$ is not a very good measure of solution simplicity. For instance, suppose that one were solving an inverse problem for density fluctuations in the ocean. One may not want to find a solution that is smallest in the sense of closest to zero but one that is smallest in the sense that it is closest to some other value, such as the average density of sea water. The obvious generalization of L is then

$$L = (\mathbf{m} - \langle\mathbf{m}\rangle)^{\mathrm{T}}(\mathbf{m} - \langle\mathbf{m}\rangle) \tag{3.41}$$

where $\langle\mathbf{m}\rangle$ is the prior value of the model parameters (a known typical value of sea water, in this case).

Sometimes the whole idea of length as a measure of simplicity is inappropriate. For instance, one may feel that a solution is simple if it is flat or if it is smooth. These measures may be particularly appropriate when the model parameters represent a discretized continuous function such as density or X-ray opacity. One may have the expectation that these parameters vary only slowly with position. Fortunately, properties such as *flatness* can be easily quantified by measures that are generalizations of length. For example, the flatness of a continuous function of space can be quantified by the norm of its first derivative, which is a measure of *steepness* (the opposite of flatness). For discrete model parameters, one can use the difference between physically adjacent model parameters as approximations of a derivative. The steepness $\mathbf{l}$ of a vector $\mathbf{m}$ is then

$$\mathbf{l} = \frac{1}{\Delta x}\begin{bmatrix} -1 & 1 & & \\ & -1 & 1 & \\ & & \ddots & \ddots \\ & & & -1 & 1 \end{bmatrix}\begin{bmatrix} m_1 \\ m_2 \\ \vdots \\ m_M \end{bmatrix} = \mathbf{Dm} \tag{3.42}$$

where $\mathbf{D}$ is the *steepness matrix*. Other methods of simplicity can also be represented by a matrix multiplying the model parameters. For instance, solution smoothness can be implemented by quantifying the *roughness* (the opposite of smoothness) by the second derivative. The matrix multiplying the model parameters would then have rows containing $(\Delta x)^{-2}[\cdots \ 1 \ -2 \ 1 \ \cdots]$. The overall steepness or roughness of the solution is then just the length

$$L = \mathbf{l}^T \mathbf{l} = [\mathbf{Dm}]^T [\mathbf{Dm}] = \mathbf{m}^T \mathbf{D}^T \mathbf{Dm} = \mathbf{m}^T \mathbf{W}_m \mathbf{m} \tag{3.43}$$

The matrix $\mathbf{W}_m = \mathbf{D}^T \mathbf{D}$ can be interpreted as a weighting factor that enters into the calculation of the length of the vector $\mathbf{m}$. Note, however, that $\|\mathbf{m}\|^2_{\text{weighted}} = \mathbf{m}^T \mathbf{W}_m \mathbf{m}$ is *not* a proper norm, since it violates the positivity condition given in Eq. (3.3a); that is, $\|\mathbf{m}\|^2_{\text{weighted}} = 0$ for some nonzero vectors (such as the constant vector). This behavior usually poses no insurmountable problems, but it can cause solutions based on minimizing this norm to be nonunique.

The measure of solution simplicity can therefore be generalized to

$$L = [\mathbf{m} - \langle \mathbf{m} \rangle]^T \mathbf{W}_m [\mathbf{m} - \langle \mathbf{m} \rangle] \tag{3.44}$$

By suitably choosing the prior model vector $\langle \mathbf{m} \rangle$ and the weighting matrix $\mathbf{W}_m$, we can quantify a wide variety of measures of simplicity.

Weighted measures of the prediction error can also be useful. Frequently, some observations are made with more accuracy than others. In this case, one would like the prediction error e_i of the more accurate observations to have a greater weight in the quantification of the overall error E than the inaccurate observations. To accomplish this weighting, we define a generalized prediction error

$$E = \mathbf{e}^T \mathbf{W}_e \mathbf{e} \tag{3.45}$$

where the matrix $\mathbf{W}_e$ defines the relative contribution of each individual error to the total prediction error. Normally we would choose this matrix to be diagonal. For example, if $N = 5$ and the third observation is known to be twice as accurately determined as the others, one might use

$$\mathbf{W}_e = \begin{bmatrix} 1 & & & & \\ & 1 & & & \\ & & 2 & & \\ & & & 1 & \\ & & & & 1 \end{bmatrix} \tag{3.46}$$

The inverse problem solutions stated earlier can then be modified to take into account these new measures of prediction error and solution simplicity. The derivations are substantially the same as for the unweighted cases but the algebra is lengthy.

3.9.1 Weighted Least Squares

If the equation $\mathbf{Gm} = \mathbf{d}$ is completely overdetermined, then one can estimate the model parameters by minimizing the generalized prediction error $E = \mathbf{e}^T \mathbf{W}_e \mathbf{e}$. This procedure leads to the solution

$$\mathbf{m}^{\text{est}} = [\mathbf{G}^T \mathbf{W}_e \mathbf{G}]^{-1} \mathbf{G}^T \mathbf{W}_e \mathbf{d} \tag{3.47}$$

3.9.2 Weighted Minimum Length

If the equation $\mathbf{Gm}=\mathbf{d}$ is completely underdetermined, then one can estimate the model parameters by choosing the solution that is simplest, where simplicity is defined by the generalized length $L=[\mathbf{m}-\langle\mathbf{m}\rangle]^{\mathrm{T}}\mathbf{W}_{\mathrm{m}}[\mathbf{m}-\langle\mathbf{m}\rangle]^{\mathrm{T}}$. This procedure leads to the solution

$$\mathbf{m}^{\mathrm{est}} = \langle\mathbf{m}\rangle + \mathbf{W}_{\mathrm{m}}^{-1}\mathbf{G}^{\mathrm{T}}\left[\mathbf{GW}_{\mathrm{m}}^{-1}\mathbf{G}^{\mathrm{T}}\right]^{-1}[\mathbf{d}-\mathbf{G}\langle\mathbf{m}\rangle] \tag{3.48}$$

3.9.3 Weighted Damped Least Squares

If the equation $\mathbf{Gm}=\mathbf{d}$ is slightly underdetermined, it can be solved by minimizing a combination of prediction error and solution length, $\Phi(\mathbf{m})=E+\varepsilon^2 L$ (Franklin, 1970; Jackson, 1972; Jordan and Franklin, 1971). The parameter ε is chosen by trial and error to yield a solution that has a reasonably small prediction error. The equation for the solution, obtained by minimizing Φ with respect to $\mathbf{m}$, is then

$$\left[\mathbf{G}^{\mathrm{T}}\mathbf{W}_{\mathrm{e}}\mathbf{G} + \varepsilon^2\mathbf{W}_{\mathrm{m}}\right]\mathbf{m}^{\mathrm{est}} = \mathbf{G}^{\mathrm{T}}\mathbf{W}_{\mathrm{e}}\mathbf{d} + \varepsilon^2\mathbf{W}_{\mathrm{m}}\langle\mathbf{m}\rangle \tag{3.49}$$

or

$$\mathbf{m}^{\mathrm{est}} = \left[\mathbf{G}^{\mathrm{T}}\mathbf{W}_{\mathrm{e}}\mathbf{G} + \varepsilon^2\mathbf{W}_{\mathrm{m}}\right]^{-1}\left[\mathbf{G}^{\mathrm{T}}\mathbf{W}_{\mathrm{e}}\mathbf{d} + \varepsilon^2\mathbf{W}_{\mathrm{m}}\langle\mathbf{m}\rangle\right] \tag{3.50}$$

This equation appears to be rather complicated. However, it can be vastly simplified by noting that it is equivalent to solving the equation

$$\mathbf{Fm}^{\mathrm{est}} = \mathbf{f} \ \text{ with } \mathbf{F}=\begin{bmatrix}\mathbf{W}_{\mathrm{e}}^{1/2}\mathbf{G}\\ \varepsilon\mathbf{D}\end{bmatrix} \ \text{ and } \ \mathbf{f}=\begin{bmatrix}\mathbf{W}_{\mathrm{e}}^{1/2}\mathbf{d}\\ \varepsilon\mathbf{D}\langle\mathbf{m}\rangle\end{bmatrix} \ \text{ and } \ \mathbf{W}_{\mathrm{m}}=\mathbf{D}^{\mathrm{T}}\mathbf{D} \tag{3.51}$$

by simple least squares; that is, the equation $\mathbf{F}^{\mathrm{T}}\mathbf{Fm}^{\mathrm{est}}=\mathbf{F}^{\mathrm{T}}\mathbf{f}$, when multiplied out, is identical to Eq. (3.50). As explained previously, the weight matrix $\mathbf{W}_{\mathrm{e}}$ is typically diagonal. In that case, its square root, $\mathbf{W}_{\mathrm{e}}^{1/2}$, is also diagonal with elements that are the square roots of the corresponding elements of $\mathbf{W}_{\mathrm{e}}$.

Eq. (3.46) has a very simple interpretation: its top row is the data equation $\mathbf{Gm}^{\mathrm{est}}=\mathbf{d}$, with both sides multiplied by the weight matrix $\mathbf{W}_{\mathrm{e}}^{1/2}$, and its bottom row is the prior equation, $\mathbf{m}^{\mathrm{est}}=\langle\mathbf{m}\rangle$, with both sides multiplied by the prior matrix, $\varepsilon\mathbf{D}$. Note that the data and prior information play completely symmetric roles in this equation.

Eq. (3.51) is extremely well suited to computations, especially if a sparse matrix is used for $\mathbf{F}$. As an example, suppose that $\mathbf{m}$ represents the values of a function $m(z)$ at evenly spaced zs, but that only a few of these ms have been observed. The data equation is then just $m_i=d_j$, where the indices i and j "match up" the observation with the corresponding model parameter. The ith row of the data kernel matrix $\mathbf{G}$ is all zero, except for a single one in the jth column. Since the observations are insufficient to determine all the model parameters, we add prior information of smoothness using a roughness matrix $\mathbf{D}$. Each row of $\mathbf{D}$ is mostly zeros, except for the sequence $[1 \ -2 \ 1]$, with the -2 centered on the model parameter whose second derivative is being computed. We can only form $M-2$ of these rows, since computing the second derivative of m_1 or m_M would require model parameters off the ends of $\mathbf{m}$. We choose to add prior information of flatness at these two points, with a steepness matrix $\mathbf{D}$ with rows containing the sequence $[-1\ 1]$. In both the roughness and steepness case, the vector $\mathbf{D}\langle\mathbf{m}\rangle$ is taken to be zero, since the solution is taken to be smooth and flat. This leads to an equation $\mathbf{Fm}=\mathbf{f}$ of the form

$$
\mathbf{F} = \begin{bmatrix}
 & 1 & & & & & & & & & \\
\cdots & \cdots & \cdots & \cdots & \cdots & \cdots & \cdots & \cdots & \cdots & \cdots & \cdots \\
 & & & 1 & & & & & & & \\
\hline
a & -2a & a & & & & & & & & \\
\cdots & \cdots & \cdots & \cdots & \cdots & \cdots & \cdots & \cdots & \cdots & \cdots & \cdots \\
 & & & & & & a & -2a & a & & \\
\hline
-b & b & & & & & & & & & \\
 & & & & & & & & -b & b &
\end{bmatrix}
\quad \text{and} \quad
\mathbf{f} = \begin{bmatrix}
d_1 \\
\vdots \\
d_N \\
\hline
0 \\
\vdots \\
0 \\
\hline
0 \\
0
\end{bmatrix}
\tag{3.52}
$$

Here $a=\varepsilon(\Delta x)^{-2}$ and $b=\varepsilon(\Delta x)^{-1}$. This equation can be solved using the biconjugate gradient algorithm, using the *MatLab* code. An example is shown in Fig. 3.11.

```
tol = 1e-6;
maxit = 3*M;
mest = bicg( @weightedleastsquaresfcn, F'*f, tol, maxit );
```

(*MatLab* script gda03_08)

The function `weightedleastsquaresfcn()`, which performs the multiplication $\mathbf{F}^{\mathrm{T}}(\mathbf{Fv})$, is similar to the `leastsquaresfcn()` discussed previously.

Prior information of flatness (small first derivative) and smoothness (small second derivative) are *not* the same, but both result in a solution that is qualitatively "smooth." The difference between them can understood by studying the *data smoothing problem*, in which a smooth estimate $m(x)$ of a function is reconstructed from noisy observations $d(x)$ of the same

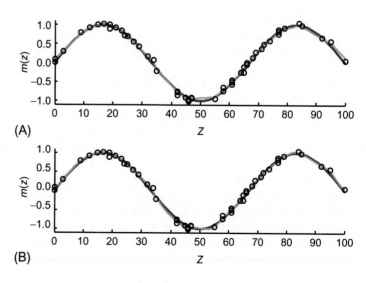

FIG. 3.11 Examples of weighted damped least squares applied to the problem of filling in data gaps, for two different types of prior information. (A) *(Red curve)* The true model is a sinusoid sampled with $\Delta z=1$. *(Black circles)* The data are noisy observations of the model at just a few points. *(Green curve)* The estimated model is reconstructed from the data using the prior information of flatness (small first derivative). (B) Same as (A) except using the prior information of smoothness (small second derivative) in the interior of the (0,100) interval and flatness at its ends. *MatLab* script gda03_09.

function. This inverse problem is a special case of the one in the previous example, with no missing data, so that the data kernel is simply $\mathbf{G}=\mathbf{I}$. The prior information has the form $\mathbf{Dm}\approx 0$, where $\mathbf{D}$ is a matrix that approximates either the first or second derivative. The weighted damped least squares solution is then:

$$\mathbf{m}=\left[\mathbf{I}+\varepsilon^2\mathbf{D}^{\mathrm{T}}\mathbf{D}\right]^{-1}\mathbf{d} \tag{3.53}$$

where ε quantifies the strength of the prior information, and where, for simplicity, we have assume that $\mathbf{W}_{\mathrm{e}}=\mathbf{I}$. Menke and Eilon (2015) derive an approximate solution to this equation when the data vector consists of a single spike at the origin; that is, $d(x)=\delta(x)$. The solution $m(x)$ then represents how the spike is smoothed by the prior information. In the case of first derivative smoothing, the solution is:

$$m(x)=\frac{\varepsilon^{-1}}{2}\exp\left(-\varepsilon^{-1}|x|\right) \tag{3.54}$$

and in the case of second derivative smoothing, it is:

$$m(x)=\left(\frac{a^3}{8\varepsilon^2}\right)\exp\left(-\frac{|x|}{a}\right)\left\{\cos\left(\frac{|x|}{a}\right)+\sin\left(\frac{|x|}{a}\right)\right\} \text{ with } a=(2\varepsilon)^{1/2} \tag{3.55}$$

Both of these *smoothing operators* have the desirable properties of being symmetric about $x=0$ and having unit area and so conserving the area under $d(x)$. The first derivative version (Fig. 3.12A) is everywhere positive (a good feature, because negative values cause overshoots in the smoothing) but has a cusp at the origin (a bad feature, since it adversely affects the power spectrum of the smoothed data). The second derivative version (Fig. 3.12B) has negative side lobes (bad) but lacks cusps (good). While both versions qualitatively smooth the data, their properties are subtly different and the choice of one over the other should take into consideration these differences.

Eq. (3.50) can be manipulated into another useful form by subtracting $[\mathbf{G}^{\mathrm{T}}\mathbf{W}_{\mathrm{e}}\mathbf{G}+\varepsilon^2\mathbf{W}_{\mathrm{m}}]\langle\mathbf{m}\rangle$ from both sides of the equation and rearranging to obtain

$$\left[\mathbf{m}^{\mathrm{est}}-\langle\mathbf{m}\rangle\right]=\left[\mathbf{G}^{\mathrm{T}}\mathbf{W}_{\mathrm{e}}\mathbf{G}+\varepsilon^2\mathbf{W}_{\mathrm{m}}\right]^{-1}\mathbf{G}^{\mathrm{T}}\mathbf{W}_{\mathrm{e}}[\mathbf{d}-\mathbf{G}\langle\mathbf{m}\rangle] \tag{3.56}$$

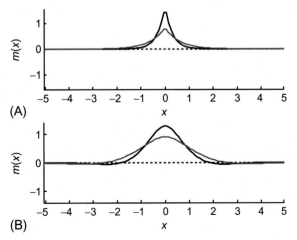

FIG. 3.12 Smoothing of a unit spike at $x=0$. (A) First derivative smoothing for $\varepsilon=0.3$ *(black)* and 0.6 *(red)*. (B) Second derivative smoothing for $\varepsilon=0.3$ *(black)* and 0.6 *(red)*. *MatLab* script gda03_10.

This equation is of the form

$$\Delta \mathbf{m} = \mathbf{M} \Delta \mathbf{d} \quad \text{with } \Delta \mathbf{d} = \mathbf{d} - \mathbf{G}\langle \mathbf{m} \rangle \quad \text{and}$$
$$\Delta \mathbf{m} = [\mathbf{m}^{\text{est}} - \langle \mathbf{m} \rangle] \quad \text{and} \quad \mathbf{M} = [\mathbf{G}^{\mathrm{T}} \mathbf{W}_{\mathbf{e}} \mathbf{G} + \varepsilon^2 \mathbf{W}_{\mathbf{m}}]^{-1} \mathbf{G}^{\mathrm{T}} \mathbf{W}_{\mathbf{e}} \tag{3.57}$$

This form emphasizes that the *deviation* $\Delta \mathbf{m}$ of the estimated solution from the prior value is a linear function of the deviation $\Delta \mathbf{d}$ of the data from the value predicted by the prior model.

Finally, we note that an alternative form of $\mathbf{M}$ in Eq. (3.57), reminiscent of the minimum-length solution, is

$$\mathbf{M} = \mathbf{W}_{\mathbf{m}}^{-1} \mathbf{G}^{\mathrm{T}} [\mathbf{G} \mathbf{W}_{\mathbf{m}}^{-1} \mathbf{G}^{\mathrm{T}} + \varepsilon^2 \mathbf{W}_{\mathbf{e}}^{-1}]^{-1} \tag{3.58}$$

The equivalence can be demonstrated by equating the two forms of $\mathbf{M}$, premultiplying by $[\mathbf{G}^{\mathrm{T}} \mathbf{W}_{\mathbf{e}} \mathbf{G} + \varepsilon^2 \mathbf{W}_{\mathbf{m}}]$ and postmultiplying by $[\mathbf{G} \mathbf{W}_{\mathbf{m}}^{-1} \mathbf{G}^{\mathrm{T}} + \varepsilon^2 \mathbf{W}_{\mathbf{e}}^{-1}]$. In both instances, one must take care to ascertain whether the inverses actually exist. Depending on the choice of the weighting matrices, sufficient prior information may or may not have been added to the problem to damp the underdeterminacy.

3.10 OTHER TYPES OF PRIOR INFORMATION

One commonly encountered type of prior information is the knowledge that some function of the model parameters equals a constant. Linear equality constraints of the form $\mathbf{Hm} = \mathbf{h}$ are particularly easy to implement. For example, one such linear constraint requires that the mean of the model parameters must equal some value h_1:

$$\mathbf{Hm} = \frac{1}{M} [1 \; 1 \; \cdots \; 1] \begin{bmatrix} m_1 \\ m_2 \\ \vdots \\ m_M \end{bmatrix} = [h_1] = \mathbf{h} \tag{3.59}$$

Another such constraint requires that a particular model parameter, m_k, equals a given value

$$\mathbf{Hm} = [0 \; \cdots \; 0 \; 1 \; 0 \; \cdots \; 0] \begin{bmatrix} m_1 \\ \vdots \\ m_k \\ \vdots \\ m_M \end{bmatrix} = [h_k] = \mathbf{h} \tag{3.60}$$

One problem that frequently arises is to solve an inverse problem $\mathbf{Gm} = \mathbf{d}$ in the least squares sense with the prior constraint that linear relationships between the model parameters of the form $\mathbf{Hm} = \mathbf{h}$ are satisfied exactly.

One way to implement this constraint is to use weighted damped least squares (Eq. 3.51), with $\mathbf{D} = \mathbf{H}$ and $\mathbf{D}\langle \mathbf{m} \rangle = \mathbf{h}$, and with the weighting factor ε chosen to be very large, so that the prior equations are given much more weight than the data equations (Lanczos, 1961). This method is well suited for computation, but it does require the value of ε to be chosen with some care—too big and the solution will suffer from numerical noise; too small and the constraints will be only very approximately satisfied.

Another method of implementing the constraints is through the use of Lagrange multipliers. One minimizes $E = \mathbf{e}^T\mathbf{e}$ with the constraint that $\mathbf{Hm} - \mathbf{h} = 0$ by forming the function

$$\Phi(m) = \sum_{i=1}^{N}\left[\sum_{j=1}^{M}G_{ij}m_j - d_i\right]^2 + 2\sum_{i=1}^{p}\lambda_i\left[\sum_{j=1}^{M}H_{ij}m_j - h_i\right] \qquad (3.61)$$

(where there are p constraints and $2\lambda_i$ are the Lagrange multipliers) and setting its derivatives with respect to the model parameters to zero as

$$\frac{\partial\Phi(\mathbf{m})}{\partial m_q} = 2\sum_{i=1}^{M}m_i\sum_{j=1}^{N}G_{jq}G_{ji} - 2\sum_{i=1}^{N}G_{iq}d_i - 2\sum_{i=1}^{p}\lambda_iH_{iq} \qquad (3.62)$$

This equation must be solved simultaneously with the constraint equations $\mathbf{Hm} = \mathbf{h}$ to yield the estimated solution. These equations, in matrix form, are

$$\begin{bmatrix} \mathbf{G}^T\mathbf{G} & \mathbf{H}^T \\ \mathbf{H} & 0 \end{bmatrix}\begin{bmatrix} \mathbf{m} \\ \boldsymbol{\lambda} \end{bmatrix} = \begin{bmatrix} \mathbf{G}^T\mathbf{d} \\ \mathbf{h} \end{bmatrix} \qquad (3.63)$$

Although these equations can be manipulated to yield an explicit formula for $\mathbf{m}^{\text{est}}$, it is often more convenient to solve directly this $M + p$ system of equations for M estimates of model parameters and p Lagrange multipliers by premultiplying by the inverse of the square matrix.

3.10.1 Example: Constrained Fitting of a Straight Line

Consider the problem of fitting the straight line $d_i = m_1 + m_2z_i$ to data, where one has prior information that the line must pass through the point (z',d') (Fig. 3.13). There are two model parameters: intercept m_1 and slope m_2. The $p = 1$ constraint is that $d' = m_1 + m_2z'$, or

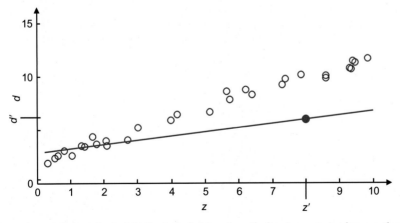

FIG. 3.13   Least squares fitting of a straight line to (z,d) data, where the line is constrained to pass through the point $(z',d') = (8,6)$. MatLab script gda03_11.

$$\mathbf{Hm} = \begin{bmatrix} 1 & z' \end{bmatrix} \begin{bmatrix} m_1 \\ m_2 \end{bmatrix} = [d'] = \mathbf{h} \tag{3.64}$$

Using the $\mathbf{G}^T\mathbf{G}$ and $\mathbf{G}^T\mathbf{d}$ computed in Section 3.5.1, the solution is

$$\begin{bmatrix} m_1^{\text{est}} \\ m_2^{\text{est}} \\ \lambda_1 \end{bmatrix} = \begin{bmatrix} N & \sum_{i=1}^{N} z_i & 1 \\ \sum_{i=1}^{N} z_i & \sum_{i=1}^{N} z_i^2 & z' \\ 1 & z' & 0 \end{bmatrix}^{-1} \begin{bmatrix} \sum_{i=1}^{N} d_i \\ \sum_{i=1}^{N} z_i d_i \\ d' \end{bmatrix} \tag{3.65}$$

Another kind of prior constraint is the *linear inequality constraint*, which we can write as $\mathbf{Hm} \geq \mathbf{h}$ (the inequality being interpreted component by component). Note that this form can also include $\leq$ inequalities by multiplying the inequality relation by -1. This kind of prior constraint has application to problems in which the model parameters are inherently positive quantities, $m_i > 0$, and to other cases when the solution is known to possess some kind of bounds. One could therefore propose a new kind of constrained least squares solution of overdetermined problems, one that minimizes the error subject to the given inequality constraints. Prior inequality constraints also have application to underdetermined problems. One can find the smallest solution that solves both $\mathbf{Gm} = \mathbf{d}$ and $\mathbf{Hm} \geq \mathbf{h}$. These problems can be solved in a straightforward fashion, which will be discussed in Chapter 7.

3.11 THE VARIANCE OF THE MODEL PARAMETER ESTIMATES

The data invariably contain noise that causes errors in the estimates of the model parameters. We can calculate how this measurement error *maps* into errors in $\mathbf{m}^{\text{est}}$ by noting that all of the formulas derived above for estimates of the model parameters are linear functions of the data, of the form $\mathbf{m}^{\text{est}} = \mathbf{Md} + \mathbf{v}$, where $\mathbf{M}$ is some matrix and $\mathbf{v}$ some vector. Therefore, if we assume that the data have a distribution characterized by some covariance matrix $[\text{cov}\,\mathbf{d}]$, the estimates of the model parameters have a distribution characterized by a covariance matrix $[\text{cov}\,\mathbf{m}] = \mathbf{M}[\text{cov}\,\mathbf{d}]\mathbf{M}^T$. The covariance of the solution can therefore be calculated in a straightforward fashion. If the data are uncorrelated and of equal variance σ_d^2, then very simple formulas are obtained for the covariance of some of the more simple inverse problem solutions.

The simple least squares solution $\mathbf{m}^{\text{est}} = [\mathbf{G}^T\mathbf{G}]^{-1}\mathbf{G}^T\mathbf{d}$ has covariance

$$[\text{cov}\,\mathbf{m}] = \left[[\mathbf{G}^T\mathbf{G}]^{-1}\mathbf{G}^T \right] \sigma_d^2 \mathbf{I} \left[[\mathbf{G}^T\mathbf{G}]^{-1}\mathbf{G}^T \right]^T = \sigma_d^2 [\mathbf{G}^T\mathbf{G}]^{-1} \tag{3.66}$$

and the simple minimum-length solution $\mathbf{m}^{\text{est}} = \mathbf{G}^T[\mathbf{GG}^T]^{-1}\mathbf{d}$ has covariance

$$[\text{cov}\,\mathbf{m}] = \left[\mathbf{G}^T[\mathbf{GG}^T]^{-1} \right] \sigma_d^2 \mathbf{I} \left[\mathbf{G}^T[\mathbf{GG}^T]^{-1} \right]^T = \sigma_d^2 \mathbf{G}^T[\mathbf{GG}^T]^{-2}\mathbf{G} \tag{3.67}$$

An important issue is how to arrive at an estimate of the variance of the data σ_d^2 that can be used in these equations. One possibility is to base it upon knowledge about the inherent accuracy of the measurement process, in which case it is termed a *prior variance*. For instance, if lengths are being measured with a ruler with 1 mm divisions, the estimate $\sigma_d \approx 1/2\,\text{mm}$

would be reasonable. Another possibility is to base the estimate upon the size distribution of prediction errors **e** determined by fitting a model to the data, in which case it is termed a *posterior* (or *a posteriori*) variance. A reasonable estimate, whose theoretical justification will be discussed in Chapter 5, is

$$\sigma_d^2 \approx \frac{1}{N-M} \sum_{i=1}^{N} e_i^2 \tag{3.68}$$

This formula is essentially the mean-squared error $N^{-1} \sum_{i=1}^{N} e_i^2$, except that N has been replaced by $N-M$ to account for the ability of a model with M parameters to exactly fit M data. Posterior estimates are usually overestimates because inaccuracies in the model contribute to the size of the prediction error.

The least squares rule for error propagation, $[\text{cov } \mathbf{m}] = \sigma_d^2 [\mathbf{G}^T \mathbf{G}]^{-1}$, indicates that the model parameters can be correlated and can be of unequal variance even when the data are uncorrelated and are of equal variance. Whether observational error is attenuated or amplified by the inversion process is critically dependent upon the structure of the data kernel **G**. In the problem for the mean of N data, discussed earlier, observational error is attenuated, but this desirable behavior is not common to all inverse problems (Fig. 3.14).

The reflection coefficient problem discussed in Section 3.5.4 has true model parameters $\mathbf{m}^{\text{true}} = [0.07, 0.10, 0.11]^T$ and prior data variance $\sigma_d^2 = 0.0050$. After solving the inverse problem, the estimated model parameters, posterior data variance, and model covariance matrix are found to be:

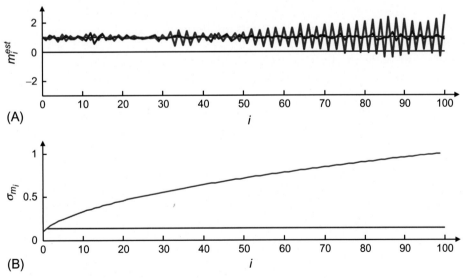

FIG. 3.14 Two hypothetical experiments to measure the weight m_i of each of 100 bricks. In experiment 1 *(red)*, the bricks are accumulated on a scale so that observation d_i is the sum of the weight of the first i bricks. In experiment 2 *(blue)*, the first brick is weighed, and then subsequently, pairs or bricks (the first and the second, the second and the third, and so forth). (A) Least squares solution for weights m_i. (B) Corresponding error σ_{m_i}. Note that the first experiment has the lower error. *MatLab* script gda03_12.

$$\mathbf{m}^{\text{est}} = \begin{bmatrix} 0.08 \\ 0.09 \\ 0.10 \end{bmatrix} \text{ and } \left(\sigma_d^{\text{est}}\right)^2 = (0.069)^2 \text{ and } [\text{cov}\,\mathbf{m}] = (10^{-5}) \begin{bmatrix} 7 & -7 & -8 \\ -7 & 8 & 8 \\ -8 & 8 & 9 \end{bmatrix} \qquad (3.69)$$

As will be discussed further in Chapter 5, the 95% confidence intervals of estimated model parameter m_i are $\pm 2\sigma_{m_i}$, where the σ_{m_i}'s are the square roots of the diagonal elements of [cov $\mathbf{m}$]. The solution, with its confidence intervals, is written as:

$$m_1^{\text{est}} = 0.08 \pm 0.02\,(95\%)$$
$$m_2^{\text{est}} = 0.09 \pm 0.02\,(95\%) \qquad (3.70)$$
$$m_3^{\text{est}} = 0.10 \pm 0.02\,(95\%)$$

Alternatively, these confidence limits can be illustrated as error bars, as in Fig. 3.7B.

3.12 VARIANCE AND PREDICTION ERROR OF THE LEAST SQUARES SOLUTION

If the prediction error $E(\mathbf{m}) = \mathbf{e}^{\text{T}}\mathbf{e}$ of an overdetermined problem has a very sharp minimum in the vicinity of the estimated solution $\mathbf{m}^{\text{est}}$, we would expect that the solution is well determined in the sense that it has small variance. Small errors in determining the shape of $E(\mathbf{m})$ due to random fluctuations in the data lead to only small errors in $\mathbf{m}^{\text{est}}$. Conversely, if $E(\mathbf{m})$ has a broad minimum, we expect that $\mathbf{m}^{\text{est}}$ has a large variance. Since the curvature of a function is a measure of the sharpness of its minimum, we expect that the variance of the solution is related to the curvature of $E(\mathbf{m})$ at its minimum, which in turn depends on the structure of the data kernel $\mathbf{G}$ (Fig. 3.15).

The curvature of the prediction error can be measured by its second derivative, as we can see by computing how small changes in the model parameters change the prediction error. Expanding the prediction error in a Taylor series about its minimum and keeping up to second-order terms give

$$\Delta E = E(\mathbf{m}) - E\left(\mathbf{m}^{\text{est}}\right) = \left[\mathbf{m} - \mathbf{m}^{\text{est}}\right]^{\text{T}} \left[\frac{1}{2}\frac{\partial^2 E}{\partial \mathbf{m}^2}\right]_{\mathbf{m}=\mathbf{m}^{\text{est}}} \left[\mathbf{m} - \mathbf{m}^{\text{est}}\right] \qquad (3.71)$$

Here the matrix $\frac{\partial^2 E}{\partial \mathbf{m}^2}$ has elements $\frac{\partial^2 E}{\partial m_i \partial m_j}$. Note that the first-order term is zero, since the expansion is made at a minimum. The second derivative can also be computed directly from the expression

$$E(\mathbf{m}) = \|\mathbf{d} - \mathbf{G}\mathbf{m}\|_2^2 = \sum_{i=1}^{N} d_i^2 - 2\sum_{i=1}^{N} d_i \sum_{j=1}^{M} G_{ij}m_j + \sum_{i=1}^{N}\sum_{j=1}^{M} G_{ij}m_j \sum_{k=1}^{M} G_{ik}m_k \qquad (3.72)$$

which gives

$$\frac{\partial^2 E}{\partial m_p \partial m_q} = 2\sum_{i=1}^{N}\sum_{j=1}^{M} G_{ij}\frac{\partial m_j}{\partial m_p}\sum_{k=1}^{M} G_{ik}\frac{\partial m_k}{\partial m_q} = 2\sum_{i=1}^{N} G_{ip}G_{iq} \text{ or } \left[\frac{1}{2}\frac{\partial^2 E}{\partial \mathbf{m}^2}\right] = \mathbf{G}^{\text{T}}\mathbf{G} \qquad (3.73)$$

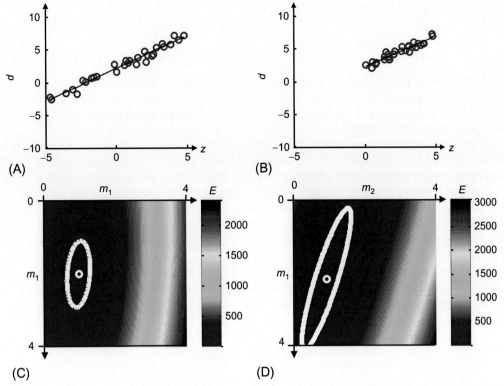

FIG. 3.15 (A) Least squares fitting of a straight line *(blue)* to (z,d) data *(red)*. (C) The best estimate of the model parameters $(m_1^{\text{est}}, m_2^{\text{est}})$ *(white circle)* occurs at the minimum of the error surface $E(m_1, m_2)$, which is a function of model parameters, intercept m_1 and slope m_2. The minimum is surrounded by a region of low error *(white ellipse)* that corresponds to lines that fit "almost as well" as the best estimate. The variance of the estimate is related to the size of the ellipse. In this example, the ellipse is narrowest in the m_2 direction, indicating that the slope m_2 is determined more accurately than intercept m_1. The geometry of the experiment, and not the overall level of observational error, determines the shape of the ellipse, as can be seen from the example in (B) and (D). The tilt of the ellipse indicates that the intercept and slope are negatively correlated. *MatLab script gda03_13.*

The covariance of the least squares solution (assuming uncorrelated data all with equal variance σ_d^2) is therefore

$$[\text{cov } \mathbf{m}] = \sigma_d^2 \left[\mathbf{G}^{\mathrm{T}}\mathbf{G}\right]^{-1} = \sigma_d^2 \left[\frac{1}{2}\frac{\partial^2 E}{\partial \mathbf{m}^2}\right]^{-1}_{\mathbf{m}=\mathbf{m}^{\text{est}}} \tag{3.74}$$

The prediction error $E = \mathbf{e}^{\mathrm{T}}\mathbf{e}$ is the sum of squares of Gaussian data minus a constant. It is, therefore, a random variable with a χ^2 distribution with $N-M$ degrees of freedom, which has mean $(N-M)\sigma_d^2$ and variance $2(N-M)\sigma_d^4$. (The degrees of freedom are reduced by M since the model can force M linear combinations of the e_i to zero.) We can use the standard deviation of E

$$\sigma_E = [2(N-M)]^{1/2}\sigma_d^2 \tag{3.75}$$

in the expression for variance as

$$[\text{cov}\,\mathbf{m}] = \sigma_d^2 \left[\mathbf{G}^T\mathbf{G}\right]^{-1} = \frac{\sigma_E}{[2(N-M)]^{1/2}} \left[\frac{1}{2}\frac{\partial^2 E}{\partial \mathbf{m}^2}\right]_{\mathbf{m}=\mathbf{m}^{est}}^{-1} \tag{3.76}$$

The covariance $[\text{cov}\,\mathbf{m}]$ can be interpreted as being controlled either by the variance of the data times a measure of how error in the data is mapped into error in the model parameters or by the standard deviation of the total prediction error times a measure of the curvature of the prediction error at its minimum.

The methods of solving inverse problems that have been discussed in this chapter emphasize the data and model parameters themselves. The method of least squares estimates the model parameters with smallest prediction length. The method of minimum-length estimates the simplest model parameters. The ideas of data and model parameters are very concrete and straightforward, and the methods based on them are simple and easily understood. Nevertheless, this viewpoint tends to obscure an important aspect of inverse problems: that the nature of the problems depends more on the *relationship* between the data and model parameters than on the data or model parameters themselves. It should, for instance, be possible to tell a well-designed experiment from a poor one without knowing what the numerical values of the data or model parameters are, or even the range in which they fall. In the next chapter, we will begin to explore this kind of problem.

3.13 PROBLEMS

3.1 Show that the equations worked out for the straight line problem in Eq. (3.6) have the solution given in Eq. (3.18).

3.2 This problem builds on Problem 1.1. Suppose that you determine the masses of 100 objects by weighing the first, then weighing the first and second together, and then weighing the rest in triplets: the first, second, and third; the second, third, and fourth; and so forth. Write a *MatLab* script that (A) randomly assigns masses m_i^{true} to the objects in the range of 0–1 kg; (B) builds the appropriate data kernel $\mathbf{G}$; (C) creates synthetic observed data $\mathbf{d}^{obs} = \mathbf{Gm} + \mathbf{n}$, where $\mathbf{n}$ is a vector of Gaussian random numbers with zero mean and $\sigma_d = 0.01$ kg; (D) solves the inverse problem by simple least squares; (E) estimates the variance of each of the estimated model parameters $\mathbf{m}^{est}$; and (F) counts up the number of estimated model parameters that are within $\pm 2\sigma_m$ of their true value. (G) Make a plot of σ_m as a function of the index of the model parameter. Does it decline, remain constant, or grow?

3.3 This problem builds on Problem 1.2. Suppose that you determine the height of 50 objects by measuring the first, and then stacking the second on top of the first and measuring their combined height, stacking the third on top of the first two and measuring their combined height, and so forth. Write a *MatLab* script that (A) randomly assigns heights m_i^{true} to the objects in the range of 0–1 m; (B) builds the appropriate data kernel $\mathbf{G}$; (C) creates synthetic observed data $\mathbf{d}^{obs} = \mathbf{Gm} + \mathbf{n}$, where $\mathbf{n}$ is a vector of Gaussian random numbers with zero mean and $\sigma_d = 0.01$ m; (D) solves the inverse problem by simple least squares; (E) estimates the variance of each of the estimated model parameters $\mathbf{m}^{est}$; and (F) counts up the number of estimated model parameters that are within $\pm 2\sigma_m$ of their true value. (G) Make a plot of σ_m as a function of the index of the model parameter. Does it decline, remain constant, or grow?

3.4 This problem builds on Problem 1.3, which considers the cubic equation, $d_i = m_1 + m_2 z_i + m_3 z_i^2 + m_4 z_i^3$. Write a *MatLab* script that (A) computes a vector $\mathbf{z}$ with $N=11$ elements equally spaced from 0 to 1; (B) randomly assigns the elements of $\mathbf{m}^{\text{true}}$ in the range of -1 to 1; (C) builds the appropriate data kernel $\mathbf{G}$; (D) creates synthetic observed data $\mathbf{d}^{\text{obs}} = \mathbf{Gm} + \mathbf{n}$, where $\mathbf{n}$ is a vector of Gaussian random numbers with zero mean and $\sigma_d = 0.05$; (E) solves the inverse problem by simple least squares; (F) calculates the predicted data, $\mathbf{d}^{\text{pre}} = \mathbf{Gm}^{\text{est}}$; and (G) plots d_i^{obs} and d_i^{pre}. Comment upon the results.

3.5 This problem builds on Problem 3.4. Modify your solution of Problem 3.4 by adding the constraint that the predicted data pass through the point $(z', d') = (5, 0)$. Comment upon the results.

3.6 Modify the reflection coefficient example of Section 3.5.4 to include the prior information that the ratio of compressional to shear wave velocities is exactly $r = 1.76$. (A) Convert the equation $(\alpha + \Delta\alpha)/(\beta + \Delta\beta)$ to a linear relationship involving model parameters $\Delta\alpha/\alpha$ and $\Delta\beta/\beta$ using small number approximations. (B) Apply this *prior* information to the estimation of the model parameters using Eq. (3.31). Omit calculation of confidence intervals. (B) Compare the results to the case where no prior information is used.

References

Aki, K., Richards, P.G., 2009. Quantitative Seismology, second ed. University Science Books, Sausalito, CA. 700 pp.

Franklin, J.N., 1970. Well-posed stochastic extensions of ill-posed linear problems. J. Math. Anal. Appl. 31, 682–716.

Jackson, D.D., 1972. Interpretation of inaccurate, insufficient and inconsistent data. Geophys. J. R. Astron. Soc. 28, 97–110.

Jackson, D.D., 1979. The use of a priori data to resolve non-uniqueness in linear inversion. Geophys. J. R. Astron. Soc. 57, 137–157.

Jordan, T.H., Franklin, J.N., 1971. Optimal solutions to a linear inverse problem in geophysics. Proc. Natl. Acad. Sci. U. S. A. 68, 291–293.

Lanczos, C., 1961. Linear Differential Operators. Van Nostrand-Reinhold, Princeton, NJ.

Menke, W., Eilon, Z., 2015. Relationship between data smoothing and the regularization of inverse problems. Pure Appl. Geophys. 172, 2711–2726.

Menke, W., Menke, J., 2011. Environmental Data Analysis with MatLab. Academic Press, Elsevier Inc., Oxford. 263 pp.

Solution of the Linear, Gaussian Inverse Problem, Viewpoint 2: Generalized Inverses

Geophysical Data Analysis
https://doi.org/10.1016/B978-0-12-813555-6.00004-6

© 2018 Elsevier Inc. All rights reserved.

4.1 SOLUTIONS VERSUS OPERATORS

In the previous chapter, we derived methods of solving the linear inverse problem $\mathbf{Gm} = \mathbf{d}$ that were based on examining two properties of its solution: prediction error and solution simplicity (or length). Most of these solutions had a form that was linear in the data, $\mathbf{m}^{est} = \mathbf{Md} + \mathbf{v}$, where $\mathbf{M}$ is some matrix and $\mathbf{v}$ is some vector, both of which are independent of the data $\mathbf{d}$. This equation indicates that the estimate of the model parameters is controlled by some matrix $\mathbf{M}$ operating on the data (i.e., multiplying the data). We therefore shift our emphasis from the estimates $\mathbf{m}^{est}$ to the operator matrix $\mathbf{M}$, with the expectation that by studying it we can learn more about the properties of inverse problems. Since the matrix $\mathbf{M}$ solves, or "inverts," the inverse problem $\mathbf{Gm} = \mathbf{d}$, it is often called the *generalized inverse* and given the symbol $\mathbf{G}^{-g}$. The exact form of the generalized inverse depends on the problem at hand. The generalized inverse of the overdetermined least squares problem is $\mathbf{G}^{-g} = [\mathbf{G}^T\mathbf{G}]^{-1}\mathbf{G}^T$, and for the minimum length underdetermined solution it is $\mathbf{G}^{-g} = \mathbf{G}^T[\mathbf{GG}^T]^{-1}$.

Note that in some ways the generalized inverse is analogous to the ordinary matrix inverse. The solution to the square (even-determined) matrix equation $\mathbf{Ax} = \mathbf{y}$ is $\mathbf{x} = \mathbf{A}^{-1}\mathbf{y}$, and the solution to the inverse problem $\mathbf{Gm} = \mathbf{d}$ is $\mathbf{m}^{est} = \mathbf{G}^{-g}\mathbf{d}$ (plus some vector, possibly). The analogy is very limited, however. The generalized inverse is not a matrix inverse in the usual sense. It is not square, and neither $\mathbf{G}^{-g}\mathbf{G}$ nor $\mathbf{GG}^{-g}$ need equal an identity matrix.

4.2 THE DATA RESOLUTION MATRIX

Suppose we have found a generalized inverse that in some sense solves the inverse problem $\mathbf{Gm} = \mathbf{d}$, yielding an estimate of the model parameters $\mathbf{m}^{est} = \mathbf{G}^{-g}\mathbf{d}$ (for the sake of simplicity we assume that there is no additive vector). We can then retrospectively ask how well this estimate of the model parameters fits the data. By plugging our estimate into the equation $\mathbf{Gm} = \mathbf{d}$ we conclude

$$\mathbf{d}^{pre} = \mathbf{Gm}^{est} = \mathbf{G}\left[\mathbf{G}^{-g}\mathbf{d}^{obs}\right] = [\mathbf{GG}^{-g}]\mathbf{d}^{obs} = \mathbf{Nd}^{obs} \qquad (4.1)$$

Here, the superscripts obs and pre mean observed and predicted, respectively. The $N \times N$ square matrix $\mathbf{N} = \mathbf{GG}^{-g}$ is called the *data resolution matrix*. This matrix describes how well the predictions match the data. If $\mathbf{N} = \mathbf{I}$, then $\mathbf{d}^{pre} = \mathbf{d}^{obs}$ and the prediction error is zero. On the other hand, if the data resolution matrix is not an identity matrix, the prediction error is nonzero.

If the elements of the data vector $\mathbf{d}$ possess a natural ordering, then the data resolution matrix has a simple interpretation. Consider, for example, the problem of fitting a curve to (z,d) points, where the data have been ordered according to the value of the auxiliary variable z. If $\mathbf{N}$ is not an identity matrix but is close to an identity matrix (in the sense that its largest elements are near its main diagonal), then the configuration of the matrix signifies that averages of neighboring data can be predicted, whereas individual data cannot. Consider the ith row of $\mathbf{N}$. If this row contained all zeros except for a one in the ith column, then d_i would be predicted exactly. On the other hand, suppose that the row contained the elements

$$[\dots\ 0\ 0\ 0\ 0.1\ 0.8\ 0.1\ 0\ 0\ 0\ \dots] \qquad (4.2)$$

where the 0.8 is in the ith column. Then the ith datum is given by

$$d_i^{\text{pre}} = \sum_{j=1}^{N} N_{ij} d_j^{\text{obs}} = 0.1 d_{i-1}^{\text{obs}} + 0.8 d_i^{\text{obs}} + 0.1 d_{i+1}^{\text{obs}} \tag{4.3}$$

The predicted value is a weighted average of three neighboring observed data. If the true data vary slowly with the auxiliary variable, then such an average might produce an estimate reasonably close to the observed value.

The rows of the data resolution matrix $\mathbf{N}$ describe how well neighboring data can be independently predicted, or *resolved*. If the data have a natural ordering, then a graph of the elements of the rows of $\mathbf{N}$ against column indices illuminates the sharpness of the resolution (Fig. 4.1A). If the graphs have a single sharp maximum centered about the main diagonal, then the data are well resolved. If the graphs are very broad, then the data are poorly resolved. Even in cases where there is no natural ordering of the data, the resolution matrix still shows how much weight each observation has in influencing the predicted value. There is then no special significance to whether large off-diagonal elements fall near to or far from the main diagonal.

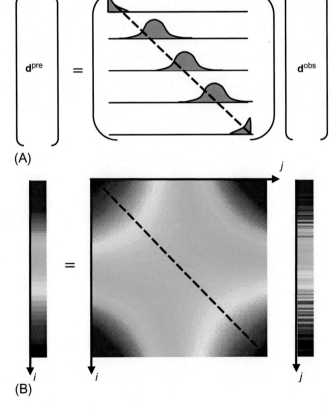

(A)

(B)

FIG. 4.1 (A) Plots of selected rows of the data resolution matrix, $\mathbf{N}$, indicate how well the data can be predicted. Narrow peaks occurring near the main diagonal of the matrix *(dashed line)* indicate that the resolution is good. (B) Actual $\mathbf{N}$ for the case of fitting a straight line to 100 data, equally spaced along the z-axis. Large values *(red colors)* occur only near the ends of the main diagonal *(dashed line)*, indicating that the resolution is poor at intermediate values of z. *MatLab* script gda04_01.

A straight line has only two parameters and so cannot accurately predict many independent data. Consequently, the data resolution matrix for the problem of fitting a straight line to data is not diagonal (Fig. 4.1B). Its largest amplitudes are at its top-right and bottom-left corners, indicating that the points at the *ends* of the line are controlling the fit.

Because the diagonal elements of the data resolution matrix indicate how much weight a datum has in its own prediction, these diagonal elements are often singled out and called the *importance* $\mathbf{n}$ of the data (Minster et al., 1974)

$$\mathbf{n} = \mathrm{diag}(\mathbf{N}) \tag{4.4}$$

The data resolution matrix is not a function of the data but only of the data kernel $\mathbf{G}$ (which embodies the model and experimental geometry) and any prior information applied to the problem. It can therefore be computed and studied without actually performing the experiment and can be a useful tool in experimental design.

4.3 THE MODEL RESOLUTION MATRIX

The data resolution matrix characterizes whether the data can be independently predicted, or resolved. The same question can be asked about the model parameters. To explore this question we imagine that there is a true, but unknown set of model parameters $\mathbf{m}^{\mathrm{true}}$ that solve $\mathbf{Gm}^{\mathrm{true}} = \mathbf{d}^{\mathrm{obs}}$. We then inquire how closely a particular estimate of the model parameters $\mathbf{m}^{\mathrm{est}}$ is to this true solution. Plugging the expression for the observed data $\mathbf{Gm}^{\mathrm{true}} = \mathbf{d}^{\mathrm{obs}}$ into the expression for the estimated model $\mathbf{m}^{\mathrm{est}} = \mathbf{G}^{-g}\mathbf{d}^{\mathrm{obs}}$ gives

$$\mathbf{m}^{\mathrm{est}} = \mathbf{G}^{-g}\mathbf{d}^{\mathrm{obs}} = \mathbf{G}^{-g}\left[\mathbf{Gm}^{\mathrm{true}}\right] = [\mathbf{G}^{-g}\mathbf{G}]\mathbf{m}^{\mathrm{true}} = \mathbf{Rm}^{\mathrm{true}} \tag{4.5}$$

(Wiggins, 1972). Here $\mathbf{R}$ is the $M \times M$ *model resolution matrix*. If $\mathbf{R} = \mathbf{I}$, then each model parameter is uniquely determined. If $\mathbf{R}$ is not an identity matrix, then the estimates of the model parameters are really weighted averages of the true model parameters. If the model parameters have a natural ordering (as they would if they represented a discretized version of a continuous function), then plots of the rows of the resolution matrix can be useful in determining to what scale features in the model can actually be resolved (Fig. 4.2A). Like the data resolution matrix, the model resolution is a function of only the data kernel and the prior information added to the problem. It is therefore independent of the actual values of the data and can therefore be another important tool in experimental design.

As an example, we examine the resolution of the discrete version of the *Laplace transform*

$$d(c) = \int_0^\infty \exp(-cz)m(z)\mathrm{d}z \rightarrow d_i = \sum_{j=1}^M \exp(-c_i z_j)m_j \tag{4.6}$$

Here, the datum d_i is a weighted average of the model parameters m_j, with weights that decline exponentially with depth z. The decay rate of the exponential is controlled by the constant, c_i, so that the smaller *is* correspond to averages over a wider range of depths and the larger *is* over a shallower range of depths. Not surprisingly, the shallow model parameters are better resolved (Fig. 4.2B).

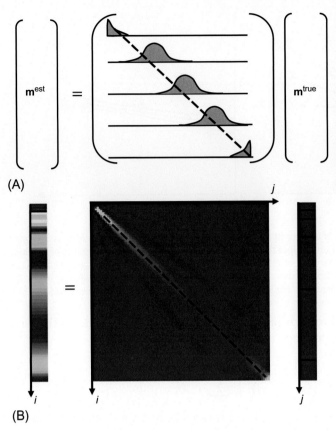

(A)

(B)

FIG. 4.2 (A) Plots of selected rows of the model resolution matrix, **R**, indicate how well the model parameters can be resolved. Narrow peaks occurring near the main diagonal of the matrix *(dashed line)* indicate that the resolution is good. (B) Actual **R** for the case where the model parameters, $m_j(z_j)$, are related to the data through the kernel, $G_{ij} = \exp(-c_i z_j)$, where the cs are constants. Large values *(red colors)* occur only near the top (small z) of the main diagonal *(dashed line)*, indicating that the resolution is poor at larger values of z. *MatLab* script gda04_02.

4.4 THE UNIT COVARIANCE MATRIX

The covariance of the model parameters depends on the covariance of the data and the way in which error is mapped from data to model parameters. This mapping is a function of only the data kernel and the generalized inverse, not of the data itself. A *unit covariance matrix* can be defined to characterize the degree of error amplification that occurs in the mapping. If the data are assumed to be uncorrelated and to have uniform variance σ^2, the unit covariance matrix is given by

$$[\text{cov}_u\,\mathbf{m}] = \sigma^{-2}\mathbf{G}^{-g}[\text{cov }\mathbf{d}]\mathbf{G}^{-gT} = \mathbf{G}^{-g}\mathbf{G}^{-gT} \tag{4.7}$$

Even if the data are correlated, one can often find some normalization of the data covariance matrix, so that one can define a *unit data covariance matrix* $[\text{cov}_u\,\mathbf{d}]$, related to the model covariance matrix by

$$[\text{cov}_u\,\mathbf{m}] = \mathbf{G}^{-g}[\text{cov}_u\,\mathbf{d}]\mathbf{G}^{-gT} \tag{4.8}$$

The unit covariance matrix is a useful tool in experimental design, especially because it is independent of the actual values and variances of the data themselves.

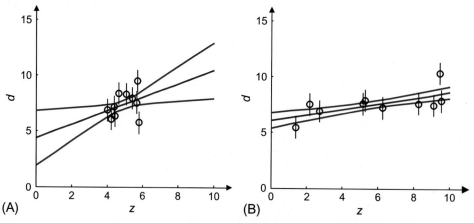

FIG. 4.3   (A) The method of least squares is used to fit a straight line *(red)* to uncorrelated data *(black circles)* with uniform variance *(vertical bars, 1σ confidence limits)*. Since the data are not well-separated in z, the variance of the slope and intercept is large, and consequently the variance of the predicted data is large as well *(blue curves, 1σ confidence limits)*. (B) Same as (A) but with the data well-separated in z. Although the variance of the data is the same as in (A), the variance of the intercept and slope, and consequently the predicted data, is much smaller. *MatLab* script gda04_03.

As an example, reconsider the problem of fitting a straight line to (z,d) data. The unit covariance matrix for intercept m_1 and slope m_2 is given by

$$[\mathrm{cov_u\,m}] = \frac{1}{N\sum z_i^2 - \left(\sum z_i\right)^2} \begin{bmatrix} \sum z_i^2 & -\sum z_i \\ -\sum z_i & N \end{bmatrix} \tag{4.9}$$

Note that the estimates of intercept and slope are uncorrelated only when the data are centered about $z=0$. The overall size of the variance is controlled by the denominator of the fraction. If all the z values are nearly equal, then the denominator of the fraction is small and the variance of the intercept and slope is large (Fig. 4.3A). On the other hand, if the z values have a large spread, the denominator is large, and the variance is small (Fig. 4.3B).

4.5 RESOLUTION AND COVARIANCE OF SOME GENERALIZED INVERSES

The data and model resolution and unit covariance matrices describe many interesting properties of the solutions to inverse problems. We therefore calculate these quantities for some of the simpler generalized inverses (with $[\mathrm{cov_u\,d}]=\mathbf{I}$).

4.5.1 Least Squares

$$\mathbf{G}^{-g} = \left[\mathbf{G}^{\mathrm{T}}\mathbf{G}\right]^{-1}\mathbf{G}^{\mathrm{T}}$$

$$\mathbf{N} = \mathbf{G}\mathbf{G}^{-g} = \mathbf{G}\left[\mathbf{G}^{\mathrm{T}}\mathbf{G}\right]^{-1}\mathbf{G}^{\mathrm{T}}$$

$$\mathbf{R} = \mathbf{G}^{-g}\mathbf{G} = \left[\mathbf{G}^{\mathrm{T}}\mathbf{G}\right]^{-1}\mathbf{G}^{\mathrm{T}}\mathbf{G} = \mathbf{I} \tag{4.10}$$

$$[\mathrm{cov_u\,m}] = \mathbf{G}^{-g}\mathbf{G}^{-g\mathrm{T}} = \left[\mathbf{G}^{\mathrm{T}}\mathbf{G}\right]^{-1}\mathbf{G}^{\mathrm{T}}\mathbf{G}\left[\mathbf{G}^{\mathrm{T}}\mathbf{G}\right]^{-1} = \left[\mathbf{G}^{\mathrm{T}}\mathbf{G}\right]^{-1}$$

4.5.2 Minimum Length

$$\mathbf{G}^{-g} = \mathbf{G}^T \left[\mathbf{GG}^T \right]^{-1}$$

$$\mathbf{N} = \mathbf{GG}^{-g} = \mathbf{GG}^T \left[\mathbf{GG}^T \right]^{-1} = \mathbf{I}$$

$$\mathbf{R} = \mathbf{G}^{-g}\mathbf{G} = \mathbf{G}^T \left[\mathbf{GG}^T \right]^{-1} \mathbf{G}$$

$$[\text{cov}_u\,\mathbf{m}] = \mathbf{G}^{-g}\mathbf{G}^{-gT} = \mathbf{G}^T \left[\mathbf{GG}^T \right]^{-1} \left[\mathbf{GG}^T \right]^{-1}\mathbf{G} = \mathbf{G}^T \left[\mathbf{G}^T\mathbf{G} \right]^{-2}\mathbf{G}$$

(4.11)

Note that there is a great deal of symmetry between the least squares and minimum length solutions. Least squares solves the completely overdetermined problem and has perfect model resolution; minimum length solves the completely underdetermined problem and has perfect data resolution. As we shall see later, generalized inverses that solve the intermediate mixed-determined problems will have data and model resolution matrices that are intermediate between these two extremes.

4.6 MEASURES OF GOODNESS OF RESOLUTION AND COVARIANCE

Just as we were able to quantify the goodness of the model parameters by measuring their overall prediction error and simplicity, we shall develop techniques that quantify the goodness of data and model resolution matrices and unit covariance matrices. Because the resolution is best when the resolution matrices are identity matrices, one possible measure of resolution is based on the size, or *spread*, of the off-diagonal elements.

$$\text{spread}(\mathbf{N}) = \|\mathbf{N} - \mathbf{I}\|_2^2 = \sum_{i=1}^{N}\sum_{j=1}^{N} \left[N_{ij} - \delta_{ij} \right]^2$$

$$\text{spread}(\mathbf{R}) = \|\mathbf{R} - \mathbf{I}\|_2^2 = \sum_{i=1}^{M}\sum_{j=1}^{M} \left[R_{ij} - \delta_{ij} \right]^2$$

(4.12)

Here δ_{ij} are the elements of the identity matrix $\mathbf{I}$. These measures of the goodness of the resolution spread are based on the L_2 norm of the difference between the resolution matrix and an identity matrix. They are sometimes called the *Dirichlet spread functions*. When $\mathbf{R} = \mathbf{I}$, spread$(\mathbf{R}) = 0$.

Since the unit standard deviation of the model parameters is a measure of the amount of error amplification mapped from data to model parameters, this quantity can be used to estimate the size of the unit covariance matrix as

$$\text{size}([\text{cov}_u\,\mathbf{m}]) = \left\| [\text{var}_u\,\mathbf{m}]^{1/2} \right\|_2^2 = \sum_{i=1}^{M} [\text{cov}_u\,\mathbf{m}]_{ii}$$

(4.13)

where the square root is interpreted element by element. Note that this measure of covariance size does not take into account the size of the off-diagonal elements in the unit covariance matrix.

4.7 GENERALIZED INVERSES WITH GOOD RESOLUTION AND COVARIANCE

Having found a way to measure quantitatively the goodness of the resolution and covariance of a generalized inverse, we now consider whether it is possible to use these measures as guiding principles for deriving generalized inverses. This procedure is analogous to that of Chapter 3, which involves first defining measures of solution prediction error and simplicity and then using those measures to derive the least squares and minimum length estimates of the model parameters.

4.7.1 Overdetermined Case

We first consider a purely overdetermined problem of the form $\mathbf{Gm} = \mathbf{d}$. We postulate that this problem has a solution of the form $\mathbf{m}^{est} = \mathbf{G}^{-g}\mathbf{d}$ and try to determine $\mathbf{G}^{-g}$ by minimizing some combination of the above measures of goodness. Since we previously noted that the overdetermined least squares solution had perfect model resolution, we shall try to determine $\mathbf{G}^{-g}$ by minimizing only the spread of the data resolution. We begin by examining the spread of the kth row of $\mathbf{N}$, say, J_k:

$$J_k = \sum_{i=1}^{N}(N_{ki} - \delta_{ki})^2 = \sum_{i=1}^{N}N_{ki}^{2} - 2\sum_{i=1}^{N}N_{ki}\,\delta_{ki} + \sum_{i=1}^{N}\delta_{ki}^{2} \tag{4.14}$$

Since each of the J_ks is positive, we can minimize the total spread $(\mathbf{N}) = \sum J_k$ by minimizing each individual J_k. We therefore insert the definition of the data resolution matrix $\mathbf{N} = \mathbf{GG}^{-g}$ into the formula for J_k and minimize it with respect to the elements of the generalized inverse matrix:

$$\frac{\partial J_k}{\partial G_{qr}^{-g}} = 0 \tag{4.15}$$

We shall perform the differentiation separately for each of the three terms of J_k. The first term is given by

$$\frac{\partial}{\partial G_{qr}^{-g}}\left[\sum_{i=1}^{N}\left[\sum_{j=1}^{M}G_{kj}G_{ji}^{-g}\right]\left[\sum_{p=1}^{M}G_{kp}G_{pi}^{-g}\right]\right] = \frac{\partial}{\partial G_{qr}^{-g}}\left[\sum_{i=1}^{N}\sum_{j=1}^{M}\sum_{p=1}^{M}G_{ji}^{-g}G_{pi}^{-g}G_{kj}G_{kp}\right]$$

$$= 2\sum_{i=1}^{N}\sum_{j=1}^{M}\sum_{p=1}^{M}\delta_{jq}\delta_{ir}G_{pi}^{-g}G_{kj}G_{kp} \tag{4.16}$$

$$= 2\sum_{p=1}^{M}G_{pr}^{-g}G_{kq}G_{kp}$$

The second term is given by

$$-2\frac{\partial}{\partial G_{qr}^{-g}}\sum_{i=1}^{N}\sum_{j=1}^{M}G_{kj}G_{ji}^{-g}\delta_{ki} = -2\sum_{i=1}^{N}\sum_{j=1}^{M}G_{kj}\,\delta_{jq}\,\delta_{ir}\,\delta_{ki} = -2G_{kq}\,\delta_{kr} \tag{4.17}$$

The third term is zero, since it is not a function of the generalized inverse. The complete equation is $\sum_{p=1}^{M}G_{kq}\,G_{kp}\,G_{pr}^{-g}=G_{kq}\,\delta_{kr}$. After summing over k and converting to matrix notation, we obtain

$$\mathbf{G}^{\mathsf{T}}\mathbf{G}\mathbf{G}^{-g}=\mathbf{G}^{\mathsf{T}} \tag{4.18}$$

Since $\mathbf{G}^{\mathsf{T}}\mathbf{G}$ is square, we can premultiply by its inverse to solve for the generalized inverse, $\mathbf{G}^{-g}=[\mathbf{G}^{\mathsf{T}}\mathbf{G}]^{-1}\mathbf{G}^{\mathsf{T}}$, which is precisely the same as the formula for the least squares generalized inverse. The least squares generalized inverse can be interpreted either as the inverse that minimizes the L_2 norm of the prediction error or as the inverse that minimizes the Dirichlet spread of the data resolution.

4.7.2 Underdetermined Case

The data can be satisfied exactly in a purely underdetermined problem. The data resolution matrix is, therefore, precisely an identity matrix and its spread is zero. We might therefore try to derive a generalized inverse for this problem by minimizing the spread of the model resolution matrix with respect to the elements of the generalized inverse. It is perhaps not particularly surprising that the generalized inverse obtained by this method is exactly the minimum length generalized inverse $\mathbf{G}^{-g}=\mathbf{G}^{\mathsf{T}}[\mathbf{G}\mathbf{G}^{\mathsf{T}}]^{-1}$. The minimum length solution can be interpreted either as the inverse that minimizes the L_2 norm of the solution length or as the inverse that minimizes the Dirichlet spread of the model resolution. This is another aspect of the symmetrical relationship between the least squares and minimum length solutions.

4.7.3 The General Case With Dirichlet Spread Functions

We seek the generalized inverse $\mathbf{G}^{-g}$ that minimizes the weighted sum of Dirichlet measures of resolution spread and covariance size.

$$\text{Minimize}: \quad \alpha_1\,\text{spread}(\mathbf{N})+\alpha_2\,\text{spread}(\mathbf{R})+\alpha_3\,\text{size}([\text{cov}_{\mathrm{u}}\,\mathbf{m}]) \tag{4.19}$$

where the αs are arbitrary weighting factors. This problem is done in exactly the same fashion as the one in Section 4.7.1, except that there are now three times as much algebra. The result is an equation for the generalized inverse:

$$\alpha_1\left[\mathbf{G}^{\mathsf{T}}\mathbf{G}\right]\mathbf{G}^{-g}+\mathbf{G}^{-g}\left[\alpha_2\left[\mathbf{G}\mathbf{G}^{\mathsf{T}}\right]+\alpha_3[\text{cov}_{\mathrm{u}}\,\mathbf{d}]\right]=[\alpha_1+\alpha_2]\mathbf{G}^{\mathsf{T}} \tag{4.20}$$

An equation of this form is called a *Sylvester equation*. It is just a set of linear equations in the elements of the generalized inverse $\mathbf{G}^{-g}$ and so could be solved by writing the elements of $\mathbf{G}^{-g}$ as a vector in a huge $NM\times NM$ matrix equation, but it has no explicit solution in terms of algebraic functions of the component matrices. Explicit solutions can be written, however, for a variety of special choices of the weighting factors. The least squares solution is recovered if $\alpha_1=1$ and $\alpha_2=\alpha_3=0$, and the minimum length solution is recovered if $\alpha_1=0$, $\alpha_2=1$, and $\alpha_3=0$. Of more interest is the case in which $\alpha_1=1$, $\alpha_2=0$, α_3 equals some constant (say, ε^2) and $[\text{cov}_{\mathrm{u}}\,\mathbf{d}]=\mathbf{I}$. The generalized inverse is then given by

$$\mathbf{G}^{-g}=\left[\mathbf{G}^{\mathsf{T}}\mathbf{G}+\varepsilon^2\mathbf{I}\right]^{-1}\mathbf{G}^{\mathsf{T}} \tag{4.21}$$

This formula is precisely the damped least squares inverse, which we derived in the previous chapter by minimizing a combination of prediction error and solution length. The damped least squares solution can also be interpreted as the inverse that minimizes a weighted combination of data resolution spread and covariance size.

Another interesting solution is obtained when a weighted combination of model resolution spread and covariance size is minimized. Setting $\alpha_1 = 0$, $\alpha_2 = 1$, $\alpha_3 = \varepsilon^2$, and $[\text{cov}_u \, \mathbf{d}] = \mathbf{I}$, we find

$$\mathbf{G}^{-g} = \mathbf{G}^T \left[\mathbf{G}\mathbf{G}^T + \varepsilon^2 \mathbf{I} \right]^{-1} \tag{4.22}$$

This solution might be termed *damped minimum length*. It will be important in the discussion later in this chapter, because it is the Dirichlet analog to the Backus-Gilbert generalized inverse that will be introduced there.

Note that it is quite possible for these generalized inverses to possess resolution matrices containing *negative* off-diagonal elements. Physically, an average makes most sense when it contained only positive weighting factors, so negative elements interfere with the interpretation of the rows of $\mathbf{R}$ as localized averages. In principle, it is possible to include nonnegativity as a constraint when choosing the generalized inverse by minimizing the spread functions. However, in practice, this constraint is never implemented because it makes the calculation of the generalized inverse very difficult. Furthermore, the more constraints that one places on $\mathbf{R}$, the less localized it tends to become.

4.8 SIDELOBES AND THE BACKUS-GILBERT SPREAD FUNCTION

The Dirichlet spread function is not a particularly appropriate measure of the goodness of resolution when the data or model parameters have a natural ordering because the off-diagonal elements of the resolution matrix are all weighted equally, regardless of whether they are close or far from the main diagonal. We would much prefer that any large elements be close to the main diagonal when there is a natural ordering (Fig. 4.4) because the rows of the resolution matrix then represent *localized* averaging functions.

If one uses the Dirichlet spread function to compute a generalized inverse, it will often have *sidelobes*, that is, large amplitude regions in the resolution matrices far from the main diagonal.

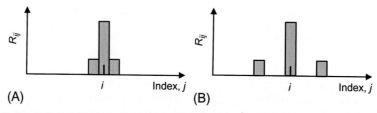

FIG. 4.4   (A, B) Resolution matrices have the same spread when measured by the Dirichlet spread function. Nevertheless, if the model parameters possess a natural ordering, then (A) is better resolved. The Backus-Gilbert spread function is designed to measure (A) as having a smaller spread than (B).

We would prefer to find a generalized inverse without sidelobes, even at the expense of widening the band of nonzero elements near the main diagonal, since a solution with such a resolution matrix is then interpretable as a localized average of physically adjacent model parameters.

We therefore add a weighting factor $w(i,j)$ to the measure of spread that weights the (i,j) element of $\mathbf{R}$ according to its physical distance from the diagonal element. This weighting preferentially selects resolution matrices that are "spiky," or "deltalike." If the natural ordering were a simple linear one, then the choice $w(i,j)=(i-j)^2$ would be reasonable. If the ordering is multidimensional, a more complicated weighting factor is needed. It is usually convenient to choose the spread function so that the diagonal elements have no weight, that is, $w(i,i)=0$, and so that $w(i,j)$ is always nonnegative and symmetric in i and j. The new spread function, often called the Backus-Gilbert spread function (Backus and Gilbert, 1967, 1968), is then given by

$$\text{spread } (\mathbf{R}) = \sum_{i=1}^{M}\sum_{j=1}^{M} w(i,j)\left[R_{ij} - \delta_{ij}\right]^2 = \sum_{i=1}^{M}\sum_{j=1}^{M} w(i,j)R_{ij}^2 \tag{4.23}$$

A similar expression holds for the spread of the data resolution. One can now use this measure of spread to derive new generalized inverses. Their sidelobes will be smaller than those based on the Dirichlet spread functions. On the other hand, they are sometimes worse when judged by other criteria. As we shall see, the Backus-Gilbert generalized inverse for the completely underdetermined problem does not exactly satisfy the data, even though the analogous minimum length generalized inverse does. These facts demonstrate that there are unavoidable trade-offs inherent in finding solutions to inverse problems.

4.9 THE BACKUS-GILBERT GENERALIZED INVERSE FOR THE UNDERDETERMINED PROBLEM

This problem is analogous to deriving the minimum length solution by minimizing the Dirichlet spread of model resolution. Since it is very easy to satisfy the data when the problem is underdetermined (so that the data resolution has small spread), we shall find a generalized inverse that minimizes the spread of the model resolution alone.

We seek the generalized inverse $\mathbf{G}^{-g}$ that minimizes the Backus-Gilbert spread of model resolution. Since the diagonal elements of the model resolution matrix are given no weight, we also require that the resulting model resolution matrix satisfy the equation

$$\sum_{j=1}^{M} R_{ij} = [1]_i \tag{4.24}$$

This constraint ensures that the diagonal of the resolution matrix is finite and that the rows are unit averaging functions acting on the true model parameters. Writing the spread of one row of the resolution matrix as J_k and inserting the expression for the resolution matrix, we have

$$
\begin{aligned}
J_k &= \sum_{l=1}^{M} w(l,k) R_{kl} R_{kl} \\
&= \sum_{l=1}^{M} w(l,k) \left[\sum_{i=1}^{N} G_{ki}^{-g} G_{il} \right] \left[\sum_{j=1}^{N} G_{kj}^{-g} G_{jl} \right] \\
&= \sum_{i=1}^{N} \sum_{j=1}^{N} G_{ki}^{-g} G_{kj}^{-g} \sum_{l=1}^{M} w(l,k) G_{il} G_{jl} \\
&= \sum_{i=1}^{N} \sum_{j=1}^{N} G_{ki}^{-g} G_{kj}^{-g} S_{ij}^{(k)}
\end{aligned}
\tag{4.25}
$$

where the quantity $S_{ij}^{(k)}$ is defined as

$$
S_{ij}^{(k)} = \sum_{l=1}^{M} w(l,k) G_{il} G_{jl}
\tag{4.26}
$$

The left-hand side of the constraint equation $\sum_j R_{ij} = [1]_i$ can also be written in terms of the generalized inverse

$$
\sum_{k=1}^{M} R_{ik} = \sum_{k=1}^{M} \left[\sum_{j=1}^{N} G_{ij}^{-g} G_{jk} \right] = \sum_{j=1}^{N} G_{ij}^{-g} \sum_{k=1}^{M} G_{jk} = \sum_{j=1}^{N} G_{ij}^{-g} u_j
\tag{4.27}
$$

Here the quantity u_j is defined as

$$
u_j = \sum_{k=1}^{M} G_{jk}
\tag{4.28}
$$

The problem of minimizing J_k with respect to the elements of the generalized inverse (under the given constraints) can be solved through the use of Lagrange multipliers. We first define a Lagrange function Φ such that

$$
\Phi = \sum_{i=1}^{N} \sum_{j=1}^{M} G_{ki}^{-g} G_{kj}^{-g} S_{ij}^{(k)} + 2\lambda \sum_{j=1}^{N} G_{kj}^{-g} u_j
\tag{4.29}
$$

where 2λ is the Lagrange multiplier. We then differentiate Φ with respect to the elements of the generalized inverse and set the result equal to zero as

$$
\frac{\partial \Phi}{\partial G_{kp}^{-g}} = 2 \sum_{i=1}^{N} S_{pi}^{(k)} G_{ki}^{-g} + 2\lambda u_p = 0
\tag{4.30}
$$

(Note that one can solve for each row of the generalized inverse separately, so that it is only necessary to take derivatives with respect to the elements in the kth row.) This equation must be solved along with the original constraint equation. Treating the kth row of $\mathbf{G}^{-g}$ as the transform of a column-vector $\mathbf{g}^{(k)}$ and the quantity $S_{ij}^{(k)}$ as a matrix $\mathbf{S}^{(k)}$, we can write these equations as the matrix equation

$$
\begin{bmatrix} \mathbf{S}^{(k)} & \mathbf{u} \\ \mathbf{u}^{\mathrm{T}} & 0 \end{bmatrix} \begin{bmatrix} \mathbf{g}^{(k)} \\ \lambda \end{bmatrix} = \begin{bmatrix} \mathbf{O} \\ 1 \end{bmatrix} \tag{4.31}
$$

This is a square $(N+1) \times (N+1)$ system of linear equations that must be solved for the N elements of the kth row of the generalized inverse and for the one Lagrange multiplier λ.

The matrix equation can be solved explicitly using a variant of the *bordering method* of linear algebra, which is used to construct the inverse of a matrix by partitioning it into submatrices with simple properties. Suppose that the inverse of the symmetric matrix in Eq. (4.31) exists and that we partition it into an $N \times N$ symmetric square matrix $\mathbf{A}$, vector $\mathbf{b}$, and scalar c. By assumption, premultiplication by the inverse yields the identity matrix

$$
\begin{bmatrix} \mathbf{A} & \mathbf{b} \\ \mathbf{b}^{\mathrm{T}} & c \end{bmatrix} \begin{bmatrix} \mathbf{S}^{(k)} & \mathbf{u} \\ \mathbf{u}^{\mathrm{T}} & 0 \end{bmatrix} = \begin{bmatrix} \mathbf{I} & \mathbf{O} \\ 0 & 1 \end{bmatrix} = \begin{bmatrix} \mathbf{A}\mathbf{S}^{(k)} + \mathbf{b}\mathbf{u}^{\mathrm{T}} & \mathbf{A}\mathbf{u} \\ \mathbf{b}^{\mathrm{T}}\mathbf{S}^{(k)} + c\mathbf{u}^{\mathrm{T}} & \mathbf{b}^{\mathrm{T}}\mathbf{u} \end{bmatrix} \tag{4.32}
$$

The unknown submatrices $\mathbf{A}$, $\mathbf{b}$, and c can now be determined by equating the submatrices

$$
\mathbf{A}\mathbf{S}^{(k)} + \mathbf{b}\mathbf{u}^{\mathrm{T}} = \mathbf{I} \text{ so that } \mathbf{A} = \left[\mathbf{S}^{(k)}\right]^{-1}\left[\mathbf{I} - \mathbf{b}\mathbf{u}^{\mathrm{T}}\right]
$$

$$
\mathbf{A}\mathbf{u} = \mathbf{O} \text{ so that } \left[\mathbf{S}^{(k)}\right]^{-1}\mathbf{u} = \mathbf{b}\mathbf{u}^{\mathrm{T}}\left[\mathbf{S}^{(k)}\right]^{-1}\mathbf{u} \text{ and } \mathbf{b} = \frac{\left[\mathbf{S}^{(k)}\right]^{-1}\mathbf{u}}{\mathbf{u}^{\mathrm{T}}\left[\mathbf{S}^{(k)}\right]^{-1}\mathbf{u}} \tag{4.33}
$$

$$
\mathbf{b}^{\mathrm{T}}\mathbf{S}^{(k)} + c\mathbf{u}^{\mathrm{T}} = 0 \text{ so that } c = \frac{-1}{\mathbf{u}^{\mathrm{T}}\left[\mathbf{S}^{(k)}\right]^{-1}\mathbf{u}}
$$

Multiplying Eq. (4.31) by the inverse matrix yields $\mathbf{g}^{(k)} = \mathbf{b}$ and $\lambda = c$. The generalized inverse, written with summations, is

$$
G_{kl}^{-g} = \frac{\sum_{i=1}^{N} u_i \left[\left(\mathbf{S}^{(k)}\right)^{-1}\right]_{il}}{\sum_{i=1}^{N}\sum_{j=1}^{N} u_i \left[\left(\mathbf{S}^{(k)}\right)^{-1}\right]_{ij} u_j} \tag{4.34}
$$

This generalized inverse is the Backus-Gilbert analog to the minimum length solution.

As an example, we compare the Dirichlet and Backus-Gilbert solutions for the Laplace transform problem discussed in Section 4.3 (Fig. 4.5). The Backus-Gilbert solution is the smoother of the two and has a corresponding model resolution matrix that consists of a single band along the main diagonal. The Dirichlet solution has more details but also more artifacts (such as negative values at $z \approx 3$). They are associated with the large-amplitude sidelobes in the corresponding model resolution matrix.

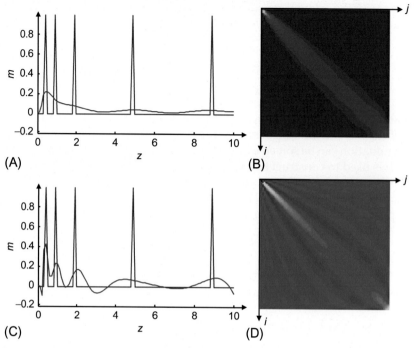

FIG. 4.5 Comparison of the Backus-Gilbert and Dirichlet solutions of the inverse problem described in Fig. 4.2. (A) The true model *(red)* contains a series of sharp spikes. The estimated model *(blue)* using the Backus-Gilbert spread function is much smoother, with the width of the smoothing increasing with z. (B) Corresponding model resolution matrix, $\mathbf{R}$. (C, D) Same, but for a Dirichlet spread function. Note that the Backus-Gilbert resolution matrix has the lower intensity sidelobes, but a wider central band. *MatLab* scripts gda04_02 and gda04_04.

4.10 INCLUDING THE COVARIANCE SIZE

The measure of goodness that was used to determine the Backus-Gilbert inverse can be modified to include a measure of the covariance size of the model parameters (Backus and Gilbert, 1970). We shall use the same measure as we did when considering the Dirichlet spread functions, so that goodness is measured by

$$\alpha \operatorname{spread}(\mathbf{R}) + (1-\alpha) \operatorname{size} \left([\operatorname{cov}_u \mathbf{m}]\right) = \alpha \sum_{i=1}^{M} \sum_{j=1}^{M} w(i,j) R_{ij}^2 + (1-\alpha) \sum_{i=1}^{M} [\operatorname{cov}_u \mathbf{m}]_{ii} \qquad (4.35)$$

where $0 \le \alpha \le 1$ is a weighting factor that determines the relative contribution of model resolution and covariance to the measure of the goodness of the generalized inverse. The goodness J'_k of the kth row is then

$$J'_k = \alpha \sum_{l=1}^{M} w(k,l) R_{kl}^2 + (1-\alpha)[\text{cov}_u \mathbf{m}]_{kk}$$

$$= \alpha \sum_{i=1}^{N} \sum_{j=1}^{N} G_{ki}^{-g} G_{kj}^{-g} [S_{ij}]_k + (1-\alpha) \sum_{i=1}^{N} \sum_{j=1}^{N} G_{ki}^{-g} G_{kj}^{-g} [\text{cov}_u \mathbf{d}]_{ij} \qquad (4.36)$$

$$= \sum_{i=1}^{N} \sum_{j=1}^{N} G_{ki}^{-g} G_{kj}^{-g} S'^{(k)}_{ij}$$

where the quantity $S'^{(k)}_{ij}$ is defined by the equation

$$S'^{(k)}_{ij} = \alpha S^{(k)}_{ij} + (1-\alpha)[\text{cov}_u \mathbf{d}]_{ij} \qquad (4.37)$$

Since the function J_k' has exactly the same form as J_k had in the previous section, the generalized inverse is just the previous result with $S^{(k)}_{ij}$ replaced by $S'^{(k)}_{ij}$:

$$G_{kl}^{-g} = \frac{\sum_{i=1}^{N} u_i \left[\left(\mathbf{S}'^{(k)} \right)^{-1} \right]_{il}}{\sum_{i=1}^{N} \sum_{j=1}^{N} u_i \left[\left(\mathbf{S}'^{(k)} \right)^{-1} \right]_{il} u_j} \qquad (4.38)$$

This generalized inverse is the Backus-Gilbert analog to the damped minimum length solution. In *MatLab*, the one-dimensional Backus-Gilbert generalized inverse GMG (i.e., for the $w(i,j) = (i-j)^2$ weight function) is calculated as

```
GMG = zeros(M,N);
u = G*ones(M,1);
for k = [1:M]
S = G * diag(([1:M]-k).^2) * G';
Sp = alpha*S + (1-alpha)*eye(N,N);
uSpinv = u'/Sp;
GMG(k,:) = uSpinv / (uSpinv*u);
end
```

(*MatLab* script gda04_05)

In higher dimensions, the definition of S is more complicated, since the weight function must represent the physical distance between model parameters. In two dimensions, a reasonable choice is

```
S = G * diag( (abs(ixofj([1:M])-ixofj(k))).^2 + ...
        (abs(iyofj([1:M])-iyofj(k))).^2 ) * G';
```

Here the index vectors ixofj(k) and iyofj(k) give the x and y values of model parameter k.

4.11 THE TRADE-OFF OF RESOLUTION AND VARIANCE

Suppose that one is attempting to determine a set of model parameters that represents a discretized version of a continuous function, such as X-ray opacity in the medical tomography

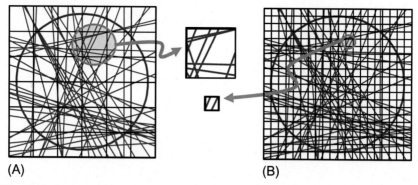

(A)   (B)

FIG. 4.6 Hypothetical tomography experiment with (A) large voxels and (B) small voxels. *MatLab* Script. The small voxel case not only has better spatial resolution but also higher variance, as fewer rays pass through each voxel, leaving less opportunity for measurement error to average out. *MatLab* script gda04_06.

problem (Fig. 4.6). If the discretization is made very fine, then the X-rays will not sample every box; the problem will be underdetermined. If we try to determine the opacity of each box individually, then estimates of opacity will tend to have rather large variance. Few boxes will have several X-rays passing through them, so that little averaging out of the errors will take place. On the other hand, the boxes are very small—and very small features can be detected (the resolution is very good). The large variance can be reduced by increasing the box size (or alternatively, averaging several neighboring boxes). Each of these larger regions will then contain several X-rays, and noise will tend to be averaged out. But because the regions are now larger, small features can no longer be detected and the resolution of the X-ray opacity has become poorer.

This scenario illustrates an important trade-off between model resolution spread and variance size. One can be decreased only at the expense of increasing the other. We can study this trade-off by choosing a generalized inverse that minimizes a weighted sum of resolution spread and covariance size:

$$\alpha \text{ spread } (\mathbf{R}) + (1 - \alpha) \text{ size } ([\text{cov}_u \mathbf{m}]) \tag{4.39}$$

If the weighting parameter α is set near 1, then the model resolution matrix of the generalized inverse will have small spread, but the model parameters will have large variance. If α is set close to 0, then the model parameters will have a relatively small variance, but the resolution will have a large spread. A *trade-off curve* can be defined by varying α on the interval (0,1) (Fig. 4.7). Such curves can be helpful in choosing a generalized inverse that has an optimum trade-off in model resolution and variance (judged by criteria appropriate to the problem at hand).

Trade-off curves play an important role in continuous inverse theory, where the discretization is (so to speak) infinitely fine, and all problems are underdetermined. It is known that in this continuous limit, the curves are monotonic and possess asymptotes in resolution and variance (Fig. 4.8). The process of approximating a continuous function by a finite set of discrete parameters somewhat complicates this picture. The resolution and variance, and indeed the solution itself, are dependent on the parameterization, so it is difficult to make any definitive statement regarding the properties of the trade-off curves. Nevertheless, if the

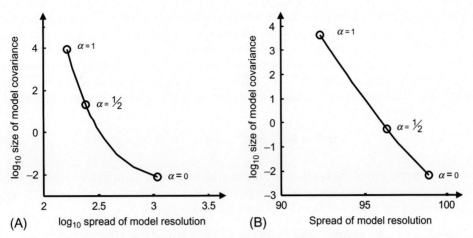

FIG. 4.7 Trade-off curves of resolution and variance for the inverse problem shown in Fig. 4.2. (A) Backus-Gilbert solution, (B) damped minimum length solution. The larger the parameter α, the more weight resolution is given (relative to variance) when forming the generalized inverse. The details of the trade-off curve depend upon the parameterization. The resolution can be no better than the smallest element in the parameterization and no worse than the sum of all the elements. *MatLab* script gda04_05.

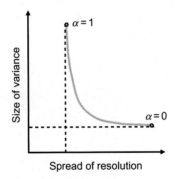

FIG. 4.8 Trade-off curve of resolution and variance has two asymptotes in the case when the model parameter is a continuous function.

discretization is sufficiently fine, the discrete trade-off curves are usually close to ones obtained with the use of continuous inverse theory. Therefore, discretizations should always be made as fine as computational considerations permit.

4.12 CHECKERBOARD TESTS

In general, the resolution matrix $\mathbf{R}$ is not symmetric, so its rows are numerically different than its columns; furthermore, their interpretation is different, too. The kth row of the resolution matrix (let's call it the row-vector $\mathbf{r}^{(k)}$) relates the kth estimated model parameter to the true model parameters, since $m_k^{est} = \sum_i R_{ki} m_i^{true} = \mathbf{r}^{(k)}\mathbf{m}$. The kth estimated model parameter is a *weighted average* of the true model parameters, with $\mathbf{r}^{(k)}$ giving the weights. (However, the elements of $\mathbf{r}^{(k)}$ do not necessarily sum to unity, as the weights of a true weighted average should). The kth *column* of the resolution matrix (let's call it the column-vector $\mathbf{c}^{(k)}$) specifies how each of the estimated model parameters is influenced by the kth true model parameter. This can be seen by setting $\mathbf{m}^{\text{true}} = \mathbf{s}^{(k)}$ with $s_i^{(k)} = \delta_{ik}$; that is, all the elements of $\mathbf{s}^{(k)}$ are zero except the kth, which is unity. Denoting the set of estimated model parameters associated with $\mathbf{s}^{(k)}$ as $\mathbf{m}^{(k)}$, we have:

$$\mathbf{c}^{(k)} = \mathbf{m}^{(\mathbf{k})} = \mathbf{R}\mathbf{s}^{(k)} \text{ or } c_j^{(k)} = m_j^{(k)} = \sum_i R_{ji}\delta_{ik} = R_{jk} \tag{4.40}$$

Thus, the kth column of the resolution matrix quantifies how a single true model parameter spreads out into many estimated model parameters. It is sometimes called a *point-spread function*. Substituting the equation $\mathbf{R} = \mathbf{G}^{-g}\mathbf{G}$ into Eq. (4.40) yields:

$$\mathbf{c}^{(k)} = \mathbf{R}\mathbf{m}^{(k)} = \mathbf{G}^{-g}\mathbf{G}\mathbf{m}^{(k)} = \mathbf{G}^{-g}\mathbf{d}^{(k)} \text{ with } \mathbf{d}^{(k)} = \mathbf{G}\mathbf{m}^{(k)} \tag{4.41}$$

Thus, the kth column of the model resolution matrix solves the inverse problem for synthetic data $\mathbf{d}^{(k)}$ corresponding to a specific model parameter vector $\mathbf{m}^{(k)}$, one that is zero except for its kth element, which is unity (i.e., a unit spike at row k). This suggests a procedure for calculating the resolution: construct the desired $\mathbf{m}^{(k)}$, solve the forward problem to generate $\mathbf{d}^{(k)}$, solve the inverse problem, and then interpret the result as the kth column of the resolution matrix (Fig. 4.9A and B). The great advantage of this technique is that $\mathbf{R}$ in its entirety need not

FIG. 4.9 Resolution of an acoustic tomography problem solved with the minimum length method. The physical model space is a 20×20 grid of pixels on an (x,y) grid. Data are measured only along rows and columns, as in Fig. 1.2. (Top row) One column of the resolution matrix, for a model parameter near the center of the (x,y) grid, calculated using two methods, (A) by computing the complete matrix $\mathbf{R}$ and extracting one column and (B) by calculating the column separately. (Bottom row) Checkerboard resolution test showing (C) true checkerboard and (D) reconstructed checkerboard. *MatLab* scripts gda04_07 and gda04_08.

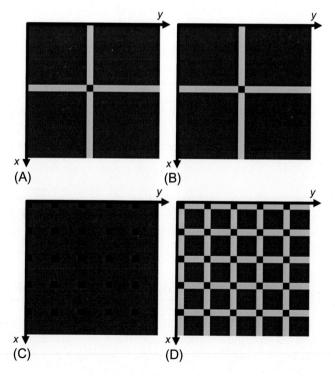

be constructed. Furthermore, the technique will work when the inverse problem is solved by an iterative method, such as the biconjugate gradient method, that does not explicitly construct $\mathbf{G}^{-g}$.

If the resolution of a problem is sufficiently good that the pattern for two well-separated model parameters does not overlap, or overlaps only minimally, then the calculation of two columns of the resolution matrix can be combined into one. One merely solves the inverse problem for synthetic data corresponding to a model parameter vector containing two unit spikes. Nor need one stop with two; a model parameter vector corresponding to a grid of spikes (i.e., a *checkerboard*) allows the resolution to be assessed throughout the model volume (Fig. 4.9C and D). If the problem has perfect resolution, this checkerboard pattern will be perfectly reproduced. If not, portions of the model volume with poor resolution will contain fuzzy spikes. The main limitation of this technique is that it makes the detection of unlocalized side lobes very difficult, since an unlocalized sidelobe associated with a particular spike will tend to be confused with a localized sidelobe of another spike.

Rows of the resolution matrix and the covariance matrix also can be calculated individually (see Menke, 2014).

4.13 PROBLEMS

4.1 Consider an underdetermined problem in which each datum is the sum of three neighboring model parameters, that is, $d_i = m_{i-1} + m_i + m_{i+1}$ for $2 \leq i \leq (M-1)$ with $M = 100$. Compute and plot both the Dirichlet and Backus-Gilbert model resolution matrices. Use the standard Backus-Gilbert weight function $w(i, j) = (i-j)^2$. Interpret the results.

4.2 This problem builds upon Problem 4.1. How does the Backus-Gilbert result change if you use the weight function $w(i,j) = |i-j|^{1/2}$, which gives less weight to distant sidelobes?

4.3 This problem is especially difficult. Consider a two-dimensional acoustic tomography problem like the one discussed in Section 1.3.3, consisting of a 20×20 rectangular array of pixels, with observations only along rows and columns. (A) Design an appropriate Backus-Gilbert weight function that quantifies the spread of resolution. (B) Write a *MatLab* script that calculates the model resolution matrix $\mathbf{R}$. (C) Plot a few representative rows of $\mathbf{R}$, but where each row is reorganized into a two-dimensional image, using the same scheme that was applied to the model parameters. Interpret the results. (Hint: You will need to switch back and forth between a 20×20 rectangular array of model parameters and a length $M = 400$ vector of model parameters, as in Fig. 10.12.)

References

Backus, G.E., Gilbert, J.F., 1967. Numerical application of a formalism for geophysical inverse problems. Geophys. J. R. Astron. Soc. 13, 247–276.

Backus, G.E., Gilbert, J.F., 1968. The resolving power of gross earth data. Geophys. J. R. Astron. Soc. 16, 169–205.

Backus, G.E., Gilbert, J.F., 1970. Uniqueness in the inversion of gross Earth data. Philos. Trans. R. Soc. Lond. A 266, 123–192.

Menke, W., 2014. Review of the generalized least squares method, Surv. Geophys. 36, 1–25.

Minster, J.F., Jordan, T.J., Molnar, P., Haines, E., 1974. Numerical modelling of instantaneous plate tectonics. Geophys. J. R. Astron. Soc. 36, 541–576.

Wiggins, R.A., 1972. The general linear inverse problem: implication of surface waves and free oscillations for Earth structure. Rev. Geophys. Space Phys. 10, 251–285.

Solution of the Linear, Gaussian Inverse Problem, Viewpoint 3: Maximum Likelihood Methods

© 2018 Elsevier Inc. All rights reserved.

5.1 THE MEAN OF A GROUP OF MEASUREMENTS

Suppose that an experiment is performed N times and that each time a single datum d_i is collected. Suppose further that these data are all noisy measurements of the same model parameter m_1. In the view of probability theory, N realizations of random variables, all of which have the same probability density function, have been measured. If these random variables are Gaussian, their joint probability density function can be characterized in terms of a variance σ^2 and a mean m_1 (see Section 2.4) as

$$p(\mathbf{d}) = \sigma^{-N}(2\pi)^{-N/2} \exp\left[-\frac{1}{2}\sigma^{-2}\sum_{i=1}^{N}[d_i - m_1]^2\right] \qquad (5.1)$$

The data $\mathbf{d}^{\text{obs}}$ can be represented graphically as a point in the N-dimensional space whose coordinate axes are $d_1, d_2, \ldots, d_N$ (Fig. 5.1). The probability density function for the data can also be graphed (Fig. 5.2). Note that the probability density function is centered about the line $d_1 = d_2 = \cdots = d_N$, since all the ds are supposed to have the same mean, and that it is spherically symmetric, since all the ds have the same variance.

Suppose that we guess a value for the unknown mean and variance, thus fixing the center and diameter of the probability density function. We can then calculate its numerical value at

FIG. 5.1 The data are represented by a single point *(black)* in a space whose dimensions equal the number of observations (in this case, 3). These data are realizations of random variables with the same mean and variance. Nevertheless, they do not necessarily fall on the line $d_1 = d_2 = d_3$ *(blue)*. MatLab script gda05_01.

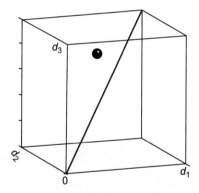

FIG. 5.2 If the data d_i are assumed to be uncorrelated with equal mean and uniform variance, their probability density function $p(\mathbf{d})$ is a spherical cloud *(red)*, centered on the line $d_1 = d_2 = d_3$ *(blue)*. MatLab script gda05_02.

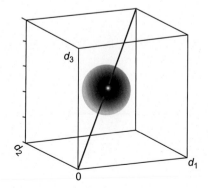

the data $p(\mathbf{d}^{obs})$. If the guessed values of mean and variance are close to being correct, then $p(\mathbf{d}^{obs})$ should be a relatively large number. If the guessed values are incorrect, then the probability, or *likelihood*, of the observed data will be small. We can imagine sliding the cloud of probability in Fig. 5.2 up along the line and adjusting its diameter until its probability at the point $\mathbf{d}^{obs}$ is maximized.

This procedure defines a method of estimating the unknown parameters in the distribution, the *method of maximum likelihood*. It asserts that the optimum values of the parameters maximize the probability that the observed data are in fact observed. In other words, the value of the probability density function at the point $\mathbf{d}^{obs}$ is made as large as possible. The maximum is located by differentiating $p(\mathbf{d}^{obs})$ with respect to mean and variance and setting the result to zero as

$$\partial p / \partial m_1 = \partial p / \partial \sigma = 0 \tag{5.2}$$

Maximizing $\log p(\mathbf{d}^{obs})$ gives the same result as maximizing $p(\mathbf{d}^{obs})$, since $\log(p)$ is a monotonic function of p. We therefore compute derivatives of the *likelihood function*, $L = \log p(\mathbf{d}^{obs})$ (Fig. 5.3). Ignoring the overall normalization of $(2\pi)^{-N/2}$ we have

$$L = \log\left(p\left(\mathbf{d}^{obs}\right)\right) = -N \log(\sigma) - \frac{1}{2}\sigma^{-2} \sum_{i=1}^{N} \left(d_i^{obs} - m_1\right)^2 \tag{5.3}$$

$$\frac{\partial L}{\partial m_1} = 0 = \frac{1}{2}\sigma^{-2} 2 \sum_{i=1}^{N} \left(d_i^{obs} - m_1\right)$$

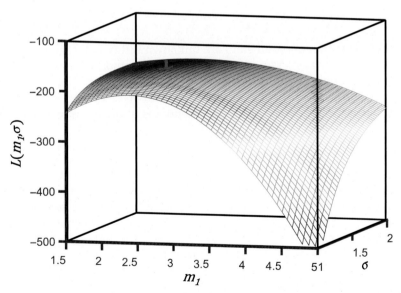

FIG. 5.3　Likelihood surface for 100 realizations of random variables with equal mean $m_1 = 2.5$ and uniform variance $\sigma^2 = (1.5)^2$. The curvature in the direction of m_1 is greater than the maximum in the direction of the σ, indicating that the former can be determined to greater certainty. *MatLab* script gda05_03.

$$\frac{\partial L}{\partial \sigma} = 0 = -\frac{N}{\sigma} + \sigma^{-3} \sum_{i=1}^{N} \left(d_i^{\text{obs}} - m_1 \right)^2$$

These equations can be solved for the estimated mean and variance as

$$m_1^{\text{est}} = \frac{1}{N} \sum_{i=1}^{N} d_i^{\text{obs}} \quad \text{and} \quad \sigma^{\text{est}} = \left[\frac{1}{N} \sum_{i=1}^{N} \left(d_i^{\text{obs}} - m_1^{\text{est}} \right)^2 \right]^{1/2} \tag{5.4}$$

The estimate for m_1 is just the usual formulas for the sample mean. The estimate for σ is the root mean squared error and also is almost the formula for the sample standard deviation, except that it has a leading factor of $1/N$, instead of $1/(N-1)$. We note that these estimates arise as a direct consequence of the assumption that the data possess a Gaussian distribution. If the data distribution were not Gaussian, then the arithmetic mean might not be an appropriate estimate of the mean of the distribution. (As we shall see in Section 8.2, the sample median is the maximum likelihood estimate of the mean of an exponential distribution.)

5.2 MAXIMUM LIKELIHOOD APPLIED TO INVERSE PROBLEM

5.2.1 The Simplest Case

Assume that the data in the linear inverse problem $\mathbf{Gm} = \mathbf{d}$ have a multivariate Gaussian probability density function, as given by

$$p(\mathbf{d}) \propto \exp \left[-\frac{1}{2} (\mathbf{d} - \mathbf{Gm})^{\text{T}} [\text{cov } \mathbf{d}]^{-1} (\mathbf{d} - \mathbf{Gm}) \right] \tag{5.5}$$

We assume that the model parameters are unknown but (for the sake of simplicity) that the data covariance is known. We can then apply the method of maximum likelihood to estimate the model parameters. The optimum values for the model parameters are the ones that maximize the probability that the observed data are in fact observed. The maximum of $p(\mathbf{d}^{\text{obs}})$ occurs when the argument of the exponential is a maximum or when the quantity given by

$$\left(\mathbf{d}^{\text{obs}} - \mathbf{Gm} \right)^{\text{T}} [\text{cov } \mathbf{d}]^{-1} \left(\mathbf{d}^{\text{obs}} - \mathbf{Gm} \right) \tag{5.6}$$

is a minimum. But this expression is just a weighted measure of prediction error. The maximum likelihood estimate of the model parameters is nothing but the weighted least squares solution, where the weighting matrix is the inverse of the covariance matrix of the data (in the notation of Chapter 3, $\mathbf{W}_e = [\text{cov } \mathbf{d}]^{-1}$). If the data happen to be uncorrelated and all have equal variance, then $[\text{cov } \mathbf{d}] = \sigma_d^2 \mathbf{I}$, and the maximum likelihood solution is the simple least squares solution. If the data are uncorrelated but their variances are all different (say, σ_{di}^2), then the prediction error is given by

$$E = \sum_{i=1}^{N} \sigma_{di}^{-2} e_i^2 \tag{5.7}$$

where $e_i = (d_i^{obs} - d_i^{pre})$ is the prediction error for each datum. Each individual error is weighted by the reciprocal of its standard deviation; the most certain data are weighted most.

We have justified the use of the L_2 norm through the application of probability theory. The least squares procedure for minimizing the L_2 norm of the prediction error makes sense if the data are uncorrelated, have equal variance, and obey Gaussian statistics. If the data are not Gaussian, then other measures of prediction error may be more appropriate.

5.2.2 Prior Distributions

The least squares solution does not exist when the linear problem is underdetermined. From the standpoint of probability theory, the probability density function of the data $p(\mathbf{d}^{obs})$ has no well-defined maximum with respect to variations of the model parameters. At best, it has a ridge of maximum probability (Fig. 5.4).

We must add prior information that causes the distribution to have a well-defined peak in order to solve an underdetermined problem. One way to accomplish this goal is to write the prior information about the model parameters as a probability density function $p_A(\mathbf{m})$, where the subscript A indicates "*a priori*" or "prior". The mean of this probability density function is then the value we expect the model parameter vector to have, and its shape reflects the certainty of this expectation.

Prior distributions for the model parameters can take a variety of forms. For instance, if we expected that the model parameters are close to $\langle \mathbf{m} \rangle$, we might use a Gaussian distribution with mean $\langle \mathbf{m} \rangle$ and variance that reflects the certainty of our knowledge (Fig. 5.5). If the prior value of one model parameter were more certain than another, we might use different variances for the different model parameters (Fig. 5.6). The general Gaussian case, with covariance $[\text{cov } \mathbf{m}]_A$, is

$$p_A(\mathbf{m}) \propto \exp\left[-\frac{1}{2}(\mathbf{m} - \langle \mathbf{m} \rangle)^T [\text{cov } \mathbf{m}]_A^{-1} (\mathbf{m} - \langle \mathbf{m} \rangle) \right] \tag{5.8}$$

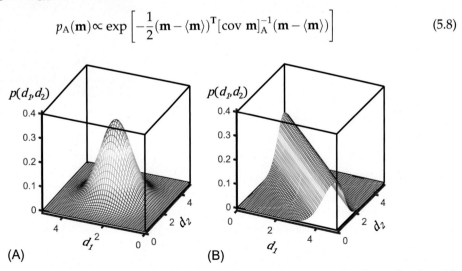

FIG. 5.4 (A) Probability density function $p(d_1, d_2)$ with a well-defined peak. (B) Probability density function with a ridge. *MatLab* script gda05_04.

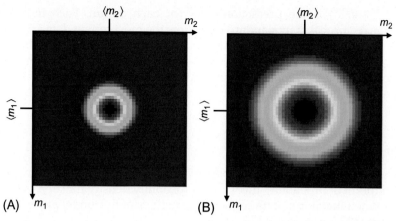

FIG. 5.5   Prior information about model parameters m_1 and m_2, represented with a probability density function $p(m_1,m_2)$. Most probable values are given by means $\langle m_1 \rangle$ and $\langle m_2 \rangle$. Width of the probability density function reflects certainty of knowledge: (A) certain; (B) uncertain. *MatLab* script gda05_05.

FIG. 5.6   Prior information about model parameters m_1 and m_2 represented with a probability density function $p(m_1,m_2)$. The model parameters are thought to be near $\langle \mathbf{m} \rangle$, with the uncertainty in m_1 less than the uncertainty of m_2. *MatLab* script gda05_06.

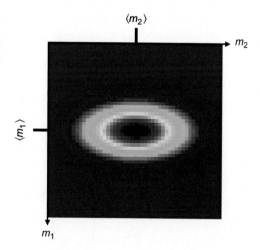

Equality constraints can be implemented with a distribution that contains a ridge (Fig. 5.7). This distribution is non-Gaussian but might be approximated by a Gaussian distribution with nonzero covariance if the expected range of the model parameters were small. Inequality constraints can also be represented by an prior distribution but are inherently non-Gaussian (Fig. 5.8).

Similarly, one can summarize the state of knowledge about the measurements with a prior probability density function $p_A(\mathbf{d})$. It simply summarizes the observations, so its mean is $\mathbf{d}^{\mathrm{obs}}$ and its covariance is the prior covariance [cov $\mathbf{d}$] of the data

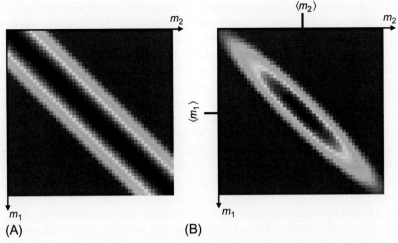

FIG. 5.7   Prior information about model parameters m_1 and m_2, represented with a probability density function $p(m_1,m_2)$. (A) Case when the values of m_1 and m_2 are unknown, but believed to be correlated. (B) Approximation of (A) with a Gaussian probability density function with finite variance. *MatLab* script gda05_07.

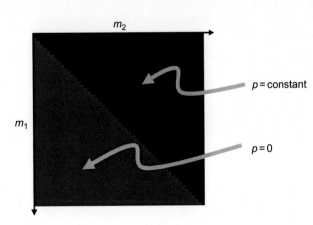

FIG. 5.8 Prior information about model parameters m_1 and m_2 represented with a probability density function $p(m_1,m_2)$. The values of the model parameters are unknown, but the relationship $m_1 \leq m_2$ is believed to hold exactly. This is a non-Gaussian probability density function. *MatLab* script gda05_08.

$$p_A(\mathbf{d}) \propto \exp\left[-\frac{1}{2}\left(\mathbf{d} - \mathbf{d}^{\text{obs}}\right)^{\mathbf{T}}[\text{cov }\mathbf{d}]^{-1}\left(\mathbf{d} - \mathbf{d}^{\text{obs}}\right)\right] \tag{5.9}$$

One important attribute of $p_A(\mathbf{m})$ and $p_A(\mathbf{d})$ is that they contain information about the model parameters $\mathbf{m}$ and the data $\mathbf{d}$, respectively. The amount of information can be quantified by the *information gain*, a scalar number S defined as

$$S[p_A(\mathbf{m})] = \int p_A(\mathbf{m}) \log\left[\frac{p_A(\mathbf{m})}{p_N(\mathbf{m})}\right] \mathrm{d}^M\mathbf{m}$$

$$S[p_A(\mathbf{d})] = \int p_A(\mathbf{d}) \log\left[\frac{p_A(\mathbf{d})}{p_N(\mathbf{d})}\right] \mathrm{d}^N\mathbf{d}$$

$$\tag{5.10}$$

Here, the *null* probability density functions $p_N(\mathbf{m})$ and $p_N(\mathbf{d})$ express the state of complete ignorance about the model parameters and data, respectively. When the range of $\mathbf{m}$ and $\mathbf{d}$ are bounded, the null probability density functions can be taken to be proportional to a constant; that is, $p_N(\mathbf{m}) \propto$ constant and $p_N(\mathbf{d}) \propto$ constant, meaning $\mathbf{m}$ and $\mathbf{d}$ "could be anything." However, when $\mathbf{m}$ and $\mathbf{d}$ are unbounded, the uniform distribution does not exist, and some other probability density function, such as a very wide Gaussian, must be used, instead. The quantity $-S$ is sometimes called the *relative entropy* between the two probability density functions. A wide distribution is "more random" than a narrow one; it has more *entropy*.

The information gain is always a nonnegative number and is only zero when $p_A(\mathbf{m}) = p_N(\mathbf{m})$ and $p_A(\mathbf{d}) = p_N(\mathbf{d})$ (Fig. 5.9). The information gain S has the following properties (Tarantola and Valette, 1982b): (1) the information gain of the null distribution is zero; (2) all distributions except the null distribution have positive information gain; (3) the more sharply peaked the probability density function becomes, the more its information gain increases; and (4) the information gain is invariant under reparameterizations.

We can summarize the state of knowledge about the inverse problem *before* it is solved by first defining a prior probability density function for the data $p_A(\mathbf{d})$ and then combining it with the prior probability density function for the model $p_A(\mathbf{m})$. The prior data probability density function simply summarizes the observations, so its mean is $\mathbf{d}^{obs}$ and its variance is equal to the prior variance of the data. Since the prior model probability density function is completely independent of the actual values of the data, we can form the joint prior probability density function simply by multiplying the two as

$$p_A(\mathbf{m}, \mathbf{d}) = p_A(\mathbf{m})p_A(\mathbf{d}) \tag{5.11}$$

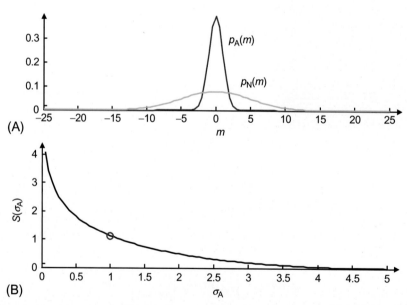

FIG. 5.9   (A) In this example, a wide Gaussian (*green*, $\sigma_N = 5$) is used for the null probability density function $p_N(\mathbf{m})$ and a narrow Gaussian (*red*, $\sigma_N = 1$) is used for the prior probability density function $p_A(\mathbf{m})$. (B) The information gain S decreases as the width of $p_A(\mathbf{m})$ is increased. The case in (A) is depicted with a red circle. *MatLab* script gda05_09.

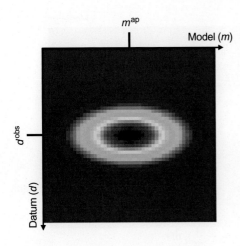

FIG. 5.10 Joint probability density function $p_A(\mathbf{m}, \mathbf{d})$ for model parameter m and datum d. The distribution is peaked at mean values m^{ap} and d^{obs}. *MatLab* script gda05_10.

This probability density function can be depicted graphically as a "cloud" of probability centered on the observed data and prior model parameters, with a width that reflects the certainty of these quantities (Fig. 5.10). If we apply the maximum likelihood method to this distribution, we simply recover the data and prior model. We have not yet applied our knowledge of the model (the relationship between data and model parameters).

5.2.3 Maximum Likelihood for an Exact Theory

Suppose that the model is the rather general equation $\mathbf{g}(\mathbf{m}) = \mathbf{d}$ (which may or may not be linear). This equation defines a surface in the space of model parameters and data along which the solution must lie (Fig. 5.11). The maximum likelihood problem then translates into finding the maximum of the joint distribution $p_A(\mathbf{m}, \mathbf{d})$ (or, equivalently, its logarithm) on the surface $\mathbf{d} = \mathbf{g}(\mathbf{m})$ (Tarantola and Valette, 1982a):

$$\text{maximize } \log[p(\mathbf{m}, \mathbf{d})] \text{ with the constraint } \mathbf{g}(\mathbf{m}) - \mathbf{d} = 0 \tag{5.12}$$

Note that if the prior probability density function for the model parameters is much more certain than that of the observed data (i.e., if $\sigma_m < \sigma_d$), then the estimate of the model parameters (the maximum likelihood point) tends to be close to the prior model parameters (Fig. 5.12). On the other hand, if the data are far more certain than the model parameters (i.e., $\sigma_d < \sigma_m$), then the estimates of the model parameters primarily reflect information contained in the data (Fig. 5.13).

In the case of Gaussian probability density function, and the linear theory $\mathbf{d} = \mathbf{Gm}$, we need only to substitute $\mathbf{Gm}$ for $\mathbf{d}$ in the expression for $p_A(\mathbf{d})$ to obtain

$$\text{minimize } \varPhi(\mathbf{m}) = L(\mathbf{m}) + E(\mathbf{m}) \text{ with respect to } \mathbf{m} \text{ with}$$

$$L(\mathbf{m}) = (\mathbf{m} - \langle \mathbf{m} \rangle)^{\mathsf{T}} [\text{cov } \mathbf{m}]_A^{-1} (\mathbf{m} - \langle \mathbf{m} \rangle) \tag{5.13}$$

$$E(\mathbf{m}) = \left(\mathbf{Gm} - \mathbf{d}^{\mathrm{obs}}\right)^{\mathsf{T}} [\text{cov } \mathbf{d}]^{-1} \left(\mathbf{Gm} - \mathbf{d}^{\mathrm{obs}}\right)$$

FIG. 5.11 (A) Prior joint probability density function $p(m,d)$ for model parameter m and datum d represents the idea that the model parameter is near its prior value m^{ap} and the datum is near its observed value d^{obs} *(white circle)*. The data and model parameters are believed to be related by an exact theory $d = g(m)$ *(white curve)*. The estimated model parameter m^{est} and predicted datum d^{pre} fall on this curve at the point of maximum probability *(black dot)*. (B) Probability density p evaluated along the curve. *MatLab* script gda05_11.

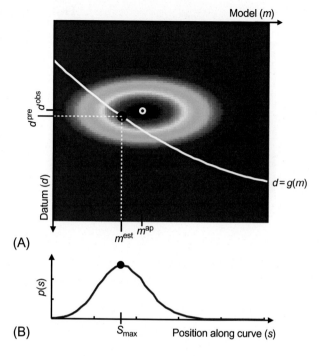

(A)

(B)

FIG. 5.12 (A) If the prior model parameter m^{ap} is much more certain than the observed datum d^{obs}, the solution is close to m^{ap} but may be far from d^{obs}. (B) The probability density function p evaluated along the curve. *MatLab* script gda05_12.

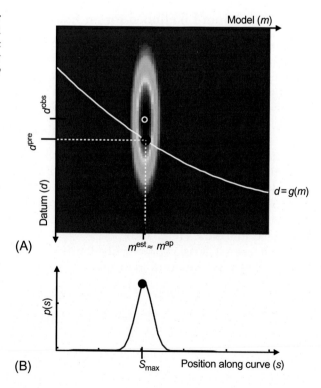

(A)

(B)

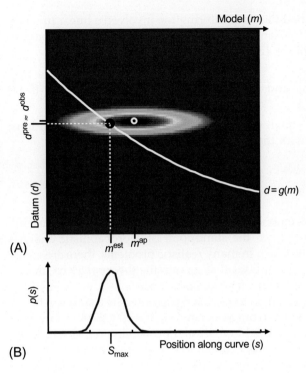

FIG. 5.13 (A) If the prior model parameter m^{ap} is much less certain than the observed datum d^{obs}, the solution is close to d^{obs} but may be far from m^{ap}. (B) The probability density function p evaluated along the curve. *MatLab* script gda05_13.

Comparison with Section 3.9.3 indicates that this is the weighted damped least squares problem, with

$$\varepsilon^2 \mathbf{W}_m = [\operatorname{cov} \mathbf{m}]_A^{-1} \quad \text{and} \quad \mathbf{W}_e = [\operatorname{cov} \mathbf{d}]^{-1} \tag{5.14}$$

so its solution is the least squares solution of $\mathbf{Fm} = \mathbf{f}$; that is, $\mathbf{F}^{\mathrm{T}}\mathbf{Fm}^{\mathrm{est}} = \mathbf{F}^{\mathrm{T}}\mathbf{f}$ with

$$\mathbf{F} = \begin{bmatrix} [\operatorname{cov} \mathbf{d}]^{-1/2}\, \mathbf{G} \\ [\operatorname{cov} \mathbf{m}]_A^{-1/2}\, \mathbf{I} \end{bmatrix} \quad \text{and} \quad \mathbf{f} = \begin{bmatrix} [\operatorname{cov} \mathbf{d}]^{-1/2}\, \mathbf{d}^{\mathrm{obs}} \\ [\operatorname{cov} \mathbf{m}]_A^{-1/2}\, \langle\mathbf{m}\rangle \end{bmatrix} \tag{5.15}$$

The matrices $[\operatorname{cov} \mathbf{d}]^{-1/2}$ and $[\operatorname{cov} \mathbf{m}]^{-1/2}$ can be interpreted as the *certainty* of $\mathbf{d}^{\mathrm{obs}}$ and $\langle\mathbf{m}\rangle$, respectively, since they are numerically large when the uncertainty of these quantities are small. Thus, the top part of the equation $\mathbf{Fm} = \mathbf{f}$ is the data equation $\mathbf{Gm} = \mathbf{d}^{\mathrm{est}}$, weighted by its certainty, and the bottom part is the prior equation $\mathbf{m} = \langle\mathbf{m}\rangle$, weighted by its certainty. Thus, the matrices $\mathbf{W}_m$ and $\mathbf{W}_e$ of weighted least squares have an important probabilistic interpretation.

The vector $\mathbf{f}$ in Eq. (5.15) has unit covariance $[\operatorname{cov} \mathbf{f}] = \mathbf{I}$, since its component quantities $[\operatorname{cov} \mathbf{d}]^{-1/2}\mathbf{d}^{\mathrm{obs}}$ and $[\operatorname{cov} \mathbf{m}]_A^{-1/2}\langle\mathbf{m}\rangle$ each have unit covariance (e.g., by the usual rules of error propagation, $[\operatorname{cov} \mathbf{d}]^{-1/2\mathrm{T}}[\operatorname{cov} \mathbf{d}][\operatorname{cov} \mathbf{d}]^{-1/2} = \mathbf{I}$). Thus, the covariance of the estimated model parameters is

$$[\operatorname{cov} \mathbf{m}^{\mathrm{est}}] = [\mathbf{F}^{\mathrm{T}}\mathbf{F}]^{-1} \tag{5.16}$$

Somewhat incidentally, we note that, had the prior information involved a linear function $\mathbf{Hm}=\mathbf{h}$ of the model parameters, with covariance $[\text{cov } \mathbf{h}]_A$, in contrast to the model parameters themselves, the appropriate form of Eq. (5.15) would be

$$\mathbf{F}=\begin{bmatrix}[\text{cov } \mathbf{d}]^{-1/2}\,\mathbf{G}\\[\text{cov } \mathbf{h}]_A^{-1/2}\,\mathbf{H}\end{bmatrix} \quad \text{and} \quad \mathbf{f}=\begin{bmatrix}[\text{cov } \mathbf{d}]^{-1/2}\,\mathbf{d}^{\text{obs}}\\[\text{cov } \mathbf{h}]_A^{-1/2}\,\mathbf{h}\end{bmatrix} \tag{5.17}$$

This form of weighted damped least squares is especially well suited for computations, especially when $\mathbf{F}^T\mathbf{Fm}^{\text{est}}=\mathbf{F}^T\mathbf{f}$ is solved with the biconjugate gradient method.

5.2.4 Inexact Theories

Weighted damped least squares, as it is embodied in Eqs. (5.15)–(5.17), is extremely useful. Nevertheless, it is somewhat unsatisfying from the standpoint of a probabilistic analysis because the theory has been assumed to be exact. In many realistic problems, there are errors associated with the theory. Some of the assumptions that go into the theory may be unrealistic, or it may be an approximate form of a clumsier but exact theory.

In this case, the model equation $\mathbf{g}(\mathbf{m})=\mathbf{d}$ can no longer be represented by a simple surface. It has become "fuzzy" because there are now errors associated with it (Fig. 5.14A; Tarantola and Valette, 1982b). Instead of a surface, one might envision a probability density function $p_g(\mathbf{m}, \mathbf{d})$ centered about $\mathbf{g}(\mathbf{m})=\mathbf{d}$, with width proportional to the uncertainty of the theory. Rather than find the maximum likelihood point of $p_A(\mathbf{m}, \mathbf{d})$ on a surface, we should instead *combine* $p_A(\mathbf{m}, \mathbf{d})$ and $p_g(\mathbf{m}, \mathbf{d})$ into a single distribution and find the maximum likelihood point in the overall volume (Fig. 5.13C). To proceed, we need a way of combining two probability density functions, each of which contains information about the data and model parameters.

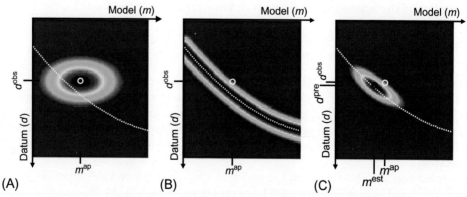

FIG. 5.14   (A) The prior probability density function $p_A(\mathbf{m}, \mathbf{d})$ represents the state of knowledge before the theory is applied. Its mean *(white circle)* is the prior model parameter m^{ap} and observed data d^{obs}. (B) An inexact theory is represented by the conditional probability density function $p_g(\mathbf{m}, \mathbf{d})$, which is centered about the exact theory *(dotted white curve)*. (C) The product $p_T(\mathbf{m}, \mathbf{d})=p_A(\mathbf{m}, \mathbf{d})p_g(\mathbf{m}, \mathbf{d})$ combines the prior information and theory. Its peak is at the estimated data m^{est} and predicted data d^{pre}. *MatLab script gda05_14.*

We have already encountered one special case of a combination in our discussion of Bayesian inference (Section 2.7). After adjusting the variable names to match the current discussion, Bayes' theorem takes the form

$$p(\mathbf{m}|\mathbf{d}) = \frac{p(\mathbf{d}|\mathbf{m})\,p(\mathbf{m})}{p(\mathbf{d})} \quad \text{or} \quad p(\mathbf{m}|\mathbf{d}) \propto p(\mathbf{d}|\mathbf{m})\,p(\mathbf{m}) \tag{5.18}$$

Note that the second form omits the denominator, which is not a function of the model parameters and hence acts only as an overall normalization. This second form can be interpreted as updating the information in $p(\mathbf{m})$ (identified now as the prior information) with $p(\mathbf{d}|\mathbf{m})$ (identified now with the data and quantitative model). Thus, in Bayesian inference, probability density functions are combined by multiplication.

In the general case, we denote the process of combining two probability density functions as $p_3 = C(p_1, p_2)$, meaning that functions 1 and 2 are combined into function 3. Then, clearly, the process of combining must have the following properties (adapted from Tarantola and Valette, 1982b):

1. The order in which two probability density functions are combined should not matter; that is, $C(p_1, p_2)$ should be commutative: $C(p_1, p_2) = C(p_2, p_1)$.
2. The order in which three probability density functions are combined should not matter; that is, $C(p_1, p_2)$ should be associative: $C(p_1, C(p_2, p_3)) = C(C(p_1, p_2), p_3)$.
3. Combining a distribution with the null distribution should return the same distribution; that is, $C(p_1, p_N) = p_1$.
4. The combination $C(p_1, p_2)$ should never be everywhere zero except if p_1 or p_2 is everywhere zero.
5. The combination $C(p_1, p_2)$ should be invariant under reparameterizations.

These conditions can be shown to be satisfied by the choice (Tarantola and Valette, 1982b):

$$p_3 = C(p_1, p_2) = \frac{p_1 p_2}{p_N} \tag{5.19}$$

(at least up to an overall normalization). Note that if the null distribution is constant (as we shall assume for the rest of this chapter), one combines distributions simply by multiplying them:

$$p_T(\mathbf{m}, \mathbf{d}) = p_A(\mathbf{m}, \mathbf{d}) p_g(\mathbf{m}, \mathbf{d}) \tag{5.20}$$

Here, the subscript T means the combined or total distribution. Note that as the error associated with the theory increases, the maximum likelihood point moves back toward the prior values of model parameters and observed data (Fig. 5.15). The limiting case in which the theory is infinitely accurate is equivalent to the case in which the distribution is replaced by a distinct surface.

The maximum likelihood point of $p_T(\mathbf{m}, \mathbf{d})$ is specified by both a set of model parameters $\mathbf{m}^{\text{est}}$ and a set of data $\mathbf{d}^{\text{pre}}$. They are determined simultaneously. This approach is different from that of the least squares problem examined in Section 5.2. In that section, we maximized the probability density function with respect to the model parameters only to determine the most probable model parameters $\mathbf{m}^{\text{est}}$ and afterward can calculate the predicted data via $\mathbf{d}^{\text{pre}} = \mathbf{G}\mathbf{m}^{\text{est}}$. The two methods do not necessarily yield the same estimates for the model

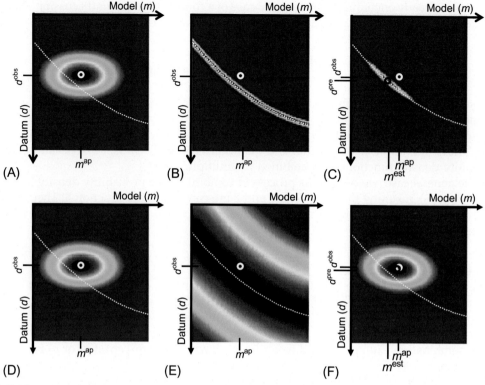

FIG. 5.15 The rows of the figure have the same format as Fig. 5.14. If the theory is made more and more inexact (compare (A–C) with (D–F)), the solution *(black circle)* moves toward the maximum likelihood point of the prior distribution. *MatLab* script gda05_15.

parameters. To find the likelihood point of $p_T(\mathbf{m}, \mathbf{d})$ with respect to model parameters only, we must sum all the probabilities along lines of equal model parameter. This summation can be thought of as projecting the distribution onto the $\mathbf{d} = 0$ plane (Fig. 5.16) and then finding the maximum. The projected distribution $p(\mathbf{m})$ is then

$$p(\mathbf{m}) = \int p_T(\mathbf{m}, \mathbf{d}) d^N d \tag{5.21}$$

where the integration is performed over the entire range of the *d*s.

5.2.5 The Simple Gaussian Case With a Linear Theory

To illustrate this method, we rederive the weighted damped least squares solution, with prior probability density function:

$$p_A(\mathbf{m}, \mathbf{d}) \propto \exp\left[-\frac{1}{2}(\mathbf{m} - \langle \mathbf{m} \rangle)^T [\text{cov } \mathbf{m}]_A^{-1}(\mathbf{m} - \langle \mathbf{m} \rangle) - \frac{1}{2}\left(\mathbf{d} - \mathbf{d}^{obs}\right)^T [\text{cov } \mathbf{d}]^{-1}\left(\mathbf{d} - \mathbf{d}^{obs}\right)\right]$$

$$\tag{5.22}$$

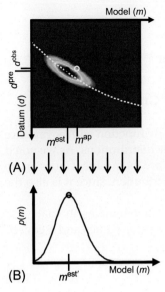

FIG. 5.16   (A) The joint probability density function $p_T(\mathbf{m}, \mathbf{d})$ can be considered the solution to the inverse problem. Its maximum likelihood point *(black circle)* gives an estimate of the model parameter m^{est} and a prediction of the data d^{pre}. (B) The function $p_T(m, \mathbf{d})$ is projected onto the m-axis, by integrating over d, to form the probability density function $p(m)$ of the model parameter irrespective of the datum. This function also has a maximum likelihood point $\mathbf{m}^{\text{est}\prime}$ which in general can be different than m^{est}. The distinction points out the difficulty of defining a unique "solution" to an inverse problem. *MatLab* script gda05_16.

If there are no errors in the theory, then its probability density function is "infinitely narrow" and can be represented by a *Dirac delta function*

$$p_g(\mathbf{m}, \mathbf{d}) = \delta(\mathbf{Gm} - \mathbf{d}) \tag{5.23}$$

where we assume that we are dealing with the linear theory $\mathbf{Gm} = \mathbf{d}$. The total distribution is then given by

$$p_T(\mathbf{m}, \mathbf{d}) = p_A(\mathbf{m}, \mathbf{d})\delta\,(\mathbf{Gm} - \mathbf{d}) \tag{5.24}$$

Performing the projection "integrates away" the delta function

$$p(\mathbf{m}) = \int p_A(\mathbf{m}, \mathbf{d})\,\delta(\mathbf{Gm} - \mathbf{d})\mathrm{d}^N\mathbf{d} \propto$$
$$\exp\left[-\frac{1}{2}(\mathbf{m} - \langle\mathbf{m}\rangle)^{\mathrm{T}}[\text{cov } \mathbf{m}]_A^{-1}(\mathbf{m} - \langle\mathbf{m}\rangle) - \frac{1}{2}\left(\mathbf{Gm} - \mathbf{d}^{\text{obs}}\right)^{\mathrm{T}}[\text{cov } \mathbf{d}]^{-1}\left(\mathbf{Gm} - \mathbf{d}^{\text{obs}}\right)\right]$$
$$\tag{5.25}$$

This projected distribution is exactly the one we encountered in the weighted damped least squares problem (the logarithm of which is shown in Eq. (5.13)).

5.2.6 The General Linear, Gaussian Case

In the general linear, Gaussian case, we assume that all the component probability density functions are Gaussian and that the theory is the linear equation $\mathbf{Gm} = \mathbf{d}$, so that

$$p_A(\mathbf{m}) \propto \exp\left[-\frac{1}{2}(\mathbf{m} - \langle\mathbf{m}\rangle)^T[\text{cov } \mathbf{m}]_A^{-1}(\mathbf{m} - \langle\mathbf{m}\rangle)\right]$$

$$p_A(\mathbf{d}) \propto \exp\left[-\frac{1}{2}\left(\mathbf{d} - \mathbf{d}^{\text{obs}}\right)^T[\text{cov } \mathbf{d}]^{-1}\left(\mathbf{d} - \mathbf{d}^{\text{obs}}\right)\right]$$

$$p_g(\mathbf{m}, \mathbf{d}) \propto \exp\left[-\frac{1}{2}(\mathbf{d} - \mathbf{Gm})^T[\text{cov } \mathbf{g}]^{-1}(\mathbf{d} - \mathbf{Gm})\right]$$

(5.26)

Here, the theory is represented by a Gaussian probability density function with covariance $[\text{cov } \mathbf{g}]$. The total distribution $p_T(\mathbf{m}, \mathbf{d})$ is the product of these three distributions. As we noted in Section 2.4, products of Gaussian probability density functions are themselves Gaussian, so $p_T(\mathbf{m}, \mathbf{d})$ is Gaussian. We need to determine the mean of this distribution, since it is also the maximum likelihood point that defines $\mathbf{m}^{\text{est}}$ and $\mathbf{d}^{\text{pre}}$.

To simplify the algebra, we first define a vector $\mathbf{x} = [\mathbf{d}^T, \mathbf{m}^T]^T$ that contains the data and model parameters, a vector $\langle\mathbf{x}\rangle = [\mathbf{d}^{\text{obs}T}, \langle\mathbf{m}\rangle^T]^T$ that contains their prior values and a covariance matrix

$$[\text{cov } \mathbf{x}] = \begin{bmatrix} [\text{cov } \mathbf{d}] & 0 \\ 0 & [\text{cov } \mathbf{m}]_A \end{bmatrix}$$

(5.27)

The first two products in the total distribution can then be combined into an exponential, with the argument given by

$$-\frac{1}{2}(\mathbf{x} - \langle\mathbf{x}\rangle)^T[\text{cov } \mathbf{x}]^{-1}(\mathbf{x} - \langle\mathbf{x}\rangle)$$

(5.28)

To express the third product in terms of $\mathbf{x}$, we define a matrix $\mathbf{F} = [\mathbf{I}, -\mathbf{G}]$ such that $\mathbf{Fx} = \mathbf{d} - \mathbf{Gm} = 0$. The argument of the third product's exponential is then given by

$$-\frac{1}{2}(\mathbf{Fx})^T[\text{cov } \mathbf{g}]^{-1}(\mathbf{Fx})$$

(5.29)

The total distribution is proportional to an exponential with argument

$$-\frac{1}{2}(\mathbf{x} - \langle\mathbf{x}\rangle)^T[\text{cov } \mathbf{x}]^{-1}(\mathbf{x} - \langle\mathbf{x}\rangle) - \frac{1}{2}(\mathbf{Fx})^T[\text{cov } \mathbf{g}]^{-1}(\mathbf{Fx}) =$$

$$-\frac{1}{2}\mathbf{x}^T\left([\text{cov } \mathbf{x}]^{-1} + \mathbf{F}^T[\text{cov } \mathbf{g}]^{-1}\mathbf{F}\right)\mathbf{x} - \mathbf{x}^T[\text{cov } \mathbf{x}]^{-1}\langle\mathbf{x}\rangle - \frac{1}{2}\langle\mathbf{x}\rangle^T[\text{cov } \mathbf{x}]^{-1}\langle\mathbf{x}\rangle$$

(5.30)

We would like to manipulate this expression into the standard form of the argument of a Gaussian probability density function; that is, an expression involving just a single vector, say $\mathbf{x}^*$, and a single covariance matrix, say $[\text{cov } \mathbf{x}^*]$, related by

$$-\frac{1}{2}(\mathbf{x} - \mathbf{x}^*)^T[\text{cov } \mathbf{x}^*]^{-1}(\mathbf{x} - \mathbf{x}^*) =$$

$$-\frac{1}{2}\mathbf{x}^T[\text{cov } \mathbf{x}^*]^{-1}\mathbf{x} + \mathbf{x}^T[\text{cov } \mathbf{x}^*]^{-1}\mathbf{x}^* - \frac{1}{2}\mathbf{x}^{*T}[\text{cov } \mathbf{x}^*]^{-1}\mathbf{x}^*$$

(5.31)

We can identify $\mathbf{x}^*$ and $[\text{cov } \mathbf{x}^*]$ by matching terms of equal powers of $\mathbf{x}$ with Eq. (5.30). Matching the quadratic terms yields

$$[\text{cov } \mathbf{x}^*]^{-1} = [\text{cov } \mathbf{x}]^{-1} + \mathbf{F}^T[\text{cov } \mathbf{g}]^{-1}\mathbf{F} \tag{5.32}$$

and matching the linear terms yields

$$[\text{cov } \mathbf{x}^*]^{-1}\mathbf{x}^* = [\text{cov } \mathbf{x}]^{-1}\langle \mathbf{x} \rangle \quad \text{or} \quad \mathbf{x}^* = [\text{cov } \mathbf{x}^*][\text{cov } \mathbf{x}]^{-1}\langle \mathbf{x} \rangle \tag{5.33}$$

These choices do not match the constant term, but such a match is not necessary, because the constant term effects only the overall normalization of the probability density function $p_T(\mathbf{x})$. The vector $\mathbf{x}^*$ corresponds to the maximum likelihood point of $p_T(\mathbf{x})$ and so can be considered the solution to the inverse problem. This solution has covariance $[\text{cov } \mathbf{x}^*]$.

Eq. (5.32) for $[\text{cov } \mathbf{x}^*]^{-1}$ can be explicitly inverted to yield an expression for $[\text{cov } \mathbf{x}^*]$, using two matrix identities that we now derive (adapted from Tarantola and Valette, 1982a, with permission). Let $\mathbf{C}_1$ and $\mathbf{C}_2$ be two symmetric matrices whose inverses exist, and let $\mathbf{M}$ be a third matrix. The expression $\mathbf{M}^T + \mathbf{M}^T\mathbf{C}_1^{-1}\mathbf{M}\mathbf{C}_2\mathbf{M}^T$ can be written in two ways by grouping terms as $\mathbf{M}^T\mathbf{C}_1^{-1}[\mathbf{C}_1 + \mathbf{M}\mathbf{C}_2\mathbf{M}^T]$ or $[\mathbf{C}_2^{-1} + \mathbf{M}^T\mathbf{C}_1^{-1}\mathbf{M}]\mathbf{C}_2\mathbf{M}^T$. Multiplying by the matrix inverses gives

$$\mathbf{C}_2\mathbf{M}^T[\mathbf{C}_1 + \mathbf{M}\mathbf{C}_2\mathbf{M}^T]^{-1} = [\mathbf{C}_2^{-1} + \mathbf{M}^T\mathbf{C}_1^{-1}\mathbf{M}]^{-1}\mathbf{M}^T\mathbf{C}_1^{-1} \tag{5.34}$$

Now consider the symmetric matrix expression $\mathbf{C}_2 - \mathbf{C}_2\mathbf{M}^T[\mathbf{C}_1 + \mathbf{M}\mathbf{C}_2\mathbf{M}^T]^{-1}\mathbf{M}\mathbf{C}_2$. By Eq. (5.34), this expression equals $\mathbf{C}_2 - [\mathbf{C}_2^{-1} + \mathbf{M}^T\mathbf{C}_1^{-1}\mathbf{M}]^{-1}\mathbf{M}^T\mathbf{C}_1^{-1}\mathbf{M}\mathbf{C}_2$. Factoring out the term in brackets gives $[\mathbf{C}_2^{-1} + \mathbf{M}^T\mathbf{C}_1^{-1}\mathbf{M}]\{[\mathbf{C}_2^{-1} + \mathbf{M}^T\mathbf{C}_1^{-1}\mathbf{M}]\mathbf{C}_2 - \mathbf{M}^T\mathbf{C}_1^{-1}\mathbf{M}\mathbf{C}_2\}$. Canceling terms gives $[\mathbf{C}_2^{-1} + \mathbf{M}^T\mathbf{C}_1^{-1}\mathbf{M}]^{-1}$ from which we conclude

$$\begin{aligned}
[\mathbf{C}_2^{-1} + \mathbf{M}^T\mathbf{C}_1^{-1}\mathbf{M}]^{-1} &= \mathbf{C}_2 - \mathbf{C}_2\mathbf{M}^T[\mathbf{C}_1 + \mathbf{M}\mathbf{C}_2\mathbf{M}^T]^{-1}\mathbf{M}\mathbf{C}_2 \\
&= \left\{ \mathbf{I} - \mathbf{C}_2\mathbf{M}^T[\mathbf{C}_1 + \mathbf{M}\mathbf{C}_2\mathbf{M}^T]^{-1}\mathbf{M} \right\}\mathbf{C}_2
\end{aligned} \tag{5.35}$$

Equating now $\mathbf{C}_2 = [\text{cov } \mathbf{x}]$, $\mathbf{C}_1 = [\text{cov } \mathbf{g}]$, and $\mathbf{M} = \mathbf{F}$, Eq. (5.32) becomes

$$\begin{aligned}
[\text{cov } \mathbf{x}^*] &= \left[[\text{cov } \mathbf{x}]^{-1} + \mathbf{F}^T[\text{cov } \mathbf{g}]^{-1}\mathbf{F} \right]^{-1} = \\
&\left\{ \mathbf{I} - [\text{cov } \mathbf{x}]\mathbf{F}^T \left[[\text{cov } \mathbf{g}] + \mathbf{F}[\text{cov } \mathbf{x}]\mathbf{F}^T \right]^{-1}\mathbf{F} \right\}[\text{cov } \mathbf{x}]
\end{aligned} \tag{5.36}$$

and Eq. (5.33) becomes

$$\mathbf{x}^* = [\text{cov } \mathbf{x}^*][\text{cov } \mathbf{x}]^{-1}\langle \mathbf{x} \rangle = \left\{ \mathbf{I} - [\text{cov } \mathbf{x}]\mathbf{F}^T \left[[\text{cov } \mathbf{g}] + \mathbf{F}[\text{cov } \mathbf{x}]\mathbf{F}^T \right]^{-1}\mathbf{F} \right\}\langle \mathbf{x} \rangle \tag{5.37}$$

Nothing in this derivation requires the special forms of $\mathbf{F}$ and $[\text{cov } \mathbf{x}]$ assumed above that made $\mathbf{Fx} = 0$ separable into an *explicit* linear inverse problem. Eq. (5.37) is in fact the solution to the completely general, *implicit*, linear inverse problem.

When $\mathbf{Fx} = 0$ is an explicit equation, the formula for $\mathbf{x}^*$ in Eq. (5.21) can be decomposed into its component vectors $\mathbf{d}^{\text{pre}}$ and $\mathbf{m}^{\text{est}}$ by substituting the definition of $\mathbf{F}$ and $[\text{cov } \mathbf{x}]$ (Eq. 5.27) into it and performing the matrix multiplications. An explicit formula for the estimated model parameters is then given by

$$\mathbf{m}^{\text{est}} = \langle \mathbf{m} \rangle + \mathbf{G}^{-g}\left(\mathbf{d}^{\text{obs}} - \mathbf{G}\langle \mathbf{m} \rangle \right) = \mathbf{G}^{-g}\mathbf{d}^{\text{obs}} + [\mathbf{I} - \mathbf{R}]\langle \mathbf{m} \rangle$$

$$\text{with} \quad \mathbf{G}^{-g} = [\text{cov } \mathbf{m}]_A\mathbf{G}^T\left\{ [\text{cov } \mathbf{d}] + [\text{cov } \mathbf{g}] + \mathbf{G}[\text{cov } \mathbf{m}]_A\mathbf{G}^T \right\}^{-1} \tag{5.38}$$

where we have used the generalized inverse $\mathbf{G}^{-g}$ and resolution matrix $\mathbf{R} = \mathbf{G}^{-g}\mathbf{G}$ notation for convenience. This generalized inverse is reminiscent of minimum-length inverse $\mathbf{G}^T[\mathbf{G}\mathbf{G}^T]^{-1}$. However, by equating $\mathbf{C}_2 = [\text{cov } \mathbf{m}]_A$, $\mathbf{C}_1 = [\text{cov } \mathbf{d}] + [\text{cov } \mathbf{g}]$, and $\mathbf{M} = \mathbf{G}$ in matrix identity Eq. (5.34), we can also write the generalized inverse as

$$\mathbf{G}^{-g} = \left\{ \mathbf{G}^T([\text{cov } \mathbf{d}] + [\text{cov } \mathbf{g}])^{-1}\mathbf{G} + [\text{cov } \mathbf{m}]_A^{-1} \right\}^{-1} \mathbf{G}^T([\text{cov } \mathbf{d}] + [\text{cov } \mathbf{g}])^{-1} \tag{5.39}$$

which is reminiscent of least squares generalized inverse $[\mathbf{G}^T\mathbf{G}]^{-1}\mathbf{G}^T$. Both forms of the generalized inverse depend only upon the sum of the covariance of the data $[\text{cov } \mathbf{d}]$ and the covariance of the theory $[\text{cov } \mathbf{g}]$; that is, they make only a combined contribution. This is the most important insight gained from this problem, for the exact-theory case (Eq. 5.15) can be made identical to the inexact-theory case (Eqs. 5.38, 5.39) with the substitution

$$[\text{cov } \mathbf{d}] \rightarrow [\text{cov } \mathbf{d}] + [\text{cov } \mathbf{g}] \tag{5.40}$$

Hence, from the point of view of computations, one continues to use Eq. (5.15) but adjusts the covariance using Eq. (5.40).

Since the estimated model parameters are a linear combination of observed data and prior model parameters, we can therefore calculate its covariance as

$$[\text{cov } \mathbf{m}^{\text{est}}] = \mathbf{G}^{-g}[\text{cov } \mathbf{d}]\mathbf{G}^{-g\text{T}} + [\mathbf{I} - \mathbf{R}][\text{cov } \mathbf{m}]_A[\mathbf{I} - \mathbf{R}]^T \tag{5.41}$$

This expression differs from those derived in Chapters 3 and 4 in that it contains a term dependent on the prior model parameter covariance $[\text{cov } \mathbf{m}]_A$.

We can examine a few interesting limiting cases of problems which have uncorrelated prior model parameters ($[\text{cov } \mathbf{m}]_A = \sigma_m^2\mathbf{I}$), data ($[\text{cov } \mathbf{d}]_A = \sigma_d^2\mathbf{I}$), and theory ($[\text{cov } \mathbf{g}]_A = \sigma_g^2\mathbf{I}$).

5.2.7 Exact Data and Theory

Suppose $\sigma_d^2 = \sigma_g^2 = 0$. The solution is then given by

$$\mathbf{m}^{\text{est}} = \mathbf{G}^T[\mathbf{G}\mathbf{G}^T]^{-1}\mathbf{d}^{\text{obs}} = [\mathbf{G}^T\mathbf{G}]^{-1}\mathbf{G}^T\mathbf{d}^{\text{obs}} \tag{5.42}$$

Note that the solution does not depend on the prior model variance, since the data and theory are infinitely more accurate than the prior model parameters. These solutions are just the minimum-length and least squares solutions, which (as we now see) are simply two different aspects of the same solution. The minimum-length form of the solution, however, exists only when the problem is purely underdetermined; the least squares form exists only when the problem is purely overdetermined.

If the prior model parameters are not equal to zero, then another term appears in the estimated solution:

$$\mathbf{m}^{\text{est}} = \mathbf{G}^{-g}\mathbf{d}^{\text{obs}} + (\mathbf{I} - \mathbf{R})\langle\mathbf{m}\rangle$$

$$= \mathbf{G}^T[\mathbf{G}\mathbf{G}^T]^{-1}\mathbf{d}^{\text{obs}} + \left\{ \mathbf{I} - \mathbf{G}^T[\mathbf{G}\mathbf{G}^T]^{-1}\mathbf{G} \right\}\langle\mathbf{m}\rangle \tag{5.43}$$

$$= [\mathbf{G}^T\mathbf{G}]^{-1}\mathbf{G}^T\mathbf{d}^{\text{obs}}$$

The minimum-length-type solution has been changed by adding a weighted amount of the prior model vector, with the weighting factor being $\{I - G^T[GG^T]^{-1}G\}$. This term is not zero, since it can also be written as $\{I - R\}$. The resolution matrix of the underdetermined problem never equals the identity matrix. On the other hand, the resolution matrix of the overdetermined least squares problem does equal the identity matrix, so the estimated model parameters of the overdetermined problem are not a function of the prior model parameters. Adding prior information with finite error to an inverse problem that features exact data and theory only affects the underdetermined part of the solution.

5.2.8 Infinitely Inexact Data and Theory

In the case of infinitely inexact data and theory, we take the $\sigma_d^2 \to \infty$ or $\sigma_g^2 \to \infty$ (or both). The solution becomes

$$m^{est} = \langle m \rangle \tag{5.44}$$

Since the data and theory contain no information, we simply recover the prior model parameters.

5.2.9 No Prior Knowledge of the Model Parameters

In this case, the limit is $\sigma_m^2 \to \infty$. The solutions are the same as in Section 5.2.7:

$$m^{est} = G^T[GG^T]^{-1}d^{obs} + \{I - G^T[GG^T]^{-1}G\}\langle m \rangle = [G^TG]^{-1}G^Td^{obs} \tag{5.45}$$

Infinitely weak prior information and finite-error data and theory produce the same results as finite-error prior information and error-free data and theory.

5.3 MODEL RESOLUTION IN THE PRESENCE OF PRIOR INFORMATION

The maximum likelihood solution in Eq. (5.17) does not distinguish the weighted data equation $C_d^{-1/2}Gm = C_d^{-1/2}d^{obs}$ from the weighted prior information equation $C_h^{-1/2}Hm = C_h^{-1/2}h$; the latter is simply appended to the bottom of the former to create the combined equation $Fm = f$, which is then solved by simple least squares. Consequently, in analogy to simple least squares, we can define a generalized inverse F^{-g} and a resolution matrix R^F as:

$$F^{-g} = [F^TF]^{-1}F^T \text{ so that } m^{est} = F^{-g}f$$

$$R^F = F^{-g}F \text{ so that } m^{est} = R^Fm^{true} \tag{5.46}$$

However, when defined in this way, the resolution is perfect, since:

$$R^F = F^{-g}F = [F^TF]^{-1}F^TF = I \tag{5.47}$$

Hence R^F is not a useful quantity; we need to construct one that is more informative.

The model parameters depend upon both $\mathbf{d}^{obs}$ and $\mathbf{h}$. This dependence is implicit in the equation $\mathbf{m}^{est} = \mathbf{F}^{-g}\mathbf{f}$, since $\mathbf{f}$ depends on $\mathbf{d}^{obs}$ and $\mathbf{h}$, but can be made explicit by rewriting this equation as:

$$\mathbf{m}^{est} = \mathbf{G}^{-g}\mathbf{d}^{obs} + \mathbf{H}^{-g}\mathbf{h}$$

with $\mathbf{G}^{-g} = \mathbf{A}^{-1}\mathbf{G}^{T}\mathbf{C}_d^{-1}$ and $\mathbf{H}^{-g} = \mathbf{A}^{-1}\mathbf{H}^{T}\mathbf{C}_h^{-1}$ and $\mathbf{A} = \mathbf{F}^{T}\mathbf{F} = \mathbf{G}^{T}\mathbf{C}_d^{-1}\mathbf{G} + \mathbf{H}^{T}\mathbf{C}_h^{-1}\mathbf{H}$ (5.48)

Consider, for the moment, the special case of $\mathbf{h} = 0$; we will relax this requirement later. This case commonly arises in practice, e.g., for the prior information of smoothness. The estimated model parameters depend only upon $\mathbf{d}^{obs}$; that is, $\mathbf{m}^{est} = \mathbf{G}^{-g}\mathbf{d}^{obs} + 0$. We can use the forward equation to predict data associated with an true model parameters, $\mathbf{d}^{pre} = \mathbf{Gm}^{true}$, and then invert these predictions back recovered model parameters, $\mathbf{m}^{est} = \mathbf{G}^{-g}\mathbf{d}^{pre} + 0$. Hence, we obtain the usual formula for resolution:

$$\mathbf{m}^{est} = \mathbf{R}^{G}\mathbf{m}^{true} \text{ with } \mathbf{R}^{G} = \mathbf{G}^{-g}\mathbf{G}$$ (5.49)

(Note, however, that the formula for $\mathbf{G}^{-g}$ in Equation depends on both $\mathbf{G}$ and $\mathbf{H}$). Unlike $\mathbf{R}^{F}$, $\mathbf{R}^{G}$ is not equal to $\mathbf{I}$ and is a useful measure of resolution.

We now relax that condition that $\mathbf{h}^{pri} = 0$. Suppose that the prior information is complete, in the sense that a unique *prior model* $\mathbf{m}^{H}$ can be obtained by a weighted least squares solution of the prior information equation alone; that is, $\mathbf{m}^{H} = [\mathbf{H}^{T}\mathbf{C}_h^{-1}\mathbf{H}]^{-1}\mathbf{H}^{T}\mathbf{C}_h^{-1}\mathbf{h}$. Not all prior information is complete in this sense, but it can always be approximated as complete by adding smallness information of the form $\mathbf{m} \approx 0$ but giving that information extremely large variance. We use this prior model $\mathbf{m}^{H}$ as a reference model, defining the deviation of a given model from it as $\Delta\mathbf{m} = \mathbf{m} - \mathbf{m}^{H}$. The generalized least squares solution can be rewritten in terms of this deviation:

$$\Delta\mathbf{m}^{est} = \mathbf{m}^{est} - \mathbf{m}^{H} = \mathbf{A}^{-1}\left(\mathbf{G}^{T}\mathbf{C}_d^{-1}\mathbf{d}^{obs} + \mathbf{H}^{T}\mathbf{C}_h^{-1}\mathbf{h}\right) - \mathbf{m}^{H}$$

$$= \mathbf{A}^{-1}\left(\mathbf{G}^{T}\mathbf{C}_d^{-1}\mathbf{d}^{obs} + \mathbf{H}^{T}\mathbf{C}_h^{-1}\mathbf{h}^{pri}\right) - \mathbf{A}^{-1}\mathbf{Am}^{H}$$

$$= \mathbf{A}^{-1}\left(\mathbf{G}^{T}\mathbf{C}_d^{-1}\mathbf{d}^{obs} + \mathbf{H}^{T}\mathbf{C}_h^{-1}\mathbf{h} - \mathbf{Am}^{H}\right)$$

$$= \mathbf{A}^{-1}\left(\mathbf{G}^{T}\mathbf{C}_d^{-1}\mathbf{d}^{obs} + \mathbf{H}^{T}\mathbf{C}_h^{-1}\mathbf{h} - \mathbf{G}^{T}\mathbf{C}_d^{-1}\mathbf{Gm}^{H} - \mathbf{H}^{T}\mathbf{C}_h^{-1}\mathbf{H}[\mathbf{H}^{T}\mathbf{C}_h^{-1}\mathbf{H}]^{-1}\mathbf{H}^{T}\mathbf{C}_h^{-1}\mathbf{h}^{pri}\right)$$

$$= \mathbf{A}^{-1}\left(\mathbf{G}^{T}\mathbf{C}_d^{-1}\mathbf{d}^{obs} + \mathbf{H}^{T}\mathbf{C}_h^{-1}\mathbf{h} - \mathbf{G}^{T}\mathbf{C}_d^{-1}\mathbf{Gm}^{H} - \mathbf{H}^{T}\mathbf{C}_h^{-1}\mathbf{h}^{pri}\right)$$

$$= \mathbf{G}^{-g}\left(\mathbf{d}^{obs} - \mathbf{Gm}^{H}\right) = \mathbf{G}^{-g}\left(\mathbf{d}^{obs} - \mathbf{d}^{H}\right)$$ (5.50)

Thus, the deviation of the model from $\mathbf{m}^{H}$ depends only on the deviation of the data from those predicted by $\mathbf{m}^{H}$:

$$\Delta\mathbf{m} = \mathbf{G}^{-g}\Delta\mathbf{d} \text{ with } \Delta\mathbf{m} = \mathbf{m} - \mathbf{m}^{H} \text{ and } \Delta\mathbf{d} = \mathbf{d} - \mathbf{d}^{H}$$ (5.51)

and furthermore

$$\mathbf{G}\Delta\mathbf{m} = \Delta\mathbf{d} \text{ since } \mathbf{G}\Delta\mathbf{m} = \mathbf{G}\left(\mathbf{m} - \mathbf{m}^{H}\right) = \mathbf{Gm} - \mathbf{Gm}^{H} = \mathbf{d} - \mathbf{d}^{H} = \Delta\mathbf{d}$$ (5.52)

We can now combine $\Delta \mathbf{d} = \mathbf{G} \Delta \mathbf{m}$ with $\Delta \mathbf{m} = \mathbf{G}^{-g} \Delta \mathbf{d}$ into the usual statement about resolution,

$$\Delta \mathbf{m}^{\text{est}} = \mathbf{R}^{G} \Delta \mathbf{m}^{\text{true}} \text{ with } \mathbf{R}^{G} = \mathbf{G}^{-g} \mathbf{G} \tag{5.53}$$

The resolution matrix $\mathbf{R}^{G}$ is a good choice for quantifying resolution. However, the quantity being resolved is the deviation of the model from the prior model and not the model, itself. The distinction, while of minor significance in cases where $\mathbf{m}^{H}$ has a simple shape, is more important when $\mathbf{m}^{H}$ is complicated.

5.4 RELATIVE ENTROPY AS A GUIDING PRINCIPLE

In Section 5.2.2, we introduced the information gain, S (Eq. 5.10), as a way of quantifying the difference in information content between two probability density functions. It can be used as a guiding principle for constructing solutions to inverse problems. The idea is to find the probability density function $p_T(\mathbf{m})$—the solution to the inverse problem—that minimizes the information gain of $p_T(\mathbf{m})$ relative to the prior probability density function $p_A(\mathbf{m})$. Thus, as little information as possible has been added to the prior information to create the solution. The quantity $-S$ is the relative entropy of the two probability density functions, so this method is called the *Maximum Relative Entropy* method and is often abbreviated MRE (Kapur, 1989). Some authors define the entropy as $+S$, in which case it is called the *Minimum Relative Entropy* method (also abbreviated MRE).

Constraints need to be added to the minimization of S or else the solution would simply be $p_T(\mathbf{m}) = p_A(\mathbf{m})$. One of these constraints must be that the area beneath $p_T(\mathbf{m})$ is unity. The choice of the other constraints depends on the particular type of inverse problem; that is, whether it is under- or overdetermined.

As an example, consider the underdetermined problem, where the equation $\mathbf{d} = \mathbf{G}\mathbf{m}$ can be assumed to hold in the mean. The minimization problem is

$$\text{minimize} : S = \int p_T(\mathbf{m}) \log \left(\frac{p_T(\mathbf{m})}{p_A(\mathbf{m})} \right) \mathrm{d}^M m \text{ with constraints} \tag{5.54}$$

$$\int p_T(\mathbf{m}) \mathrm{d}^M m = 1 \text{ and } \int p_T(\mathbf{m})(\mathbf{d} - \mathbf{G}\mathbf{m}) \mathrm{d}^M m = 0$$

Here, the final distribution $p_T(\mathbf{m})$ is unknown and the prior distribution $p_A(\mathbf{m})$ is prescribed. Note that the second constraint indicates that the mean (expected) value of the error $\mathbf{e} = \mathbf{d} - \mathbf{G}\mathbf{m}$ is zero.

This minimization problem can be solved by using the *Euler-Lagrange method*. It states that the integral $\int F[f(\mathbf{m}), \mathbf{m}] \mathrm{d}^M m$ is minimized subject to the integral constraint $\int G[f(\mathbf{m}), \mathbf{m}] \mathrm{d}^M m$ when $\Phi = F + \lambda G$ is minimized with respect to f. Here, λ is a Lagrange multiplier. In our case, we introduce one Lagrange multiplier λ_0 associated with the first constraint and a vector $\boldsymbol{\lambda}$ of Lagrange multipliers associated with the second

$$\Phi(\mathbf{m}) = p_T \log(p_T) - p_T \log(p_A) + \lambda_0 p_T + \boldsymbol{\lambda}^{\mathrm{T}} (\mathbf{d} - \mathbf{G}\mathbf{m}) p_T \tag{5.55}$$

Differentiating with respect to p_T yields

$$\frac{\partial \Phi}{\partial p_T} = 0 = \log(p_T) + 1 - \log(p_A) + \lambda_0 + \boldsymbol{\lambda}^T(\mathbf{d} - \mathbf{Gm}) \tag{5.56}$$

or

$$p_T(\mathbf{m}) = p_A(\mathbf{m}) \exp\left\{-(1 + \lambda_0) - \boldsymbol{\lambda}^T(\mathbf{d} - \mathbf{Gm})\right\}. \tag{5.57}$$

Now suppose that the prior probability density function is Gaussian in form, with prior value $\langle \mathbf{m} \rangle$ and prior covariance $[\text{cov } \mathbf{m}]_A$

$$p_A(\mathbf{m}) \propto \exp\left\{-\frac{1}{2}(\mathbf{m} - \langle \mathbf{m} \rangle)^T[\text{cov } \mathbf{m}]_A^{-1}(\mathbf{m} - \langle \mathbf{m} \rangle)\right\} \tag{5.58}$$

Then

$$p_T(\mathbf{m}) \propto \exp\{-A(\mathbf{m})\} \quad \text{with}$$
$$A(\mathbf{m}) = \frac{1}{2}(\mathbf{m} - \langle \mathbf{m} \rangle)^T[\text{cov } \mathbf{m}]_A^{-1}(\mathbf{m} - \langle \mathbf{m} \rangle) - (1 + \lambda_0) - \boldsymbol{\lambda}^T(\mathbf{d} - \mathbf{Gm}) \tag{5.59}$$

We now assert that the best estimate of the model parameter $\mathbf{m}^{\text{est}}$ is the mean of this distribution, which is also its maximum likelihood point. This point occurs where $A(\mathbf{m})$ is minimum:

$$\frac{\partial A}{\partial m_q} = 0 \quad \text{or} \quad 0 = [\text{cov } \mathbf{m}]_A^{-1}(\mathbf{m}^{\text{est}} - \langle \mathbf{m} \rangle) + \mathbf{G}^T\boldsymbol{\lambda} \tag{5.60}$$

Premultiplying by $\mathbf{G}[\text{cov } \mathbf{m}]_A$, substituting in the constraint equation $\mathbf{d} = \mathbf{Gm}$ (which is assumed to hold in the mean) and rearranging yields

$$\boldsymbol{\lambda} = \left\{\mathbf{G}[\text{cov } \mathbf{m}]_A\mathbf{G}^T\right\}^{-1}\{\mathbf{d} - \mathbf{G}\langle \mathbf{m} \rangle\} \tag{5.61}$$

Substituting this expression for $\boldsymbol{\lambda}$ into Eq. (5.60) for $\partial A/\partial \mathbf{m}$ yields the solution

$$\mathbf{m}^{\text{est}} - \langle \mathbf{m} \rangle = [\text{cov } \mathbf{m}]_A\mathbf{G}^T\left\{\mathbf{G}[\text{cov } \mathbf{m}]_A\mathbf{G}^T\right\}^{-1}\{\mathbf{d} - \mathbf{G}\langle \mathbf{m} \rangle\}. \tag{5.62}$$

Thus, the principle of maximum relative entropy, when applied to the underdetermined problem, yields the weighted minimum-length solution (compare Eqs. (5.62) with Eq. (3.43) when $\mathbf{W}_m^{-1} = [\text{cov } \mathbf{m}]_A$). Many of the other inverse theory solutions that were developed in this chapter using maximum likelihood techniques can also be derived using the MRE principle (Woodbury, 2011).

5.5 EQUIVALENCE OF THE THREE VIEWPOINTS

We can arrive at the same general solution to the linear inverse problem by three distinct routes.

Viewpoint 1. The solution is obtained by minimizing a weighted sum of L_2 prediction error and L_2 solution simplicity

$$\text{Minimize}: \ e^{\mathrm{T}}W_e e + \varepsilon^2 [\mathbf{m} - \langle \mathbf{m} \rangle]^{\mathrm{T}} W_m [\mathbf{m} - \langle \mathbf{m} \rangle] \tag{5.63}$$

where ε^2 is a weighting factor.

Viewpoint 2. The solution is obtained by minimizing a weighted sum of three terms: the Dirichlet spreads of model resolution and data resolution and the size of the model covariance.

$$\text{Minimize}: \ \alpha_1 \ \text{spread}(\mathbf{R}) + \alpha_2 \ \text{spread}(\mathbf{N}) + \alpha_3 \ \text{size}([\text{cov}_u \mathbf{m}]) \tag{5.64}$$

Viewpoint 3. The solution is obtained by maximizing the likelihood of the joint Gaussian distribution of data, prior model parameters, and theory.

$$\text{Maximize}: \ L = \log p_{\mathrm{T}}(\mathbf{m}, \mathbf{d}) \tag{5.65}$$

These derivations emphasize the close relationship among the L_2 norm, the Dirichlet spread function, and the Gaussian probability density function.

5.6 CHI-SQUARE TEST FOR THE COMPATIBILITY OF THE PRIOR AND POSTERIOR ERROR

The least squares solution in Eq. (5.17) is based upon the minimization of the total error $\varPhi$, as defined in Eq. (5.13). It is the sum of an error E associated with the observations and the error L associated with the prior information:

$$\varPhi = E + L$$

$$E = \mathbf{e}^{\mathrm{T}} \mathbf{e} \ \text{with} \ \mathbf{e} = [\text{cov} \ \mathbf{d}]^{-\frac{1}{2}} \left(\mathbf{d}^{\text{obs}} - \mathbf{G} \mathbf{m} \right)$$

$$L = \boldsymbol{\ell}^{\mathrm{T}} \boldsymbol{\ell} \ \text{with} \ \boldsymbol{\ell} = [\text{cov} \ \mathbf{h}]_A^{-\frac{1}{2}} (\mathbf{h} - \mathbf{H} \mathbf{m}) \tag{5.66}$$

Each type of error is weighted by its certainty, as quantified by the inverse of the square root of its covariance matrix: $[\text{cov} \, \mathbf{d}]^{-\frac{1}{2}}$ in the case of the prediction error $\mathbf{e}$; and $[\text{cov} \, \mathbf{h}]_A^{-\frac{1}{2}}$ in the case of $\boldsymbol{\ell}$. These are prior covariances, meaning that they are established independently of the outcome of the inverse problem, say from a broad knowledge of the accuracy of the measurement technique and the quality of the prior information. As discussed in Section 5.2.3, the individual errors $\mathbf{e}$ and $\boldsymbol{\ell}$ are uncorrelated random variables with zero mean and unit variance; the weighting normalizes their amplitudes. Thus, E, L, and $\varPhi$ are chi-squared distributed random variables (see Eq. 2.28), with N, K, and $N+K$ degrees of freedom, respectively, where N is the number of data and K is the number of number of pieces of prior information. Because they are chi-squared distributed, they have mean values of $\bar{E} = N$, $\bar{L} = K$, and $\bar{\varPhi} = N + L$.

Once the inverse problem is solved, estimates of the posterior errors, E^{est}, L^{est}, and $\varPhi^{\text{est}}$ can be determined by inserting $\mathbf{m}^{\text{est}}$ into Eq. (5.33). An important issue is whether these estimates are compatible with the means stated earlier; that is, whether their values are compatible with the *null hypothesis* that any difference between the estimated and expected value is due to random variation. The error is not behaving as expected when the null hypothesis is determined to be unlikely to be true. The model might be incorrect, or the prior convariances may be poorly estimated.

The estimated error decreases as more model parameters are added to the inverse problem, since they are able to fit increasingly more of the variability of the data. In the $N+K=M$ case, the estimated error can be zero, even when the true error is greater than zero. Thus, while Φ^{est} is a chi-squared distrbuted random variable, it has only $\nu_\Phi = N+K-M$ degrees of freedom, where M is the number of model pararameters. Consequently, it has mean ν_Φ and variance $2\nu_\Phi$. The null hypothesis is unlikely to be true when Φ^{est} falls outside the 95% confidence interval:

$$\nu_\Phi - 2(2\nu_\Phi)^{\frac{1}{2}} < \Phi < \nu_\Phi + 2(2\nu_\Phi)^{\frac{1}{2}} \tag{5.67}$$

Errors that are to the right of the interval correspond to models that fit the data more poorly than can be expected by random variation alone. This case can occur when the model is too simple; too few model parameters are available to capture the true variability of the data and prior information. Errors to the left of the interval correspond to models that *overfit* the data; that is, the error is smaller than can be expected by random variation alone. The overfit case can occur when the model is too complex. So many model parameters are available that even noise is being fit.

The individual compatibility of the errors E^{est} and of L^{est} with their respective prior variances can also be tested (although with a caviat described later). These errors are chi-squared distributed, but with fewer degrees of error than the total error Φ^{est}. An important issue is that the loss of degrees of freedom associated with the M model parameters should be partitioned between E^{est} and L^{est}. The Welch-Satterthwaite approximation spreads it in proportion to the number of component errors, so that E^{est} is assigned $\nu_E = \nu_\Phi N/(N+K)$ degrees of freedom and L^{est} is assigned $\nu_L = \nu_\Phi K/(N+K)$ degrees of freedom. The 95% confidence intervals are approximately:

$$\nu_E - 2(2\nu_E)^{\frac{1}{2}} < E < \nu_E + 2(2\nu_E)^{\frac{1}{2}} \text{ and } \nu_L - 2(2\nu_L)^{\frac{1}{2}} < L < \nu_L + 2(2\nu_L)^{\frac{1}{2}} \tag{5.68}$$

The caviat is that the accuracy of the Welch-Satterthwaite approximation varies from one inverse problem to another and is not always sufficient for a correct conclusion to be drawn from the test. This problem can be avoided generating empirical distributions for E^{est} and L^{est} (and Φ^{est}, too) by solving the inverse problem many times with noisy synthetic data, computing the error associated with every solution, and constructing an empirical probability distribution (histogram) from them.

The test for the compatibility of the estimated error is illustrated in the following example. Suppose that an unknown time series $\mathbf{m}$, with sampling interval $\Delta t = 1$ and length $M = 101$, is measured by three different observers, so that:

$$\mathbf{d} = \mathbf{Gm} = \begin{bmatrix} \mathbf{I} \\ \mathbf{I} \\ \mathbf{I} \end{bmatrix} \mathbf{m} \tag{5.69}$$

where $\mathbf{I}$ is a $M \times M$ identity matrix. The vector of observations $\mathbf{d}^{obs}$ is of length $N = 3M = 303$ and contains measurement noise with zero mean and a variance of, say, $\sigma_d^2 = (0.1)^2$. Furthermore, suppose that the time series is believed to be smoothly varying, corresponding to the prior information $\mathbf{Hm} = \mathbf{h}$ with:

$$\mathbf{H} = \frac{1}{(\Delta t)^2} \begin{bmatrix} 1 & -2 & 1 & & & \\ & 1 & -2 & 1 & & \\ & & & \ddots & & \\ & & & 1 & -2 & 1 \end{bmatrix} \quad \text{and} \quad \mathbf{h} = 0 \tag{5.70}$$

Here $\mathbf{H}$ is a $K \times M$ matrix of second differences that approximates the second derivative, with $K = M - 2 = 99$. If we also have prior information that $\mathbf{m}$ is approximately sinusoidal, say with an angular frequency of about $\omega_0 = 0.3$ radians/s and an amplitude of about $A = 2$, then a good choice for the variance of the prior information is $\sigma_A^2 = (\omega_0^2 A)^2/2$, since the second derivative of $m(t) = A \sin \omega_0 t$ is $\ddot{m}(t) = -\omega_0^2 A \sin \omega_0 t$ and the average value of $[\ddot{m}(t)]^2$, which is a measure of its variance, is $(\omega_0^2 A)^2/2$. We start with an $\mathbf{m}^{\text{true}}$ (Fig. 5.17A) with the derived properties and create synthetic data $\mathbf{d}^{\text{obs}}$ by adding Normally distributed random noise to $\mathbf{d}^{\text{true}} = \mathbf{G}\mathbf{m}^{\text{true}}$. We then solve the inverse problem and compare the errors with the confidence ranges from Eqs. (5.67), (5.68):

$$\Phi^{\text{est}} = 303 \text{ with a } 95\% \text{ confidence range of } 252 < \Phi < 350$$

$$E^{\text{est}} = 227 \text{ with a } 95\% \text{ confidence range of } 184 < E < 269$$

$$L^{\text{est}} = 57 \text{ with a } 95\% \text{ confidence range of } 50 < L < 98$$

Because the errors are all within the 95% confidence ranges, the null hypothesis cannot be rejected. The differences between the observed values and expected values may be due to random variation. The results of the inversion are compatible with the initial assumptions about the prior variances.

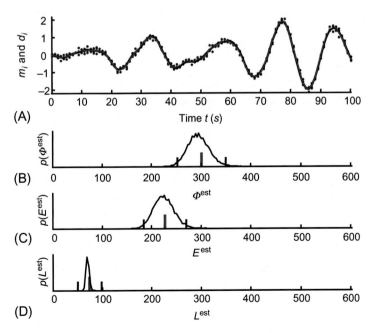

(A)

(B)

(C)

(D)

FIG. 5.17 Test for the compatibility of the prior and posterior errors. (A) The smooth model *(black curve)* is measured by three different observers at a suite of closely spaced points *(blue dots)*. The model is reconstructed *(red)* using the data together with the prior information that its second derivative is small. (B) Histogram *(black curve)* of the total error Φ^{est} for 10,000 realizations of the data, compared to mean *(red bar)* and 95% confidence interval *(blue bars)* of the corresponding chi-squared distribution. (C) Same as (B), but for the observational error E^{est}. (D) Same as (C), but for a priori error L^{est}. *MatLab* script gda05_17.

An alternative (and arguably better) set of confidence limits can be created by solving the inverse problem many times, each with a different realization of observational noise, tabulating the resulting errors and then constructing empirical probability density functions from them (Fig. 5.17B–D). These p.d.f.'s give 95% confidence ranges that are similar to, but not exactly the same, those calculated from Eqs. (5.67) and (5.68). In this particular example, they lead to the same conclusion; that the null hypothesis cannot be rejected.

5.7 THE F-TEST OF THE ERROR IMPROVEMENT SIGNIFICANCE

We sometimes have *two* candidate models for describing an inverse problem, one of which is more complicated than the other (in the sense that it possesses a greater number of model parameters). Suppose that Model A is more complicated than Model B and that the total posterior error Φ_B^{est} for Model A is less than the total posterior error Φ_A^{est} for Model B: $\Phi_B^{est} < \Phi_A^{est}$. Does Model A really fit the data and prior information better than Model B?

The answer to this question depends on the magnitude of the difference. Almost any complicated model will fit better than a less complicated one. The relevant question is whether the fit is *significantly* better, that is, whether the improvement is too large to be accounted for by random fluctuations in the data. For statistical reasons that will be cited, we pretend, in this case, that the two inverse problems are solved with two different realizations of the data.

As discussed in the previous section, Φ^{est} is chi-squared distributed with $\nu = N + K - M$ degrees of freedom. Consequently, Φ^{est} has a mean of ν and the quantity Φ^{est}/ν has a mean of unity. The ratio:

$$F = \frac{\Phi_A^{est}/\nu_B}{\Phi_B^{est}/\nu_A} \tag{5.71}$$

will tend to be less than unity when model A is the better fit and greater than unity when model B is the better fit. If the ratio is only slightly less than or greater than unity, the difference in fit may be entirely the result of random fluctuations in the data and therefore may not be significant. A ratio outside of this range is significant in the sense that it is unlikely to have arisen from random variation. The null hypothesis is that deviation from the value $F = 1$ is due to random variation. One model is said to fit *significantly* better than the other only when the null hypothesis is unlikely to be true.

The probability density function $p(F, \nu_A, \nu_B)$ is called the *Fisher-Snedecor* distribution (or the F-distribution, for short). Its functional form is known, but cannot be written in terms of elementary functions, so we omit it here. It is unimodal, with mean and variance given by

$$\bar{F} = \frac{\nu_B}{\nu_B - 2} \quad \text{and} \quad \sigma_F^2 = \frac{2\nu_B^2(\nu_A + \nu_B - 2)}{\nu_A(\nu_B - 2)^2(\nu_B - 4)} \tag{5.72}$$

For large degrees of freedom, $\bar{F} \approx 1$. The quantities F^{est} and $1/F^{est}$ play symmetrical roles, in the sense that the first quantifies the improvement of fit of model A with respect to model B, and the latter, the improvement of fit of model B with respect to model A. Thus, in testing the null hypothesis that any difference between the two estimated variances is due to random variation, we should compute the probability that F is smaller than $1/F^{est}$ or larger than F^{est}:

$$P\left(F < 1/F^{\text{est}} \text{ or } F > F^{\text{est}}\right) \tag{5.73}$$

The null hypothesis can be rejected if this probability is less than 5%. This probability is computed as:

```
Fobs = (PhiA/vA) / (PhiB/vB);
if( Fobs<1 )
      Fobs=1/Fobs;
end
Pval = 1 - abs(fcdf(1/Fobs,vA,vB)-fcdf(Fobs,vA,vB));
```

(*MatLab* script gda05_18.)

Here, the *MatLab* function fcdf() computes the cumulative probability of F.

In the special case of no prior information, $\Phi^{\text{est}} = E^{\text{est}}$, $K = 0$, and $\nu_\Phi = \nu_E = N - M$. Furthermore, in the common case where the data are uncorrelated and with equal variance, $[\text{cov } \mathbf{d}] = \sigma_d^2 \mathbf{I}$ and the F-ratio simplifies to:

$$F = \left[\frac{\mathbf{e}_A^T \mathbf{e}_A}{\sigma_d^2 (N - M_A)}\right] \bigg/ \left[\frac{\mathbf{e}_B^T \mathbf{e}_B}{\sigma_d^2 (N - M_B)}\right] = \frac{\left(\sigma_{dA}^{\text{est}}\right)^2 / \sigma_d^2}{\left(\sigma_{dB}^{\text{est}}\right)^2 / \sigma_d^2} = \frac{\left(\sigma_{dA}^{\text{est}}\right)^2}{\left(\sigma_{dB}^{\text{est}}\right)^2} \tag{5.74}$$

In this case, the F-ratio is just the ratio of the estimated data variances, and the null hypothesis is that any departure of that ratio from unity is due to random variation. An example is shown in Fig. 5.18.

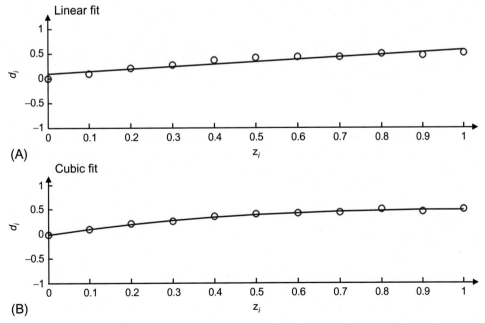

FIG. 5.18 Hypothetical data set (*red circles*) fit (*blue curve*) with (A) a straight line and (B) a cubic polynomial. Although the cubic fit appears superior, an F-test reveals that this level of improvement of fit will be obtained 6.4% of the time under the null hypothesis that the improvement is due to random variation. The improvement of fit is not significant at the 95% level. *MatLab* script gda05_18.

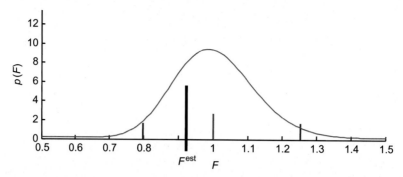

FIG. 5.19   The F-distribution *(red curve)* for the second example of Section 5.7 has $\nu_A = 301$ and $\nu_B = 301$ degrees of freedom, and mean *(red bar)* and 95% confidence interval *(blue bars)* as shown. The value $F^{est} = 0.969$ *(black bar)* estimated from the data (treated as one realization of the inverse problem) lies within the interval, indicating that the null hypothesis cannot be rejected. *MatLab* script gda05_19.

As an example of a case that includes prior information, we return to the time series reconstruction problem described at the end of the previous section and performed it in two different ways, with densely sampled Case A with M model parameters (with M odd), and with sparsely sampled Case B with $(M+1)/2$ model parameters, formed by selecting every other Case A model parameter, and by linearly interpolating between them. The data kernels are then:

$$\mathbf{G}_A = \begin{bmatrix} \mathbf{I} \\ \mathbf{I} \\ \mathbf{I} \end{bmatrix} \text{ and } \mathbf{G}_B = \begin{bmatrix} \mathbf{C} \\ \mathbf{C} \\ \mathbf{C} \end{bmatrix} \text{ where } \mathbf{C} = \begin{bmatrix} 1 & 0 & 0 & 0 & 0 \\ \tfrac{1}{2} & \tfrac{1}{2} & 0 & 0 & 0 \\ 0 & 1 & 0 & 0 & 0 \\ 0 & \tfrac{1}{2} & \tfrac{1}{2} & 0 & 0 \\ 0 & 0 & 1 & 0 & 0 \\ 0 & 0 & \tfrac{1}{2} & \tfrac{1}{2} & 0 \\ & & \vdots & & \end{bmatrix} \cdots \tag{5.75}$$

Here $\mathbf{C}$ is a matrix that takes a sparsely sampled vector of length $(M+1)/2$ into a densely sampled vector of length M by linearly interpolating between values. The solution, for a particular realization of the noise, gives $F^{est} = 0.969$ (Fig. 5.19), which corresponds to a case where Case A (the densely sampled case) fits better than case B (the sparsely sampled). However, the probability that $F^{est} < 1/0.969$ or $F^{est} > 0.969$ is 78.5%, much higher than the 5% needed to reject the null hypothesis. The estimated value of 0.969 is within the central part of the F-distribution and not near its tails (Fig. 5.19).

5.8 PROBLEMS

5.1 Suppose that a random variable m is defined on the interval $[-1,1]$. A reasonable choice for the null probability density function is $p_N(\mathbf{m}) = 1/2$, meaning that m can be anywhere within the interval with equal probability. (A) Calculate (either analytically or numerically) the information gain S of the prior probability density function

$p_A(\mathbf{m}) = 1/2 + cm$, where $0 < c < 1/2$. Note that the larger the constant c, the higher the probability that m will fall in the positive half of the interval. (B) Make and interpret a plot of $S(c)$.

5.2 Suppose that a Gaussian probability density function with mean $\langle m \rangle$ and variance σ_m^2 is used to represent prior information about the following types of model parameters: (A) the density of sea water; (B) the shear velocity at 100-km depth in the earth; (C) the O^{18} to O^{16} ratio in glacial ice. Propose plausible values of $\langle m \rangle$ and σ_m^2 in each case, citing references that justify your values.

5.3 Suppose that you are fitting a cubic polynomial to data, $d_i = m_1 + m_2 z_i + m_3 z_i^2 + m_4 z_i^3$, but have prior information that $m_1 = 2m_2 = 4m_3 = 8m_4$. Write a *MatLab* script to solve this problem using Eq. (3.63). Set up the problem so that $N = 50$, $0 < z_i < 1$, and $m_1^{\text{true}} = 1$, and generate synthetic data with $\sigma_d^2 = (0.1)^2$. How well do the data alone (i.e., without the prior information) constrain the ratios between the ms?

5.4 This problem builds upon Problem 5.3. Suppose that you are fitting a cubic polynomial to data, $d_i = m_1 + m_2 z_i + m_3 z_i^2 + m_4 z_i^3$, but have prior information that $m_1 = 2m_2 = 4m_3 = 8m_4$. Write a *MatLab* script to solve this problem using Eq. (5.17). Use a range of values for the variance σ_m^2 of the prior information, from very uncertain to very certain. Set up the problem so that $N = 50$, $0 < z_i < 1$, and $m_1^{\text{true}} = 1$ and generate synthetic data with $\sigma_d^2 = (0.1)^2$. How well do the data alone (i.e., without the prior information) constrain the ratios between the ms?

5.5 This problem builds upon Problems 5.3 and 5.4. Modify your script in Problem 5.4 in the case where the model of the data fitting a cubic polynomial is thought to be inexact, with a variance $\sigma_d^2 = (0.05)^2$. How does this modification change your results?

5.6 Write a *MatLab* function to empirically generate the $p(F)$ probability density function with $\nu_1 = \nu_2 = 20$ by (A) using the `random('Normal', ...)` function to generate batches of 20 Gaussian-distributed random numbers with zero mean and unit variance. (B) Calculating χ_{20}^2 by summing the squares of each batch. (C) Calculating F for pairs of batches. (D) Repeat many times, creating a histogram of the resulting Fs, and normalize to unit area to produce an empirical estimate of $p(F)$. (E) Compare your result with *MatLab's* `fpdf()` function.

5.7 This problem expands upon Problem 5.6. Suppose that the random numbers in step (A) are drawn from a uniform distribution with zero mean and unit variance, but that F is calculated the same as before (let us call it F'). (A) How different is $p(F')$ from $p(F)$? (B) How would treating data with uniformly distributed error as if they were Gaussian-distributed affect the results of an F-test? Hint: A distribution that is uniform between a and b has a variance of $(b-a)^2/12$.

References

Kapur, J.N., 1989. Maximum Entropy Models in Science and Engineering. Wiley, New York. 636 pp.

Tarantola, A., Valette, B., 1982a. Generalized non-linear inverse problems solved using the least squares criterion. Rev. Geophys. Space Phys. 20, 219–232.

Tarantola, A., Valette, B., 1982b. Inverse problems = quest for information. J. Geophys. 50, 159–170.

Woodbury, A.D., 2011. Minimum relative entropy, Bayes and Kapur. Geophys. J. Int. 185, 181–189.

Nonuniqueness and Localized Averages

6.1 NULL VECTORS AND NONUNIQUENESS

In Chapters 3–5, we presented the basic method of finding estimates of the model parameters in a linear inverse problem. We showed that we could always obtain such estimates but that sometimes in order to do so we had to add prior information to the problem. We shall now consider the meaning and consequences of nonuniqueness in linear inverse problems and show that it is possible to devise solutions that do not depend at all on prior information. As we shall show, however, these solutions are not estimates of the model parameters themselves but estimates of *weighted averages* (linear combinations) of the model parameters.

When the linear inverse problem $\mathbf{Gm} = \mathbf{d}$ has nonunique solutions, there exist nontrivial solutions (i.e., solutions with some nonzero m_i) to the homogeneous equation $\mathbf{Gm} = 0$. These solutions are called the *null vectors* of the inverse problem as premultiplying them by the data

© 2018 Elsevier Inc. All rights reserved.

kernel yields zero. To see why nonuniqueness implies null vectors, suppose that the inverse problem has two distinct solutions $\mathbf{m}^{(1)}$ and $\mathbf{m}^{(2)}$ as

$$\begin{aligned} \mathbf{Gm}^{(1)} &= \mathbf{d} \\ \mathbf{Gm}^{(2)} &= \mathbf{d} \end{aligned} \tag{6.1}$$

Subtracting these two equations yields

$$\mathbf{G}\left(\mathbf{m}^{(1)} - \mathbf{m}^{(2)}\right) = 0 \tag{6.2}$$

Since the two solutions are by assumption distinct, their difference $\mathbf{m}^{\text{null}} = \mathbf{m}^{(1)} - \mathbf{m}^{(2)}$ is nonzero. The converse is also true; any linear inverse problem that has null vectors is nonunique. Note that the equation $\mathbf{Gm}^{\text{null}} = 0$ can be interpreted to mean that $\mathbf{m}^{\text{null}}$ is perpendicular to every row of $\mathbf{G}$ (as its dot product with every row is zero). Consequently, no linear combination of the rows of $\mathbf{G}$ can be a null vector. If $\mathbf{m}^{\text{par}}$ (where par stands for "particular") is any nonnull solution to $\mathbf{Gm} = \mathbf{d}$ (for instance, the minimum length solution), then $\mathbf{m}^{\text{par}} + \alpha \mathbf{m}^{\text{null}}$ is also a solution with the same error for any choice of α. Note that as $\alpha \mathbf{m}^{\text{null}}$ is a null vector for any nonzero α, null vectors are only distinct if they are linearly independent. If a given inverse problem has q distinct null solutions, then the most general solution is

$$\mathbf{m}^{\text{gen}} = \mathbf{m}^{\text{par}} + \sum_{i=1}^{q} \alpha_i \mathbf{m}^{\text{null}(i)} \tag{6.3}$$

where gen stands for "general." We shall show in Section 7.6 that $0 \leq q \leq M$, that is, that there can be no more linearly independent null vectors than there are unknowns.

6.2 NULL VECTORS OF A SIMPLE INVERSE PROBLEM

As an example, consider the following very simple equations:

$$\mathbf{Gm} = \begin{bmatrix} \frac{1}{4} & \frac{1}{4} & \frac{1}{4} & \frac{1}{4} \end{bmatrix} \begin{bmatrix} m_1 \\ m_2 \\ m_3 \\ m_4 \end{bmatrix} = [d_1] \tag{6.4}$$

This equation implies that only the mean value of a set of four model parameters has been measured. One obvious solution to this equation is $\mathbf{m} = [d_1,\ d_1,\ d_1,\ d_1]^{\mathrm{T}}$ (in fact, this is the minimum length solution).

Three linearly independent null solutions can be determined by inspection as

$$\mathbf{m}^{\text{null}(1)} = \begin{bmatrix} 1 \\ -1 \\ 0 \\ 0 \end{bmatrix} \quad \mathbf{m}^{\text{null}(2)} = \begin{bmatrix} 1 \\ 0 \\ -1 \\ 0 \end{bmatrix} \quad \mathbf{m}^{\text{null}(3)} = \begin{bmatrix} 1 \\ 0 \\ 0 \\ -1 \end{bmatrix} \tag{6.5}$$

The most general solution is then

$$\mathbf{m}^{\text{gen}} = \begin{bmatrix} d_1 \\ d_1 \\ d_1 \\ d_1 \end{bmatrix} + \alpha_1 \begin{bmatrix} 1 \\ -1 \\ 0 \\ 0 \end{bmatrix} + \alpha_2 \begin{bmatrix} 1 \\ 0 \\ -1 \\ 0 \end{bmatrix} + \alpha_3 \begin{bmatrix} 1 \\ 0 \\ 0 \\ -1 \end{bmatrix} \tag{6.6}$$

where the αs are arbitrary parameters.

Finding a particular solution to this problem now consists of choosing values for the parameters α_i. If one chooses these parameters so that $||\mathbf{m}||_2$ is minimized, one obtains the minimum length solution. Since the first vector is orthogonal to all the others, this minimum occurs when $\alpha_i = 0$, $i = 1, 2, 3$. We shall show in Chapter 7 that this is a general result: the minimum length solution never contains any null vectors. Note, however, that if other definitions of solution simplicity are used (e.g., flatness or smoothness), those solutions will contain null vectors.

6.3 LOCALIZED AVERAGES OF MODEL PARAMETERS

In Chapters 3–5, we have sought to estimate the elements of the solution vector $\mathbf{m}$. Another approach is to estimate some average of the model parameter $\langle m \rangle = \mathbf{a}^T\mathbf{m}$, where $\mathbf{a}$ is some averaging vector. The average is said to be localized if this averaging vector consists mostly of zeros (except for some group of nonzero elements that multiplies model parameters centered about one particular model parameter). This definition makes particular sense when the model parameters possess some natural ordering in space and time, such as acoustic velocity as a function of depth in the earth. For instance, if $M = 8$, the averaging vector $\mathbf{a} = [0, 0, \frac{1}{4}, \frac{1}{2}, \frac{1}{4}, 0, 0, 0]^T$ could be said to be localized about the fourth model parameter. The averaging vectors are usually normalized so that the sum of their elements is unity.

The advantage of estimating averages of the model parameters rather than the model parameters themselves is that quite often it is possible to identify unique averages even when the model parameters themselves are not unique. To examine when uniqueness can occur, we compute the average of the general solution as

$$\langle m \rangle = \mathbf{a}^T\mathbf{m}^{\text{gen}} = \mathbf{a}^T\mathbf{m}^{\text{par}} + \sum_{i=1}^{q} \alpha_i \mathbf{a}^T\mathbf{m}^{\text{null}(i)} \tag{6.7}$$

If $\mathbf{a}^T\mathbf{m}^{\text{null}(i)}$ is zero for all i, then $\langle m \rangle$ is unique. The process of averaging has completely removed the nonuniqueness of the problem. Since $\mathbf{a}$ has M elements and there are $q \leq M$ constraints placed on $\mathbf{a}$, one can always find at least one vector that cancels (or "annihilates") the null vectors. One cannot, however, always guarantee that the averaging vector is localized around some particular model parameter. But, if $q < M$, one has some freedom in choosing $\mathbf{a}$ and there is some possibility of making the averaging vector at least somewhat localized. Whether this can be done depends on the structure of the null vectors, which, in turn, depends on the structure of the data kernel $\mathbf{G}$. Since the small-scale features of the model are unresolvable in many problems, unique localized averages can often be found.

6.4 RELATIONSHIP TO THE RESOLUTION MATRIX

During the discussion of the resolution matrix $\mathbf{R}$ (Section 4.3), we encountered in a somewhat different form the problem of determining averaging vectors. We showed that any estimate $\mathbf{m}^{est}$ computed from a generalized inverse $\mathbf{G}^{-g}$ was related to the true model parameters by

$$\mathbf{m}^{est} = \mathbf{R}\mathbf{m}^{true} = \mathbf{G}^{-g}\mathbf{G}\mathbf{m}^{true} \quad \text{or} \quad m_i^{est} = \sum_{j=1}^{M} R_{ij}m_j^{true} = \sum_{j=1}^{M}\sum_{k=1}^{N} G_{ik}^{-g}G_{kj}m_j^{true} \tag{6.8}$$

The ith row of $\mathbf{R}$ (or rather its transpose) can be interpreted as a unique averaging vector $\mathbf{a}$ that is centered about m_i, with the averaging vector $\mathbf{a}$ being built up from linear combinations of the rows of the data kernel $\mathbf{G}$

$$m_i^{est} = \langle m \rangle^{(i)} = \sum_{j=1}^{M} a_j^{(i)} m_j^{true} \quad \text{with} \quad a_j^{(i)} = \sum_{k=1}^{N} c_k^{(i)} G_{kj} \quad \text{and} \quad c_k^{(i)} = G_{ik}^{-g} \tag{6.9}$$

Note that the resolution matrix in Eq. (6.8) is composed of the product of the generalized inverse and the data kernel. We can interpret this product as meaning that a row of the resolution matrix is composed of a weighted sum of the rows of the data kernel $\mathbf{G}$ (where the elements of the generalized inverse are the weighting factors) $c_k^{(i)} = G_{ik}^{-g}$ regardless of the generalized inverse's particular form. An averaging vector $\mathbf{a}$ produces a unique average $\langle m \rangle$ if and only if $\mathbf{a}^T$ can be represented as a linear combination of the rows of the data kernel $\mathbf{G}$, since the rows of $\mathbf{G}$ are guaranteed to be perpendicular to every null vector.

Whether or not the average is truly localized depends on the structure of $\mathbf{R}$. The spread function discussed previously in Section 4.6 is a measure of the degree of localization.

The process of forming the generalized inverse is equivalent to "shuffling" the rows of the equation $\mathbf{Gm} = \mathbf{d}$ by forming linear combinations until the data kernel is as close as possible to an identity matrix. Each row of the data kernel can then be viewed as a localized averaging vector, and each corresponding row of the shuffled data vector is the estimated value of the average.

6.5 AVERAGES VERSUS ESTIMATES

We can, therefore, identify a type of dualism in inverse theory. Given a generalized inverse $\mathbf{G}^{-g}$ that in some sense solves $\mathbf{Gm} = \mathbf{d}$, we can speak either of estimates of model parameters $\mathbf{m}^{est} = \mathbf{G}^{-g}\mathbf{d}$ or of localized averages $\langle \mathbf{m} \rangle = \mathbf{G}^{-g}\mathbf{d}$. The numerical values are the same but the interpretation is quite different. When the solution is interpreted as a localized average, it can be viewed as a unique quantity that exists independently of any prior information applied to the inverse problem. Examination of the resolution matrix may reveal that the average is not especially localized and the solution may be difficult to interpret. When the solution is viewed as an estimate of a model parameter, the location of what is being solved for is clear. The estimate can be viewed as unique only if one accepts as appropriate whatever prior information was used to remove the inverse problem's underdeterminacy. In most instances, the choice of prior information is somewhat ad hoc, so the solution may still be difficult to interpret.

In the sample problem stated earlier, the data kernel has only one row. There is therefore only one averaging vector that will annihilate all the null vectors: one proportional to that row

$$\mathbf{a} = \begin{bmatrix} \frac{1}{4} & \frac{1}{4} & \frac{1}{4} & \frac{1}{4} \end{bmatrix}^{T} \tag{6.10}$$

This averaging vector is clearly unlocalized. In this problem, the structure of $\mathbf{G}$ is just too poor to form good averages. The generalized inverse to this problem is by inspection

$$\mathbf{G}^{-g} = [1 \ 1 \ 1 \ 1]^{T} \tag{6.11}$$

The resolution matrix is therefore

$$\mathbf{R} = \frac{1}{4} \begin{bmatrix} 1 & 1 & 1 & 1 \\ 1 & 1 & 1 & 1 \\ 1 & 1 & 1 & 1 \\ 1 & 1 & 1 & 1 \end{bmatrix} \tag{6.12}$$

which is very unlocalized and equivalent to Eq. (6.8).

6.6 NONUNIQUE AVERAGING VECTORS AND PRIOR INFORMATION

There are instances in which even nonunique averages of model parameters can be of value, especially when they are used in conjunction with other *prior* knowledge of the nature of the solution (Wunsch and Minster, 1982). Suppose that one simply picks a localized averaging vector that does not necessarily annihilate all the null vectors and that, therefore, does not lead to a unique average. In the earlier problem, the vector $\mathbf{a} = (\frac{1}{3})[1,1,1,0]^{T}$ might be such a vector. It is somewhat localized, being centered about the second model parameter. Note that it does not lead to a unique average, since

$$\langle m \rangle = \mathbf{a}^{T}\mathbf{m}^{gen} = d_1 + 0 + 0 + \frac{1}{3}\alpha_3 \tag{6.13}$$

is still a function of one of the arbitrary parameters α_i. Suppose, however, that there is prior knowledge that every m_i must satisfy $0 \le m_i \le 2d_1$. Then from the equation for $\mathbf{m}^{gen}$, α_3 must be no greater than d_1 and no less than $-d_1$. Since $-d_1 \le \alpha_3 \le d_1$, the average has bounds $(\frac{2}{3})d_1 \le \langle m \rangle \le (\frac{4}{3})d_1$. These constraints are considerably tighter than the prior bounds on m_i, which demonstrates that this technique has indeed produced some useful information. This approach works because even though the averaging vector does not annihilate all the null vectors, $\mathbf{a}^{T}\mathbf{m}^{null(i)}$ is small compared with the elements of the null vector. Localized averaging vectors often lead to small products since the null vectors often fluctuate rapidly about zero, indicating that small-scale features of the model are the most poorly resolved. A slightly more complicated example of this type is solved in Fig. 6.1.

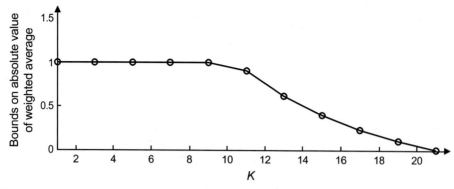

FIG. 6.1   Bounds on weighted averages of model parameters, m_i, in a problem in which the only datum is that the sum of all model parameters is zero. When this observation is combined with the prior information that each model parameter must satisfy $|m_i| \leq 1$, bounds can be placed on the weighted averages of the model parameter. The bounds shown here are for averages of K neighboring model parameters. Note that the bounds are tighter than the prior bounds only when $K > 10$. *MatLab* script gda06_01.

This approach can be generalized as follows (Oldenburg, 1983):

$$\text{maximize/minimize } \langle m \rangle = \mathbf{a}^T \mathbf{m} \text{ with respect to } \mathbf{m}$$
$$\text{with the constraints } \mathbf{Gm} = \mathbf{d}^{\text{obs}} \text{ and } \mathbf{m}^{(l)} \leq \mathbf{m} \leq \mathbf{m}^{(u)} \tag{6.14}$$

Here, $\mathbf{m}^l$ and $\mathbf{m}^u$ are the prior lower and upper bounds on $\mathbf{m}$, respectively. Note that this formulation does not explicitly include null vectors; the constraint $\mathbf{Gm} = \mathbf{d}^{\text{obs}}$ is sufficient to ensure that the model parameters satisfy the data. This problem is a special case of the *linear programming problem*

$$\text{find } \mathbf{x} \text{ that minimizes } z = \mathbf{f}^T \mathbf{x}$$
$$\text{with the constraints } \mathbf{Ax} \leq \mathbf{b} \text{ and } \mathbf{Cx} = \mathbf{d} \text{ and } \mathbf{x}^{(l)} \leq \mathbf{x} \leq \mathbf{x}^{(u)} \tag{6.15}$$

Note that the minimization problem can be converted into a maximization problem by flipping the sign of $\mathbf{f}$. Furthermore, "≥ type" inequality constraints can be converted to "≤ type" by multiplication by -1.

The linear programming problem was first studied by economists and business operations analysts. For example, z might represent the profit realized by a factory producing a product line, where the number of each product is given by $\mathbf{x}$ and the profit on each item given by $\mathbf{f}$. The problem is to maximize the total profit $\mathbf{f}^T \mathbf{x}$ without violating the constraint that one can produce only a positive amount of each product, or any other linear inequality constraints that might represent labor laws, union regulations, physical limitation of machines, etc.

MatLab provides a `linprog()` function for solving the linear programming problem. Eq. (6.13) is solved by calling it twice, once to minimize $\langle m \rangle$ and the other to maximize it

```
[mest1, amin] = linprog(a,[],[],G,dobs,mlb,mub);
[mest2, amax] = linprog(- a,[],[],G,dobs,mlb,mub);
amax = - amax;
```

(*MatLab* script gsa06_01)

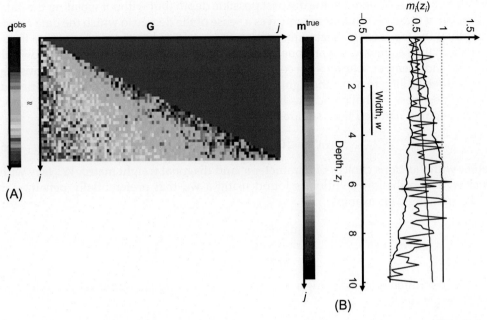

FIG. 6.2 (A) This underdetermined inverse problem, $\mathbf{d} = \mathbf{Gm}$, has $M = 100$ model parameters m_i and $N = 40$ data d_i. The data are weighted averages of the model parameters, from the surface down to a depth, z, that increases with index, i. The observed data include additive noise. (B) The true model parameters (*red curve*) increase linearly with depth z. The estimated model parameters (*blue curve*), computed using the minimum length method, scatter about the true model at shallow depths ($z < 6$) but decline toward zero at deeper depths due to poor resolution. Bounds on localized averages of the model parameters, with an averaging width, $w = 2$ (*black curves*), are for prior information, $0 < m_i < 1$ (*gray-dotted lines*). *MatLab* script gda06_02.

Here, amin and amax are its lower and upper bounds on $\langle m \rangle$, respectively. While the corresponding model parameter vectors mest1 and mest2 are also calculated, they are not usually of interest. The arguments of linprog() include the averaging vector a, the data kernel G, the observed data dobs, and the prior upper and lower bounds mlb and mub on the model parameters. An example using a Laplace transform-like data kernel is shown in Fig. 6.2.

6.7 END-MEMBER SOLUTIONS AND SQUEEZING

The nonuniqueness of an inverse problem can be explored by constructing *end-member solutions*, which satisfy the data to equal degrees, but which obey different kinds of prior information. For example, "shallow" and "deep" solutions might be useful end-members in a one-dimensional inversion for model parameters $m(z_i)$, where the auxiliary variable z represents depth, especially if $m(z_i)$ is thought to contain a single feature centered at unknown depth $z = z_0$. The shallow solution is one for which the feature is centered at the smallest possible depth (but without violating the data) and the deep solution is the one for which the feature is centered at the deepest possible depth (but without violating the data). A comparison of these two end-members gives a sense of the degree to which the data constrain the depth of the feature. This technique is sometimes referred to as *squeezing*, since the solution is alternately squeezed toward one prior notion of an end-member and then toward another. The end-member solutions are constructed by starting with a solution that satisfies the data, and then adding to it just the right combination of null vectors to satisfy the prior information of deepness or shallowness as best as possible. This process can be implemented by using a weighted damped least squares solution of the form (see Section 3.9.3):

$$\mathbf{m}^{\text{est}} = \left[\mathbf{G}^{\text{T}}\mathbf{G} + \varepsilon^2 \mathbf{W}_m\right]^{-1} \mathbf{G}^{\text{T}} \mathbf{d}^{\text{obs}} \tag{6.16}$$

together with judicious choices of parameter ε^2 and diagonal weight matrix $\mathbf{W}_m$ (say with diagonal $\mathbf{w}_m$). The shallow solution is found using a $\mathbf{w}_m$ that preferentially penalizes deep model parameters; for example

$$\mathbf{w}_m = \begin{bmatrix} 1 & 1 & \cdots & 1 & \begin{matrix} \text{smoothly} \\ \text{ramping} \\ \text{up to} \end{matrix} & w_0 & w_0 & \cdots & w_0 \end{bmatrix}^{\text{T}} \text{ with } w_0 \gg 1 \tag{6.17}$$

The deep solution is found using a $\mathbf{w}_m$ that preferentially penalizes shallow model parameters; for example

$$\mathbf{w}_m = \begin{bmatrix} w_0 & w_0 & \cdots & w_0 & \begin{matrix} \text{smoothly} \\ \text{ramping} \\ \text{down to} \end{matrix} & 1 & 1 & \cdots & 1 \end{bmatrix}^{\text{T}} \text{ with } w_0 \gg 1 \tag{6.18}$$

The two parameters ε^2 and w_0 are adjusted by trial and error, so that the corresponding solutions satisfy the data to an acceptable degree and are as different from one another as possible (Fig. 6.3).

Squeezing can be used to implement many other prior notion end-members, in addition to spatial localization. The general procedure is to represent the solution as a sum of patterns $\mathbf{p}^{(j)}$ with unknown coefficients $\hat{m}_j$:

$$m_i = \sum_{j=1}^{M} \hat{m}_j p_i^{(j)} = \mathbf{P}\hat{\mathbf{m}} \text{ with } P_{ij} = p_i^{(j)} \tag{6.19}$$

Two of these patterns, say $\mathbf{p}^{(A)}$ and $\mathbf{p}^{(B)}$, represent the end-members and the other patterns fill out the solution to make it complete. The inverse problem is then recast into one where the coefficients $\hat{\mathbf{m}}$ are the model parameters with the substitution $\mathbf{m} = \mathbf{P}\hat{\mathbf{m}}$, so that $\mathbf{Gm} = \mathbf{d}$ becomes $(\mathbf{GP})\hat{\mathbf{m}} = \mathbf{d}$, and the solution of this new equation is squeezed toward large $\hat{m}_A$ and small $\hat{m}_B$ (and vice versa). For instance, the model parameters $m_i = m(z_i)$ (where z is an auxiliary variable) could be squeezed toward the end-members of polynomials of mostly low/high order by identifying Eq. (6.16) as a Taylor series (i.e., with $p^{(j)}(z_i) = (z_i)^{j-1}$) and by squeezing toward $\hat{m}_1$ and away from $\hat{m}_M$ (and vice versa). If the solution cannot be squeezed close to $\mathbf{p}^{(A)}$ without causing unacceptably large error, one is justified in concluding that, even though the data do not uniquely determine the solution, it requires that the solution *cannot* look like $\mathbf{p}^{(A)}$ (and similarly for $\mathbf{p}^{(B)}$).

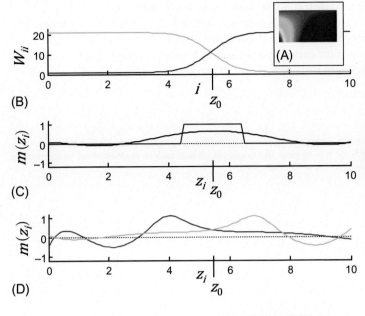

FIG. 6.3 Squeezed solutions for a simple inverse problem for model parameters $m(z_i)$, where z is depth. (A) The data kernel represents smooth averages of the data and has the form $G_{ij}=c_j\exp(-c_jz_i)$, where the c_j s are constants. (B) The true model (*blue curve*) is a boxcar centered at $z_0=5.4$. The ordinary damped least squares solution (*black curve*) is smoother than the boxcar, but has negligible error. (C) The shallow solution (*red curve*) and deep solution (*green curve*) also have negligible error, but are concentrated at shallow and deep depths, respectively, than the ordinary damped least squares solution. Taken together, they indicate that the data are sufficient to constrain the central part of the solution to the $4<z<7$ depth range. *MatLab* script gda06_03.

6.8 PROBLEMS

6.1. What is the general solution to the problem $\mathbf{Gm}=\mathbf{d}$, with

$$\mathbf{G} = \begin{bmatrix} 1 & 1 & 0 & 0 \\ 0 & 0 & 1 & 1 \end{bmatrix}$$

6.2. Give some examples of physical problems where the model parameters can be assumed, with reasonable certainty, to fall between lower and upper bounds (and identify the bounds).

6.3. Suppose that $M=21$, model parameters are known to have bounds $-1\leq m_i\leq 1$. Suppose that the unweighted average of each three adjacent model parameters are observed so that the data kernel has the form

$$\mathbf{G} = \frac{1}{3}\begin{bmatrix} 1 & 1 & 1 & 0 & \cdots & 0 & 0 & 0 & 0 & 0 & 0 \\ 0 & 1 & 1 & 1 & 0 & \cdots & 0 & 0 & 0 & 0 & 0 \\ & & & & & \cdots & & & & & \\ 0 & 0 & 0 & 0 & 0 & 0 & 0 & 0 & 1 & 1 & 1 \end{bmatrix}$$

Can useful bounds be placed on the weighted average of the three adjacent model parameters, where the weights are [¼, ½, ¼]? Adapt *MatLab* script gda06_01 to explore this problem.

6.4. Consider an inverse problem with $M=8$ model parameters $m_i=m(z_i)=\sin(\pi z_i)$, where the auxiliary depth variable z is in the range $0\leq z\leq 1$. (A) Construct a synthetic data set in which the observations are a triplicate set of measurements of m_i equally spaced depths, all made with an accuracy of $\sigma_d=0.1$ (so that $N=3M$). (B) Estimate a reference solution using simple least squares. (C) Use squeezing to develop end-member solutions for the case where the solution is represented as a Taylor series and where the end-members represent polynomials with mostly low-order or mostly high-order terms. (D) Show that the solution cannot be squeezed to one in which the low-order terms have negligible amplitude but for which the error is acceptably low.

References

Oldenburg, D.W., 1983. Funnel functions in linear and nonlinear appraisal. J. Geophys. Res. 88, 7387–7398.
Wunsch, C., Minster, J.F., 1982. Methods for box models and ocean circulation tracers: mathematical programming and non-linear inverse theory. J. Geophys. Res. 87, 5647–5662.

7

Applications of Vector Spaces

7.1 MODEL AND DATA SPACES

So far, we have used vector notation for the data $\mathbf{d}$ and model parameters $\mathbf{m}$ mainly because it facilitates the algebra. The interpretations of vectors as geometrical quantities can also be used to gain an insight into the properties of inverse problems. We therefore introduce the idea of vector spaces containing $\mathbf{d}$ and $\mathbf{m}$, which we shall denote $S(\mathbf{d})$ and $S(\mathbf{m})$. Any particular choice of $\mathbf{m}$ and $\mathbf{d}$ is then represented as a vector in these spaces (Fig. 7.1).

The linear equation $\mathbf{d} = \mathbf{Gm}$ can be interpreted as a mapping of vectors from $S(\mathbf{m})$ to $S(\mathbf{d})$ and its solution $\mathbf{m}^{est} = \mathbf{G}^{-g}\mathbf{d}$ as a mapping of vectors from $S(\mathbf{d})$ to $S(\mathbf{m})$.

One important property of a vector space, such as $S(\mathbf{m})$, is that its coordinate axes are arbitrary. Thus far we have been using axes parallel to the individual model parameters, but we recognize that we are by no means required to do so. Any set of vectors that spans the space

© 2018 Elsevier Inc. All rights reserved.

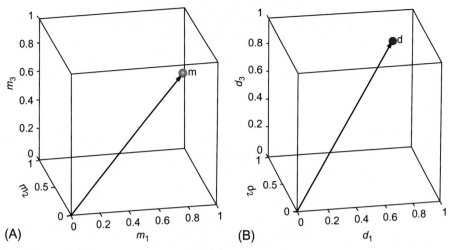

FIG. 7.1   (A) The model parameters represented as a vector **m** in the M-dimensional space $S(\mathbf{m})$ of all possible model parameters. (B) The data represented as a vector **d** in the N-dimensional space $S(\mathbf{d})$ of all possible data. *MatLab* script gda07_01.

will serve as coordinate axes. The Mth dimensional space $S(\mathbf{m})$ is spanned by any M vectors, say, $\mathbf{m}^{(i)}$, as long as these vectors are linearly independent. An arbitrary vector lying in $S(\mathbf{m})$, say, $\mathbf{m}^*$, can be expressed as a sum of these M basis vectors, written as

$$\mathbf{m}^* = \sum_{i=1}^{M} \alpha_i \mathbf{m}^{(i)} \tag{7.1}$$

where the αs are the components of the vector $\mathbf{m}^*$ in the new coordinate system. If the $\mathbf{m}^{(i)}$s are linearly dependent, then the vectors $\mathbf{m}^{(i)}$ lie in a subspace, or *hyperplane*, of $S(\mathbf{m})$ and the expansion cannot be made (Fig. 7.2).

   We shall consider, therefore, transformations of the coordinate systems of the two spaces $S(\mathbf{m})$ and $S(\mathbf{d})$. Using $S(\mathbf{m})$ as an example, if **m** is the representation of a vector in one coordinate system and $\mathbf{m}'$ its representation in another, we can write the transformation as

$$\mathbf{m}' = \mathbf{Tm} \quad \text{and} \quad \mathbf{m} = \mathbf{T}^{-1}\mathbf{m}' \tag{7.2}$$

where **T** is the transformation matrix. If the new basis vectors are still mutually orthogonal unit vectors, then **T** represents simple rotations or reflections of the coordinate axes. As we shall show later, however, it is sometimes convenient to choose a new set of basis vectors that are not unit vectors.

7.2 HOUSEHOLDER TRANSFORMATIONS

   In this section, we shall show that the minimum length, least squares, and constrained least squares solutions can be found through simple transformations of the equation $\mathbf{Gm} = \mathbf{d}$. While the results are identical to the formulas derived in Chapter 3, the approach and emphasis are quite different and will enable us to visualize these solutions in a new way. We shall

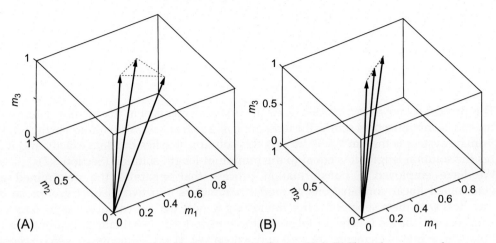

FIG. 7.2 (A) These three vectors span the three-dimensional space $S(\mathbf{m})$. (B) These three vectors do not span the space as they all lie on the same plane. *MatLab* script gda07_02.

begin by considering a purely underdetermined linear problem $\mathbf{Gm} = \mathbf{d}$ with $M > N$. Suppose we want to find the minimum-length solution (the one that minimizes $L = \mathbf{m}^T\mathbf{m}$). We shall show that it is easy to find this solution by transforming the model parameters into a new coordinate system $\mathbf{m}' = \mathbf{Tm}$. The inverse problem becomes

$$\mathbf{d} = \mathbf{Gm} = \mathbf{GIm} = \{\mathbf{GT}^{-1}\}\{\mathbf{Tm}\} = \mathbf{G}'\mathbf{m}' \tag{7.3}$$

where $\mathbf{G}' = \mathbf{GT}^{-1}$ is the data kernel in the new coordinate system. The solution length becomes

$$L = \mathbf{m}^T\mathbf{m} = \{\mathbf{T}^{-1}\mathbf{m}'\}^T\{\mathbf{T}^{-1}\mathbf{m}'\} = \mathbf{m}'^T\{\mathbf{T}^{-1T}\mathbf{T}^{-1}\}\mathbf{m}' \tag{7.4}$$

Suppose that we could choose $\mathbf{T}$ so that $\{\mathbf{T}^{-1T}\mathbf{T}^{-1}\} = \mathbf{I}$. The solution length would then have the same form in both coordinate systems, namely, the sum of squares of the vector elements. Minimizing $\mathbf{m}'^T\mathbf{m}'$ would be equivalent to minimizing $\mathbf{m}^T\mathbf{m}$. Transformations of this type that do not change the length of the vector components are called *unitary transformations*. They may be interpreted as rotations and reflections of the coordinate axes. We can see from Eq. (7.4) that unitary transformations satisfy $\mathbf{T}^T = \mathbf{T}^{-1}$.

Now suppose that we could also choose the transformation so that $\mathbf{G}'$ is the lower triangular matrix

$$\begin{bmatrix} G'_{11} & 0 & 0 & 0 & \cdots & 0 & 0 & \cdots & 0 \\ G'_{21} & G'_{22} & 0 & 0 & \cdots & 0 & 0 & \cdots & 0 \\ G'_{31} & G'_{32} & G'_{33} & 0 & \cdots & 0 & 0 & \cdots & 0 \\ \vdots & \vdots & \vdots & \vdots & \cdots & & & & \\ G'_{N1} & G'_{N2} & G'_{N3} & G'_{N4} & \cdots & G'_{NN} & 0 & \cdots & 0 \end{bmatrix} \begin{bmatrix} m'_1 \\ m'_2 \\ m'_3 \\ \vdots \\ m'_M \end{bmatrix} = \begin{bmatrix} d_1 \\ d_2 \\ d_3 \\ \vdots \\ d_N \end{bmatrix} \tag{7.5}$$

Notice that no matter what values we pick for $\{m'_i, i = N+1, M\}$, we cannot change the value of $\mathbf{G}'\mathbf{m}'$ as the last $M - N$ columns of $\mathbf{G}'$ are all zero. On the other hand, we can solve for the first N elements of $\mathbf{m}'^{\text{est}}$ uniquely as

$$m'_1 = [d_1]/G'_{11}$$
$$m'_2 = [d_2 - G'_{21}m'_1]/G'_{22}$$
$$m'_3 = [d_3 - G'_{31}m'_1 - G'_{32}m'_2]/G'_{33} \tag{7.6}$$
$$\vdots$$

This process is known as *back-solving*. As the first N elements of $\mathbf{m}'$ are thereby determined, $\mathbf{m}'^T\mathbf{m}'$ can be minimized by setting the remaining m'_i equal to zero. The solution in the original coordinate system is then $\mathbf{m}^{est} = \mathbf{T}^{-1}\mathbf{m}'$. As this solution satisfies the data exactly and minimizes the solution length, it is equal to the minimum-length solution (Section 3.7).

We have employed a transformation process that separates the determined and undetermined linear combinations of model parameters into two distinct groups so that we can deal with them separately. It is interesting to note that we can now easily determine the null vectors for the inverse problem. In the transformed coordinates they are the set of vectors whose first N elements are zero and whose last $M - N$ elements are zero except for one element. There are clearly $M - N$ such vectors, so we have established that there are never more than M null vectors in a purely underdetermined problem. The null vectors can easily be transformed back into the original coordinate system by premultiplication by $\mathbf{T}^{-1}$. As all but one element of the transformed null vectors are zero, this operation just selects a column of $\mathbf{T}^{-1}$ (or, equivalently, a row of $\mathbf{T}$).

One transformation that can triangularize a matrix is called a *Householder transformation*. We shall discuss in Section 7.3 how it can be constructed. As we shall see later, Householder transformations have application to a wide variety of methods that employ the L_2 norm as a measure of size. These transformations provide an alternative method of solving such problems and additional insight into their structure.

The overdetermined linear inverse problem $\mathbf{Gm} = \mathbf{d}$ with $N > M$ can also be solved through the use of Householder transformations. In this case, we seek a solution that minimizes the prediction error $E = \mathbf{e}^T\mathbf{e}$. We seek a transformation with two properties: it must operate on the transformed prediction error $\mathbf{e}' = \mathbf{Te}$ in such a way that minimizing $\mathbf{e}'^T\mathbf{e}'$ is the same as minimizing $\mathbf{e}^T\mathbf{e}$, and it must transform the data kernel into the upper-triangularized form. The transformed prediction error is

$$\mathbf{e}' = \mathbf{Te} = \mathbf{T}\{\mathbf{d} - \mathbf{Gm}\} = \mathbf{Td} - \mathbf{TGm} = \mathbf{d}' - \mathbf{G}'\mathbf{m} \tag{7.7}$$

where $\mathbf{d}'$ is the transformed data and $\mathbf{G}'$ is the transformed and triangularized data kernel

$$
\begin{bmatrix} e'_1 \\ e'_2 \\ e'_3 \\ \vdots \\ e'_M \\ e'_{M+1} \\ \vdots \\ e'_N \end{bmatrix}
= -
\begin{bmatrix}
G'_{11} & G'_{12} & G'_{13} & G'_{14} & \cdots & G'_{1M} \\
0 & G'_{22} & G'_{23} & G'_{24} & \cdots & G'_{2M} \\
0 & 0 & G'_{33} & G'_{34} & \cdots & G'_{3M} \\
\vdots & \vdots & \vdots & \vdots & \vdots & \vdots \\
0 & 0 & 0 & 0 & \cdots & G'_{MM} \\
0 & 0 & 0 & 0 & 0 & 0 \\
\vdots & \vdots & \vdots & \vdots & \vdots & \vdots \\
0 & 0 & 0 & 0 & \cdots & 0
\end{bmatrix}
\begin{bmatrix} m_1 \\ m_2 \\ m_3 \\ \vdots \\ \\ m_M \end{bmatrix}
+
\begin{bmatrix} d'_1 \\ d'_2 \\ d'_3 \\ \vdots \\ d'_M \\ d'_{M+1} \\ \vdots \\ d'_N \end{bmatrix}
\tag{7.8}
$$

We note that no matter what values we choose for $\mathbf{m}^{\text{est}}$, we cannot alter the last $N - M$ elements of $\mathbf{e}'$ as the last $N - M$ rows of the transformed data kernel are zero. We can, however, set the first M elements of $\mathbf{e}'$ equal to zero by satisfying the first M equations $\mathbf{e}' = \mathbf{d}' - \mathbf{G}'\mathbf{m} = 0$ exactly. As the top part of $\mathbf{G}'$ is triangular, we can use the back-solving technique described earlier. The total error is then the length of the last $N - M$ elements of $\mathbf{e}'$, written as

$$E = \sum_{i=M+1}^{N} \mathbf{e}_i'^2 \tag{7.9}$$

Again we use the Householder transformations to separate the problem into two parts: data that can be satisfied exactly and data that cannot be satisfied at all. The solution is chosen so that it minimizes the length of the prediction error, and the least square solution is thereby obtained.

Finally, we note that the constrained least squares problem can also be solved with Householder transformations. Suppose that we want to solve $\mathbf{Gm} = \mathbf{d}$ in the least squares sense except that we want the solution to obey p linear equality constraints of the form $\mathbf{Hm} = \mathbf{h}$. Because of the constraints, we do not have complete freedom in choosing the model parameters. We therefore employ Householder transformations to separate those linear combinations of $\mathbf{m}$ that are completely determined by the constraints from those that are completely undetermined. This process is precisely the same as the one used in the underdetermined problem and consists of finding a transformation, say, $\mathbf{T}$, that triangularizes $\mathbf{Hm} = \mathbf{h}$ as

$$\mathbf{h} = \mathbf{Hm} = \{\mathbf{HT}^{-1}\}\{\mathbf{Tm}\} = \mathbf{H}'\mathbf{m}' \tag{7.10}$$

The first p elements of $\mathbf{m}'^{\text{est}}$ are now completely determined and can be computed by back-solving the triangular system. The same transformation can be applied to $\mathbf{Gm} = \mathbf{d}$ to yield the transformed inverse problem $\mathbf{d} = \mathbf{Gm} = \{\mathbf{GT}^{-1}\}\{\mathbf{Tm}\} = \mathbf{G}'\mathbf{m}'$. But $\mathbf{G}'$ will not be triangular as the transformation was designed to triangularize $\mathbf{H}$, not $\mathbf{G}$. As the first p elements of $\mathbf{m}'^{\text{est}}$ have been determined by the constraints, we can partition $\mathbf{G}'$ into two submatrices $\mathbf{G}' = [\mathbf{G}'_1, \mathbf{G}'_2]$, where $\mathbf{G}'_1$ multiplies the p-determined model parameters and $\mathbf{G}'_2$ multiplies the as yet unknown $M - p$ model parameters

$$[\mathbf{G}'_1, \mathbf{G}'_2] \left[\left[m_1' \text{est} \cdots m_p' \text{est} \right], \left[m_{p+1}' \text{est} \cdots m_M' \text{est} \right] \right]^{\text{T}} = \mathbf{d} \tag{7.11}$$

The equation can be rearranged into standard form by subtracting the part involving the already-determined model parameters:

$$\mathbf{G}'_2 \left[m_{p+1}'^{\text{est}} \cdots m_M'^{\text{est}} \right]^{\text{T}} = \mathbf{d} - \mathbf{G}'_1 \left[m_1'^{\text{est}} \cdots m_p'^{\text{est}} \right]^{\text{T}} \tag{7.12}$$

The equation is now a completely overdetermined one in the $M - p$ unknown model parameters and can be solved as described earlier. Finally, the solution is transformed back into the original coordinate system by $\mathbf{m}^{\text{est}} = \mathbf{T}^{-1}\mathbf{m}'^{\text{est}}$.

7.3 DESIGNING HOUSEHOLDER TRANSFORMATIONS

For a transformation to preserve length, it must be a unitary transformation (i.e., it must satisfy $\mathbf{T}^T = \mathbf{T}^{-1}$). Any transformation of the form

$$\mathbf{T} = \mathbf{I} - \frac{2\mathbf{v}\mathbf{v}^T}{\mathbf{v}^T\mathbf{v}} \tag{7.13}$$

(where $\mathbf{v}$ is any vector) is a unitary transformation as

$$\mathbf{T}^{-1}\mathbf{T} = \mathbf{T}^T\mathbf{T} = \left[\mathbf{I} - \frac{2\mathbf{v}\mathbf{v}^T}{\mathbf{v}^T\mathbf{v}}\right]^2 = \mathbf{I} - \frac{4\mathbf{v}\mathbf{v}^T}{\mathbf{v}^T\mathbf{v}} + \frac{4\mathbf{v}\mathbf{v}^T}{\mathbf{v}^T\mathbf{v}} = \mathbf{I} \tag{7.14}$$

This can be shown to be the most general form of a unitary transformation. The problem is to find the vector $\mathbf{v}$ such that the transformation triangularizes a given matrix. To do so, we shall begin by finding a sequence of transformations, each of which converts to zeros either the elements beneath the main diagonal of one *column* of the matrix (for premultiplication of the transformation) or the elements to the right of the main diagonal of one *row* of the matrix (for postmultiplication by the transformation). The first i columns are converted to zeros by

$$\mathbf{T} = \mathbf{T}^{(i)}\mathbf{T}^{(i-1)}\mathbf{T}^{(i-2)}...\mathbf{T}^{(1)} \tag{7.15}$$

In the overdetermined problem, applying M of these transformations produces an upper-triangularized matrix. In the $M=3$, $N=4$ case, the transformations proceed as

$$
\begin{array}{ccccccc}
\mathbf{G} & \rightarrow & \mathbf{T}^{(1)}\mathbf{G} & \rightarrow & \mathbf{T}^{(2)}\mathbf{T}^{(1)}\mathbf{G} & \rightarrow & \mathbf{T}^{(3)}\mathbf{T}^{(2)}\mathbf{T}^{(1)}\mathbf{G} \\
\begin{bmatrix} x & x & x \\ x & x & x \\ x & x & x \\ x & x & x \end{bmatrix} & \rightarrow &
\begin{bmatrix} x & x & x \\ 0 & x & x \\ 0 & x & x \\ 0 & x & x \end{bmatrix} & \rightarrow &
\begin{bmatrix} x & x & x \\ 0 & x & x \\ 0 & 0 & x \\ 0 & 0 & x \end{bmatrix} & \rightarrow &
\begin{bmatrix} x & x & x \\ 0 & x & x \\ 0 & 0 & x \\ 0 & 0 & 0 \end{bmatrix}
\end{array} \tag{7.16}
$$

where the xs symbolize nonzero matrix elements. Consider the ith column of $\mathbf{G}$, denoted by the vector $\mathbf{g} = [G_{1i}, G_{2i}, ..., G_{Ni}]^T$. We want to construct a transformation such that

$$\mathbf{g}' = \mathbf{T}^{(i)}\mathbf{g} = [G'_{1i} \; G'_{2i} \; \cdots \; G'_{ii} \; 0 \; 0 \; \cdots \; 0]^T \tag{7.17}$$

Substituting the expression for the transformation yields

$$\mathbf{I}\mathbf{g} - \left(\frac{2\mathbf{v}\mathbf{v}^T}{\mathbf{v}^T\mathbf{v}}\right)\mathbf{g} = \begin{bmatrix} G_{1i} \\ G_{2i} \\ \vdots \\ G_{Ni} \end{bmatrix} - \frac{2\mathbf{v}^T\mathbf{g}}{\mathbf{v}^T\mathbf{v}} \begin{bmatrix} v_1 \\ v_2 \\ \vdots \\ v_N \end{bmatrix} = \begin{bmatrix} G'_{1i} \\ G'_{2i} \\ \vdots \\ G'_{ii} \\ 0 \\ \vdots \\ 0 \end{bmatrix} \tag{7.18}$$

As the term $2\mathbf{v}^T\mathbf{g}/\mathbf{v}^T\mathbf{v}$ is a scalar, we can only zero the last $N-i$ elements of $\mathbf{g}'$ if $[2\mathbf{v}^T\mathbf{g}/\mathbf{v}^T\mathbf{v}]$ $[v_{i+1}, ..., v_N] = [G_{i+1,i}, ..., G_{Ni}]$. We therefore set $[v_{i+1}, ..., v_N] = [G_{i+1,i}, ..., G_{Ni}]$ and choose the other i elements of $\mathbf{v}$ so that the normalization is correct. As this is but a single constraint on i elements of $\mathbf{v}$, we have considerable flexibility in making the choice. It is convenient to choose the first $i-1$ elements of $\mathbf{v}$ as zero (this choice is both simple and causes the transformation to leave the first $i-1$ elements of $\mathbf{g}$ unchanged). This leaves the ith element of $\mathbf{v}$ to be determined by the condition that $2\mathbf{v}^T\mathbf{g}/\mathbf{v}^T\mathbf{v}=1$. It is easy to show that $v_i=G_{ii}-\alpha_i$, where

$$\alpha_i^2 = \sum_{j=i}^{N} G_{ji}^2 \tag{7.19}$$

(Although the sign of α_i is arbitrary, choosing it to have a sign opposite that of G_{ii} is best, since it improves the numerical stability of computer algorithms.) The Householder transformation is then defined by the vector as

$$\mathbf{v} = \begin{bmatrix} 0 & 0 & \cdots & 0 & (G_{ii}-\alpha_i) & G_{i+1,i} & G_{i+2,i} & \cdots & G_{N,i} \end{bmatrix}^T \tag{7.20}$$

Finally, we note that the $(i+1)$st Householder transformation does not destroy any of the zeros created by the ith, as long as we apply them in order of decreasing number of zeros. We can thus apply a succession of these transformations to triangularize an arbitrary matrix. The inverse transformations are also trivial to construct as they are just the transforms of the forward transformations.

7.4 TRANSFORMATIONS THAT DO NOT PRESERVE LENGTH

Suppose we want to solve the linear inverse problem $\mathbf{Gm}=\mathbf{d}$ in the sense of finding a solution $\mathbf{m}^{\text{est}}$ that minimizes a weighted combination of prediction error and solution simplicity as

$$\text{minimize}: \quad E+L=\mathbf{e}^T\mathbf{W}_e\mathbf{e}+\mathbf{m}^T\mathbf{W}_m\mathbf{m} \tag{7.21}$$

It is possible to find transformations $\mathbf{m}'=\mathbf{T}_m\mathbf{m}$ and $\mathbf{e}'=\mathbf{T}_e\mathbf{e}$ which, although they do not preserve the length, nevertheless change it in precisely such a way that $E+L=\mathbf{e}'^T\mathbf{e}'+\mathbf{m}'^T\mathbf{m}'$ (Wiggins, 1972). The weighting factors are identity matrices in the new coordinate system.

Consider the weighted measure of length $L=\mathbf{m}^T\mathbf{W}_m\mathbf{m}$. If we could factor the weighting matrix into the product $\mathbf{W}_m=\mathbf{T}_m^T\mathbf{T}_m$, where $\mathbf{T}_m$ is some matrix, then we could identify $\mathbf{T}_m$ as a transformation of coordinates

$$L=\mathbf{m}^T\mathbf{W}_m\mathbf{m}=\mathbf{m}^T\{\mathbf{T}_m^T\mathbf{T}_m\}\mathbf{m}=\{\mathbf{T}_m\mathbf{m}\}^T\{\mathbf{T}_m\mathbf{m}\}=\mathbf{m}'^T\mathbf{m}' \tag{7.22}$$

This factorization can be accomplished in several ways. If we have built $\mathbf{W}_m$ up from the $\mathbf{D}$ matrix, then $\mathbf{W}_m=\mathbf{D}^T\mathbf{D}$ and $\mathbf{T}_m=\mathbf{D}$. If not, then we can rely upon the fact that $\mathbf{W}_m$, being symmetric, has a symmetric square root, so that $\mathbf{W}_m=\mathbf{W}_m^{1/2}\mathbf{W}_m^{1/2}=\mathbf{W}_m^{1/2T}\mathbf{W}_m^{1/2}$, in which case $\mathbf{T}=\mathbf{W}_m^{1/2}$. Then

$$\mathbf{m}'=\mathbf{W}_m^{1/2}\mathbf{m} \ \text{ or } \ \mathbf{m}'=\mathbf{Dm} \qquad\qquad \mathbf{m}=\mathbf{W}_m^{-1/2}\mathbf{m}' \ \text{ or } \ \mathbf{m}=\mathbf{D}^{-1}\mathbf{m}'$$

$$\mathbf{d}'=\mathbf{W}_e^{1/2}\mathbf{d} \qquad\qquad \text{and} \qquad\qquad \mathbf{d}=\mathbf{W}_e^{-1/2}\mathbf{d}' \tag{7.23}$$

$$\mathbf{G}'=\mathbf{W}_e^{1/2}\mathbf{G}\mathbf{W}_m^{-1/2} \ \text{ or } \ \mathbf{G}'=\mathbf{W}_e^{1/2}\mathbf{G}\mathbf{D}^{-1} \qquad \mathbf{G}=\mathbf{W}_e^{-1/2}\mathbf{G}'\mathbf{W}_m^{1/2} \ \text{ or } \ \mathbf{G}=\mathbf{W}_e^{-1/2}\mathbf{G}'\mathbf{D}$$

We have encountered this weighting before, in Eqs. (3.46), (5.15). In fact, the damped least squares solution (Eq. 3.35)

$$\mathbf{m}'^{\mathbf{est}} = \left[\mathbf{G}'^T\mathbf{G}' + \varepsilon^2\mathbf{I}\right]^{-1}\mathbf{G}'^T\mathbf{d}' \qquad (7.24)$$

transforms into the weighted damped least squares solution (Eq. 3.45 with $\langle\mathbf{m}\rangle$ set to zero) under this transformation:

$$\mathbf{W}_m^{1/2}\mathbf{m}^{\mathbf{est}} = \left[\mathbf{W}_m^{-1/2}\mathbf{G}^T\mathbf{W}_e^{1/2}\mathbf{W}_e^{1/2}\mathbf{G}\mathbf{W}_m^{-1/2} + \varepsilon^2\mathbf{I}\right]^{-1}\mathbf{W}_m^{-1/2}\mathbf{G}^T\mathbf{W}_e^{1/2}\mathbf{W}_e^{1/2}\mathbf{d}$$

$$\text{or} \qquad (7.25)$$

$$\mathbf{m}^{\mathbf{est}} = \left[\mathbf{G}^T\mathbf{W}_e\mathbf{G} + \varepsilon^2\mathbf{W}_m\right]^{-1}\mathbf{G}^T\mathbf{W}_e\mathbf{d}$$

The square root of a symmetric matrix $\mathbf{W}$ is defined via its eigenvalue decomposition. Let $\mathbf{\Lambda}$ and $\mathbf{U}$ be the eigenvalues and eigenvectors, respectively, of $\mathbf{W}$ so that $\mathbf{W} = \mathbf{U}\mathbf{\Lambda}\mathbf{U}^T$. Then

$$\mathbf{T} = \mathbf{W}^{1/2} = \mathbf{U}\mathbf{\Lambda}^{1/2}\mathbf{U}^T \quad \text{so that} \quad \mathbf{W}^{1/2T}\mathbf{W}^{1/2} = \mathbf{U}\mathbf{\Lambda}^{1/2}\mathbf{U}^T\mathbf{U}\mathbf{\Lambda}^{1/2}\mathbf{U}^T$$

$$= \mathbf{U}\mathbf{\Lambda}^{1/2}\mathbf{\Lambda}^{1/2}\mathbf{U}^T = \mathbf{U}\mathbf{\Lambda}\mathbf{U}^T = \mathbf{W} \qquad (7.26)$$

Here we have used the fact that $\mathbf{U}^T\mathbf{U} = \mathbf{I}$. The *MatLab* command `sqrt(Wm)` correctly computes the square root of a square matrix, so the eigenvalue decomposition need not be used explicitly. Yet another possible definition of the transformation $\mathbf{T}$ is

$$\mathbf{T} = \mathbf{\Lambda}^{1/2}\mathbf{U}^T \qquad (7.27)$$

as then $\mathbf{T}^T\mathbf{T} = \mathbf{U}\mathbf{\Lambda}^{1/2}\mathbf{\Lambda}^{1/2}\mathbf{U}^T = \mathbf{U}\mathbf{\Lambda}\mathbf{U}^T = \mathbf{W}$. The transformed inverse problem is then $\mathbf{G}'\mathbf{m}' = \mathbf{d}'$, where

$$\mathbf{m}' = \left\{\mathbf{\Lambda}_m^{1/2}\mathbf{U}_m^T\right\}\mathbf{m} \qquad\qquad \mathbf{m} = \left\{\mathbf{U}_m\mathbf{\Lambda}_m^{-1/2}\right\}\mathbf{m}'$$

$$\mathbf{d}' = \left\{\mathbf{\Lambda}_e^{1/2}\mathbf{U}_e^T\right\}\mathbf{d} \qquad \text{and} \qquad \mathbf{d} = \left\{\mathbf{U}_e\mathbf{\Lambda}_e^{-1/2}\right\}\mathbf{d}' \qquad (7.28)$$

$$\mathbf{G}' = \left\{\mathbf{\Lambda}_e^{1/2}\mathbf{U}_e^T\right\}\mathbf{G}\left\{\mathbf{U}_m^T\mathbf{\Lambda}_m^{-1/2}\right\} \qquad \mathbf{G} = \left\{\mathbf{U}_e\mathbf{\Lambda}_e^{-1/2}\right\}\mathbf{G}'\left\{\mathbf{\Lambda}_m^{1/2}\mathbf{U}_m^T\right\}$$

where $\mathbf{W}_e = \mathbf{U}_e\mathbf{\Lambda}_e\mathbf{U}_e^T$ and $\mathbf{W}_m = \mathbf{U}_m\mathbf{\Lambda}_m\mathbf{U}_m^T$. It is sometimes convenient to transform weighted L_2 problems into one or another of these two forms before proceeding with their solution.

7.5 THE SOLUTION OF THE MIXED-DETERMINED PROBLEM

The concept of vector spaces is particularly helpful in understanding the mixed-determined problem, in which some linear combinations of the model parameters are over-determined and some are underdetermined.

If the problem is to some degree underdetermined, then the equation $\mathbf{Gm} = \mathbf{d}$ contains information about only some of the model parameters. We can think of these combinations as lying in a subspace $S_p(\mathbf{m})$ of the model parameters space. No information is provided about the part of the solution that lies in the rest of the space, which we shall call the *null space* $S_0(\mathbf{m})$.

The part of the $\mathbf{m}$ that lies in the null space is completely "unilluminated" by $\mathbf{Gm}=\mathbf{d}$, as the equation contains no information about these linear combinations of the model parameters.

On the other hand, if the problem is to some degree overdetermined, the product $\mathbf{Gm}$ may not be able to span $S(\mathbf{d})$ no matter what one chooses for $\mathbf{m}$. At best $\mathbf{Gm}$ may span a subspace $S_p(\mathbf{d})$ of the data space. Then no part of the data lying outside of this space, say, in $S_0(\mathbf{d})$, can be satisfied for any choice of the model parameters.

If the model parameters and data are divided into parts with subscript p that lie in the p spaces and parts with subscript 0 that lie in the null spaces, we can write $\mathbf{Gm}=\mathbf{d}$ as

$$G\left[\mathbf{m}_p + \mathbf{m}_0\right] = \left[\mathbf{d}_p + \mathbf{d}_0\right] \tag{7.29}$$

The solution length is then

$$L=\mathbf{m}^T\mathbf{m}= \left[\mathbf{m}_p + \mathbf{m}_0\right]^T\left[\mathbf{m}_p + \mathbf{m}_0\right]=\mathbf{m}_p^T\mathbf{m}_p + \mathbf{m}_0^T\mathbf{m}_0 \tag{7.30}$$

(The cross terms $\mathbf{m}_p^T\mathbf{m}_0$ and $\mathbf{m}_0^T\mathbf{m}_p$ are zero as the vectors lie in different spaces.) Similarly, the prediction error is

$$E= \left[\mathbf{d}_p + \mathbf{d}_0 - \mathbf{Gm}_p\right]^T\left[\mathbf{d}_p + \mathbf{d}_0 - \mathbf{Gm}_p\right] = \left[\mathbf{d}_p - \mathbf{Gm}_p\right]^T\left[\mathbf{d}_p - \mathbf{Gm}_p\right] + \mathbf{d}_0^T\mathbf{d}_0 \tag{7.31}$$

(as $\mathbf{Gm}_0=0$ and $\mathbf{d}_p$ and $\mathbf{d}_0$ lie in different spaces). We can now define precisely what we mean by a solution to the mixed-determined problem that minimizes prediction error while adding a minimum of prior information: prior information is added to specify only those linear combinations of the model parameters that reside in the null space $S_0(\mathbf{m})$, and the prediction error is reduced to only the portion in the null space $S_0(\mathbf{d})$ by satisfying $\mathbf{e}_p=[\mathbf{d}_p - \mathbf{Gm}_p]=0$ exactly. One possible choice of prior information is $\mathbf{m}_0^{\text{est}}=0$, which is sometimes called the "natural solution" of the mixed-determined problem. We note that when $\mathbf{Gm}=\mathbf{d}$ is purely underdetermined the natural solution is just the minimum-length solution, and when $\mathbf{Gm}=\mathbf{d}$ is purely overdetermined it is just the least squares solution.

One might be tempted to view the natural solution as *better* than, say, the damped least squares solution, as the prior information is applied only to the part of the solution that is in the null space and does not increase the prediction error E. However, such an assessment is not clear-cut. If the prior information is indeed accurate, it should be fully utilized, even if its use leads to a slightly larger prediction error. Anyway, given noisy measurements, two slightly different prediction errors will not be statistically distinguishable. This analysis emphasizes that the choice of the inversion method must be tailored carefully to the problem at hand.

7.6 SINGULAR-VALUE DECOMPOSITION AND THE NATURAL GENERALIZED INVERSE

The p and null subspaces of the linear problem can be easily identified through a type of eigenvalue decomposition of the data kernel that is called the *singular-value decomposition*. We shall derive this decomposition in Section 7.7, but first we shall state the result and demonstrate its usefulness.

Any $N \times M$ square matrix can be written as the product of three matrices (Penrose, 1955; Lanczos, 1961):

$$G = U \Lambda V^T \tag{7.32}$$

The matrix U is an $N \times N$ matrix of eigenvectors that span the data space $S(d)$:

$$U = \begin{bmatrix} u^{(1)} & u^{(2)} & u^{(3)} & \cdots & u^{(N)} \end{bmatrix} \tag{7.33}$$

where the $u^{(i)}$s are the individual vectors. The vectors are orthogonal to one another and can be chosen to be of unit length so that $UU^T = U^T U = I$ (where the identity matrix is $N \times N$). Similarly, V is an $M \times M$ matrix of eigenvectors that span the model parameter space $S(m)$ as

$$V = \begin{bmatrix} v^{(1)} & v^{(2)} & v^{(3)} & \cdots & v^{(M)} \end{bmatrix} \tag{7.34}$$

Here the $v^{(i)}$s are the individual orthonormal vectors so that $VV^T = V^T V = I$ (the identity matrix being $M \times M$). The matrix Λ is an $N \times M$ diagonal eigenvalue matrix whose diagonal elements are nonnegative and are called *singular values*. In the $N=4$, $M=3$ case

$$\Lambda = \begin{bmatrix} \lambda_1 & 0 & 0 \\ 0 & \lambda_2 & 0 \\ 0 & 0 & \lambda_3 \\ 0 & 0 & 0 \end{bmatrix} \tag{7.35}$$

The singular values are usually arranged in order of decreasing size. Some of the singular values may be zero. We therefore partition Λ into a submatrix Λ_p of p nonzero singular values and several zero matrices as

$$\Lambda = \begin{bmatrix} \Lambda_p & 0 \\ 0 & 0 \end{bmatrix} \tag{7.36}$$

where Λ_p is a $p \times p$ diagonal matrix. The decomposition then becomes $U \Lambda V^T = U_p \Lambda_p V_p^T$, where U_p and V_p consist of the first p columns of U and V, respectively. The other portions of the eigenvector matrices are canceled by the zeros in Λ. The matrix G contains no information about the subspaces spanned by these portions of the data and model eigenvectors, which we shall call V_0 and U_0, respectively. As we shall soon prove, these are precisely the same spaces as the p and null spaces defined in the previous section.

The data kernel is not a function of the null eigenvectors V_0 and U_0. The equation $d = Gm = U_p \Lambda_p V_p^T m$ contains no information about the part of the model parameters in the space spanned by V_0 as the model parameters m are multiplied by V_p (which is orthogonal to everything in V_0). The eigenvector V_p, therefore, lies completely in $S_p(m)$, and V_0 lies completely in $S_0(m)$. Similarly, no matter what value $\{\Lambda_p V_p^T m\}$ attains, it can have no component in the space spanned by U_0 as it is multiplied by U_p (and U_0 and U_p are orthogonal). Therefore, U_p lies completely in $S_p(d)$ and U_0 lies completely in $S_0(d)$.

We have demonstrated that the p and null spaces can be identified through the singular-value decomposition of the data kernel. The full spaces $S(m)$ and $S(d)$ are spanned by V and U, respectively. The p spaces are spanned by the parts of the eigenvector matrices that have nonzero eigenvalues: $S_p(m)$ is spanned by V_p and $S_p(d)$ is spanned by U_p. The remaining eigenvectors V_0 and U_0 span the null spaces $S_0(m)$ and $S_0(d)$. The p and null matrices are orthogonal and are normalized in the sense that $V_p^T V_p = U_p^T U_p = I$, where I is $p \times p$ in size. However, as these matrices do not in general span the complete data and model spaces, $V_p V_p^T$ and $U_p U_p^T$ are not in general identity matrices.

The natural solution to the inverse problem can be constructed from the singular-value decomposition. This solution must have an m^{est} that has no component in $S_0(m)$ and a prediction error e that has no component in $S_p(d)$. We therefore consider the solution

$$m^{est} = V_p \Lambda_p^{-1} U_p^T d \tag{7.37}$$

which is picked in analogy to the square matrix case. To show that m^{est} has no component in $S_0(m)$, we take the dot product of the equation with V_0, which lies completely in $S_0(m)$, as

$$V_0^T m^{est} = V_0^T V_p \Lambda_p^{-1} U_p^T d = 0 \tag{7.38}$$

as V_0^T and V_p are orthogonal. To show that e has no component in $S_p(d)$, we take the dot product with U_p as

$$U_p^T e = U_p^T [d - G m^{est}] = U_p^T \left[d - U_p \Lambda_p V_p^T V_p \Lambda_p^{-1} U_p^T d \right]$$
$$= U_p^T \left[d - U_p U_p^T d \right] = U_p^T d - U_p^T d = 0 \tag{7.39}$$

as $V_p^T V_p = U_p^T U_p = \Lambda_p \Lambda_p^{-1} = I$. The natural solution of the inverse problem is therefore shown to be

$$m^{est} = V_p \Lambda_p^{-1} U_p^T d \tag{7.40}$$

We note that we can define a generalized inverse operator for the mixed-determined problem, the *natural generalized inverse* $G^{-g} = V_p \Lambda_p^{-1} U_p^T$. (This generalized inverse is so useful that it is sometimes referred to as *the* generalized inverse, although of course there are other generalized inverses that can be designed for the mixed-determined problem that embody other kinds of prior information.)

In *MatLab*, the singular-value decomposition is computed as

```
[U, L, V] = svd(G);
lambda = diag(L);
```

(*MatLab* script gda07_03)

Here we have copied the diagonal elements of matrix L of singular values into the vector lambda. The submatrices U_p and V_p (with p an integer) are extracted via:

```
Up = U(:,1:p);
Vp = V(:,1:p);
lambdap = lambda(1:p);
```

(*MatLab* script gda07_03)
and the model parameters are estimated as

```
mest = Vp*((Up'*dobs)./lambdap);
```

(*MatLab* script gda07_03)

Note that the calculation is organized so that the matrices $\mathbf{U}_p^T$ and $\mathbf{V}_p$ multiply vectors (as contrasted to matrices); the generalized inverse is never explicitly formed. The value of the integer p must be chosen so that zero eigenvalues are excluded from the estimate; else a division-by-zero error will occur. As discussed later, one often excludes near-zero eigenvalues as well, as the resulting solution (while not quite the natural solution) is often better behaved.

The natural generalized inverse has model resolution

$$\mathbf{R} = \mathbf{G}^{-g}\mathbf{G} = \left\{\mathbf{V}_p\mathbf{\Lambda}_p^{-1}\mathbf{U}_p^T\right\}\left\{\mathbf{U}_p\mathbf{\Lambda}_p\mathbf{V}_p^T\right\} = \mathbf{V}_p\mathbf{V}_p^T \tag{7.41}$$

The model parameters will be perfectly resolved only if $\mathbf{V}_p$ spans the complete space of model parameters, that is, if there are no zero eigenvalues and $p \geq M$. The data resolution matrix is

$$\mathbf{N} = \mathbf{G}\mathbf{G}^{-g} = \left\{\mathbf{U}_p\mathbf{\Lambda}_p\mathbf{V}_p^T\right\}\left\{\mathbf{V}_p\mathbf{\Lambda}_p^{-1}\mathbf{U}_p^T\right\} = \mathbf{U}_p\mathbf{U}_p^T \tag{7.42}$$

The data are only perfectly resolved if $\mathbf{U}_p$ spans the complete space of data and $p = N$. Finally, we note that if the data are uncorrelated with uniform variance σ_d^2, the model covariance is

$$\begin{aligned}
\left[\operatorname{cov} \mathbf{m}^{\text{est}}\right] &= \mathbf{G}^{-g}[\operatorname{cov}\mathbf{d}]\mathbf{G}^{-g\mathrm{T}} = \sigma_d^2\left\{\mathbf{V}_p\mathbf{\Lambda}_p^{-1}\mathbf{U}_p^T\right\}\left\{\mathbf{V}_p\mathbf{\Lambda}_p^{-1}\mathbf{U}_p^T\right\}^T \\
&= \sigma_d^2\mathbf{V}_p\mathbf{\Lambda}_p^{-2}\mathbf{V}_p^T
\end{aligned} \tag{7.43}$$

The covariance of the estimated model parameters is very sensitive to the smallest nonzero eigenvalue. (Note that forming the natural inverse corresponds to assuming that linear combinations of the prior model parameters in the p space have infinite variance and that combinations in the null space have zero variance and zero mean.) The covariance of the estimated model parameters, therefore, does not explicitly contain [cov m]. If one prefers a solution based on the natural inverse (but with the null vectors chosen to minimize the distance to a set of prior model parameters with mean $\langle\mathbf{m}\rangle$ and covariance $[\operatorname{cov}\mathbf{m}]_A$), it is appropriate to use the formula $\mathbf{m}^{\text{est}} = \mathbf{G}^{-g}\mathbf{d} + [\mathbf{I} - \mathbf{R}]\langle\mathbf{m}\rangle$, where $\mathbf{G}^{-g}$ is the natural inverse. The covariance of this estimate is now

$$\left[\operatorname{cov} \mathbf{m}^{\text{est}}\right] = \mathbf{G}^{-g}[\operatorname{cov}\mathbf{d}]\mathbf{G}^{-g\mathrm{T}} + [\mathbf{I} - \mathbf{R}][\operatorname{cov}\mathbf{m}]_A[\mathbf{I} - \mathbf{R}]^T \tag{7.44}$$

which is based on the usual rule for computing covariances.

To use the natural inverse one must be able to identify the number p, that is, to count the number of nonzero singular values. Plots of the sizes of the singular values against their index numbers (the *spectrum* of the data kernel) can be useful in this process. The value of p can be easily determined if the singular values fall into two clearly distinguishable groups, one nonzero and one zero (Fig. 7.3A and B). In realistic inverse problems, however, the singular values often smoothly decline in size (Fig. 7.3C and D), making it difficult to distinguish ones that are actually nonzero from ones that are zero but computed somewhat inaccurately owing to round-off error by the computer. Furthermore, if one chooses p so as to include these very small singular values, the solution variance will be very large, as it contains the factor $\mathbf{\Lambda}_p^{-2}$.

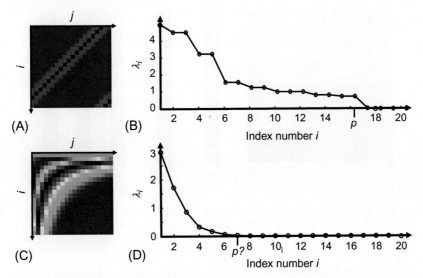

FIG. 7.3 (A) Hypothetical data kernel G_{ij} for an inverse problem with $M=20$ model parameters and $N=20$ observations. (B) Corresponding singular values λ_i have a clear cutoff at $p=16$. (C) Another hypothetical data kernel, also with $N=20$ and $M=20$. (D) Corresponding singular values do not have a clear cutoff, so the parameter, p, must be chosen in a more arbitrary fashion near $p \approx 7$. *MatLab* Script gda07_04.

One solution to this problem is to pick some cutoff size for singular values and then consider any values smaller than this as equal to zero. This process artificially reduces the dimensions of $\mathbf{V}_p$ and $\mathbf{U}_p$ that are included in the generalized inverse. The resulting estimates of the model parameters are no longer exactly the natural solution. But, if only small singular values are excluded, the solution is generally close to the natural solution and possesses better variance. On the other hand, its model and data resolution are worse. We recognize that this trade-off is just another manifestation of the trade-off between resolution and variance discussed in Chapter 4.

As an example, we consider the problem $\mathbf{Gm}=\mathbf{d}$ illustrated in Fig. 7.4A, which has the same data kernel as in Fig. 7.3A. Although $\mathbf{G}$ is 20×20, the problem is mixed-determined, as $p=16$ and four of the singular values are zero. In this case, the natural solution is very close to the true solution, presumably because, by coincidence, the true solution did not have much of its power in the null space. The natural solution is also close to the damped minimum-length solution (Eq. 4.22), which is not surprising, for both arise from similar minimization principles. The damped minimum-length solution minimizes a weighted sum of prediction error and solution length; the natural solution minimizes the prediction error and the component of the solution length in the null space.

Instead of choosing a sharp cutoff for the singular values, it is possible to include all the singular values while damping the smaller ones. We let $p=M$, but replace the reciprocals of all the singular values by $\lambda_i/(\varepsilon^2+\lambda_i^2)$, where ε is some small number. This change has little effect on the larger singular values but prevents the smaller ones from leading to large variances. Of course, the solution is no longer the natural solution. While its variance is

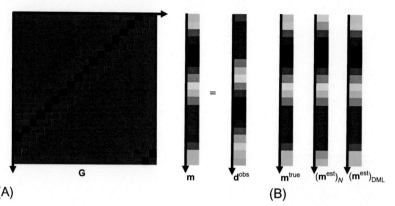

FIG. 7.4 (A) Same linear problem as in Fig. 7.3A, where $\mathbf{d}^{obs} = \mathbf{Gm}^{true} + \mathbf{n}$, with $\mathbf{n}$ uncorrelated Gaussian noise. (B) Corresponding solutions, true solutions $\mathbf{m}^{true}$, natural solution $(\mathbf{m}^{est})_N$, and damped minimum-length solution $(\mathbf{m}^{est})_{DML}$. *MatLab* Script gda07_03.

improved, its model and data resolution are degraded. In fact, this solution is precisely the damped least squares solution discussed in Chapter 3:

$$\begin{aligned}
\left[\mathbf{G}^T\mathbf{G} + \varepsilon^2\mathbf{I}\right]^{-1}\mathbf{G}^T &= \left[\mathbf{V}\mathbf{\Lambda}\mathbf{U}^T\mathbf{U}\mathbf{\Lambda}\mathbf{V}^T + \varepsilon^2\mathbf{I}\right]^{-1}\mathbf{V}\mathbf{\Lambda}\mathbf{U}^T \\
&= \left[\mathbf{V}\mathbf{\Lambda}^2\mathbf{V}^T + \varepsilon^2\mathbf{V}\mathbf{I}\mathbf{V}^T\right]^{-1}\mathbf{V}\mathbf{\Lambda}\mathbf{U}^T \\
&= \mathbf{V}\left[\mathbf{\Lambda}^2 + \varepsilon^2\mathbf{I}\right]^{-1}\mathbf{V}^T\mathbf{V}\mathbf{\Lambda}\mathbf{U}^T \\
&= \mathbf{V}\left\{\left[\mathbf{\Lambda}^2 + \varepsilon^2\mathbf{I}\right]^{-1}\mathbf{\Lambda}\right\}\mathbf{U}^T
\end{aligned} \tag{7.45}$$

Here, we rely upon the fact that if $\mathbf{M} = \mathbf{V}\mathbf{D}\mathbf{V}^T$ is the eigenvalue decomposition of $\mathbf{M}$, then $\mathbf{M}^{-1} = \mathbf{V}\mathbf{D}^{-1}\mathbf{V}^T$.

The damping of the singular values corresponds to the addition of prior information that the model parameters are small. The precise value of the number used as the cutoff or damping parameter must be chosen by a trial-and-error process which weighs the relative merits of having a solution with small variance against those of having one that fits the data and is well resolved.

In Section 6.6, we discussed the problem of bounding nonunique averages of model parameters by incorporating prior inequality constraints into the solution of the inverse problem. We see that the singular-value decomposition provides a simple way of identifying the null vectors of $\mathbf{Gm} = \mathbf{d}$. The general solution to the inverse problem (Wunsch and Minster, 1982)

$$\mathbf{m}^{gen} = \mathbf{m}^{par} + \sum_{i=1}^{p}\alpha_i\mathbf{m}^{null(i)} \tag{7.46}$$

can be thought of as having the natural solution as its particular solution and a sum over the null eigenvectors as its null solution:

$$\mathbf{m}^{gen} = \mathbf{V}_p\mathbf{\Lambda}_p^{-1}\mathbf{U}_p^T\mathbf{d} + \mathbf{V}_0\boldsymbol{\alpha} \tag{7.47}$$

There are $q = M - p$ null vectors in the general solution, with the coefficients given by $\boldsymbol{\alpha}$. In Section 6.6, upper and lower bounds on localized averages of the solution were found by determining the α that maximizes (for the upper bound) or minimizes (for the lower bound) the localized average $\langle m \rangle = \mathbf{a}^T \mathbf{m}$ with the constraint that $\mathbf{m}^l \leq \mathbf{m} \leq \mathbf{m}^u$, where $\mathbf{m}^l$ and $\mathbf{m}^u$ are prior bounds. The use of the natural solution guarantees that the prediction error is minimized in the L_2 sense.

The bounds on the localized average $\langle m \rangle = \mathbf{a}^T \mathbf{m}$ should be treated with some skepticism, as they depend on a particular choice of the solution (in this case one that minimizes the L_2 prediction error). If the total error E increases only slightly as this solution is perturbed and if this additional error can be considered negligible, then the true bounds of the localized average will be larger than those given earlier. In principle, one can handle this problem by forsaking the eigenvalue decomposition and simply determining the $\mathbf{m}$ that extremizes $\langle m \rangle$ with the constraints that $\mathbf{m}^l \leq \mathbf{m} \leq \mathbf{m}^u$ and that the total prediction error is less than some tolerable amount, say, E_M. For L_2 measures of the prediction error, this is a very difficult nonlinear problem. However, if the error is measured under the L_1 norm, it can be transformed into a linear programming problem.

7.7 DERIVATION OF THE SINGULAR-VALUE DECOMPOSITION

The singular-value decomposition can be derived in many ways. We follow here the treatment of Lanczos (1961). For an alternate derivation, see Menke and Menke (2011, Section 8.2). We first form an $(N+M) \times (N+M)$ square symmetric matrix $\mathbf{S}$ from $\mathbf{G}$ and $\mathbf{G}^T$ as

$$\mathbf{S} = \begin{bmatrix} 0 & \mathbf{G} \\ \mathbf{G}^T & 0 \end{bmatrix} \tag{7.48}$$

From elementary linear algebra, we know that this matrix has $N+M$ real eigenvalues λ_i and a complete set of eigenvectors $\mathbf{w}^{(i)}$ which solve $\mathbf{S}\mathbf{w}^{(i)} = \lambda_i \mathbf{w}^{(i)}$. Partitioning $\mathbf{w}$ into a part $\mathbf{u}$ of length N and a part $\mathbf{v}$ of length M, we obtain

$$\mathbf{S}\mathbf{w}^{(i)} = \lambda_i \mathbf{w}^{(i)} \rightarrow \begin{bmatrix} 0 & \mathbf{G} \\ \mathbf{G}^T & 0 \end{bmatrix} \begin{bmatrix} \mathbf{u}^{(i)} \\ \mathbf{v}^{(i)} \end{bmatrix} = \lambda_i \begin{bmatrix} \mathbf{u}^{(i)} \\ \mathbf{v}^{(i)} \end{bmatrix} \tag{7.49}$$

We shall now show that $\mathbf{u}^{(i)}$ and $\mathbf{v}^{(i)}$ are the same vectors as those defined in the previous section. We first note that this equation implies that $\mathbf{G}\mathbf{v}^{(i)} = \lambda_i \mathbf{u}^{(i)}$ and $\mathbf{G}^T \mathbf{u}^{(i)} = \lambda_i \mathbf{v}^{(i)}$. Suppose that there is a positive eigenvalue λ_i with eigenvector $[\mathbf{u}^{(i)}, \mathbf{v}^{(i)}]^T$. Then we note that $-\lambda_i$ is also an eigenvalue with eigenvector $[-\mathbf{u}^{(i)}, \mathbf{v}^{(i)}]^{T^T}$. If there are p positive eigenvalues, then there are $N+M-2p$ zero eigenvalues. Now by manipulating the earlier equations we obtain

$$\mathbf{G}^T \mathbf{G} \mathbf{v}^{(i)} = \lambda_i^2 \mathbf{v}^{(i)} \quad \text{and} \quad \mathbf{G} \mathbf{G}^T \mathbf{u}^{(i)} = \lambda_i^2 \mathbf{u}^{(i)} \tag{7.50}$$

As a symmetric matrix can have no more distinct eigenvectors than its dimension, we note that $p \leq \min(N, M)$. As both matrices are square and symmetric, there are M vectors $\mathbf{v}^{(i)}$ that form a complete orthogonal set $\mathbf{V}$ spanning $S(\mathbf{m})$ and N vectors $\mathbf{u}^{(i)}$ that form a complete orthogonal set $\mathbf{U}$ spanning $S(\mathbf{d})$. These include p of the $\mathbf{w}$ eigenvectors with distinct nonzero eigenvalues and remaining ones chosen from the eigenvectors with zero eigenvalues.

The equation $\mathbf{Gv}^{(i)} = \lambda_i \mathbf{u}^{(i)}$ can be written in matrix form as $\mathbf{GV} = \mathbf{U\Lambda}$, where $\mathbf{\Lambda}$ is a diagonal matrix of the eigenvalues. Postmultiplying by $\mathbf{V}^T$ gives the singular-value decomposition $\mathbf{G} = \mathbf{U\Lambda V}^T$.

7.8 SIMPLIFYING LINEAR EQUALITY AND INEQUALITY CONSTRAINTS

The singular-value decomposition can be used to simplify linear constraints.

7.8.1 Linear Equality Constraints

Consider the problem of solving $\mathbf{Gm} = \mathbf{d}$ in the sense of finding a solution that minimizes the L_2 prediction error subject to the $K < M$ constraints that $\mathbf{Hm} = \mathbf{h}$. This problem can be reduced to the unconstrained problem $\mathbf{G'm'} = \mathbf{d'}$ in $M' \leq M$ new model parameters. We first find the singular-value decomposition of the constraint matrix $\mathbf{H} = \mathbf{U}_p\mathbf{\Lambda}_p\mathbf{V}_p^T$. If $p = K$, the constraints are consistent and determine p linear combinations of the unknowns. The general solution is then $\mathbf{m} = \mathbf{V}_p\mathbf{\Lambda}_p^{-1}\mathbf{U}_p^T\mathbf{h} + \mathbf{V}_0\boldsymbol{\alpha}$, where $\boldsymbol{\alpha}$ is an arbitrary vector of length $M - p$ and is to be determined by minimizing the prediction error. Substituting this equation for $\mathbf{m}$ into $\mathbf{Gm} = \mathbf{d}$ and rearranging terms yields

$$\mathbf{GV}_0\boldsymbol{\alpha} = \mathbf{d} - \mathbf{GV}_p\mathbf{\Lambda}_p^{-1}\mathbf{U}_p^T\mathbf{h} \tag{7.51}$$

This equation in the unknown coefficients $\boldsymbol{\alpha}$ now can be solved as an unconstrained least squares problem. We note that we have encountered this problem in a somewhat different form during the discussion of Householder transformations (Section 7.2). The main advantage of using the singular-value decomposition is that it provides a test of the constraint's consistency.

7.8.2 Linear Inequality Constraints

Consider the L_2 problem

$$\text{minimize } \|\mathbf{d} - \mathbf{Gm}\|_2^2 \text{ subject to } \mathbf{Hm} \geq \mathbf{h} \tag{7.52}$$

We shall show that as long as $\mathbf{Gm} = \mathbf{d}$ is in fact overdetermined, this problem can be reduced to the simpler problem (Lawson and Hanson, 1974):

$$\text{minimize } \|\mathbf{m'}\|_2^2 \text{ subject to } \mathbf{H'm'} \geq \mathbf{h'} \tag{7.53}$$

To demonstrate this transformation, we form the singular-value decomposition of the data kernel $\mathbf{G} = \mathbf{U}_p\mathbf{\Lambda}_p\mathbf{V}_p^T$. We first divide the data $\mathbf{d}$ into two parts $\mathbf{d} = [\mathbf{d}_p^T, \mathbf{d}_0^T]^T$ where $\mathbf{d}_p = \mathbf{U}_p^T\mathbf{d}$ is in the $S_p(\mathbf{m})$ subspace and $\mathbf{d}_0 = \mathbf{U}_0^T\mathbf{d}$ is in the $S_0(\mathbf{m})$ subspace. The prediction error is then $E = \|\mathbf{d}^{\text{obs}} - \mathbf{d}^{\text{pre}}\|_2^2$ or

$$E = \left\| \begin{bmatrix} \mathbf{U}_p^{\mathrm{T}}\mathbf{d} \\ \mathbf{U}_0^{\mathrm{T}}\mathbf{d} \end{bmatrix} - \begin{bmatrix} \mathbf{U}_p^{\mathrm{T}}\mathbf{U}_p\mathbf{\Lambda}_p\mathbf{V}_p^{\mathrm{T}}\mathbf{m} \\ 0 \end{bmatrix} \right\|_2^2 = \left\| \mathbf{d}_p - \mathbf{\Lambda}_p\mathbf{V}_p^{\mathrm{T}}\mathbf{m} \right\|_2^2 + \|\mathbf{d}_0\|_2^2 \tag{7.54}$$

$$= \|\mathbf{m}'\|_2^2 + \|\mathbf{d}_0\|_2^2$$

where $\mathbf{m}' = \mathbf{d}_p - \mathbf{\Lambda}_p\mathbf{V}_p^{\mathrm{T}}\mathbf{m}$. We note that minimizing $\|\mathbf{m}'\|_2^2$ is the same as minimizing E as the other term is a constant. Inverting this expression for the unprimed model parameters gives $\mathbf{m} = \mathbf{V}_p\mathbf{\Lambda}_p^{-1}[\mathbf{d}_p - \mathbf{m}'] = \mathbf{V}_p\mathbf{\Lambda}_p^{-1}\left[\mathbf{U}_p^{\mathrm{T}}\mathbf{d} - \mathbf{m}'\right]$. Substituting this expression into the constraint equation $\mathbf{Hm} \geq \mathbf{h}$ and rearranging terms yields

$$\left\{ -\mathbf{HV}_p\mathbf{\Lambda}_p^{-1} \right\}\mathbf{m}' \geq \left\{ \mathbf{h} - \mathbf{HV}_p\mathbf{\Lambda}_p^{-1}\mathbf{U}_p^{\mathrm{T}}\mathbf{d} \right\} \quad \text{or} \quad \mathbf{H}'\mathbf{m}' \geq \mathbf{h}' \tag{7.55}$$

which is in the desired form.

7.9 INEQUALITY CONSTRAINTS

We shall now consider the solution of L_2 minimization problems with inequality constraints of the form

$$\text{minimize} \quad \|\mathbf{d} - \mathbf{Gm}\|_2^2 \quad \text{subject to} \quad \mathbf{Hm} \geq \mathbf{h} \tag{7.56}$$

We first note that problems involving $=$ and $\leq$ constraints can be reduced to this form. Equality constraints can be removed by the transformation described in Section 7.8.1, and $\leq$ constraints can be removed by multiplication by -1 to change them into $\geq$ constraints. For this minimization problem to have any solution, the constraints $\mathbf{Hm} \geq \mathbf{h}$ must be consistent; there must be at least one $\mathbf{m}$ that satisfies all the constraints. We can view these constraints as defining a volume in $S(\mathbf{m})$. Each constraint defines a hyperplane that divides $S(\mathbf{m})$ into two half-spaces, one in which that constraint is satisfied (the *feasible half-space*) and the other in which it is violated (the *infeasible half-space*). The set of inequality constraints, therefore, defines a volume in $S(\mathbf{m})$ which might be zero, finite, or infinite in extent. If the region has zero volume, then no feasible solution exists (Fig. 7.5B). If it has nonzero volume, then there is at least one solution that minimizes the prediction error (Fig. 7.5A). This volume has the shape of a polyhedron as its boundary surfaces are planes. It can be shown that the polyhedron must be convex: it can have no reentrant angles or grooves.

The starting point for solving the L_2 minimization problem with inequality constraints is the Kuhn-Tucker theorem, which describes the properties that any solution to this problem must possess. For any $\mathbf{m}$ that minimizes $\|\mathbf{d} - \mathbf{Gm}\|_2^2$ subject to p constraints $\mathbf{Hm} \geq \mathbf{h}$, it is possible to find a vector $\mathbf{y}$ of length p such that

$$\frac{1}{2}\nabla E = \nabla[\mathbf{Hm}] \cdot \mathbf{y} \quad \text{or} \quad -\mathbf{G}^{\mathrm{T}}[\mathbf{d} - \mathbf{Gm}] = \mathbf{H}^{\mathrm{T}}\mathbf{y} \tag{7.57}$$

This equation states that the gradient of the error ∇E can be written as a linear combination of hyperplane normals $\mathbf{H}^{\mathrm{T}}$, with the combination specified by the vector $\mathbf{y}$. The Kuhn-Tucker theorem goes on to state that the elements of $\mathbf{y}$ are nonnegative; that $\mathbf{y}$ can be partitioned into two parts $\mathbf{y}_E$ and $\mathbf{y}_S$ (possibly requiring reordering of the constraints) that satisfy

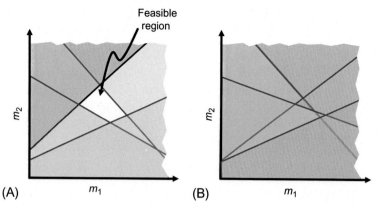

FIG. 7.5 (A) Each linear inequality constraint divides $S(\mathbf{m})$ into two half-spaces, one infeasible *(shaded)* and the other feasible *(white)*. Consistent constraints combine to form a convex, polygonal region within $S(\mathbf{m})$ *(white)* of feasible $\mathbf{m}$. (B) Inconsistent constraints do not form a feasible region.

$$\mathbf{y}=\begin{bmatrix}\mathbf{y}_E>0\\\mathbf{y}_S=0\end{bmatrix} \quad \text{and} \quad \begin{matrix}\mathbf{H}_E\mathbf{m}-\mathbf{h}_E=0\\\mathbf{H}_S\mathbf{m}-\mathbf{h}_S>0\end{matrix} \tag{7.58}$$

The first group of constraints is satisfied in the equality sense (thus the subscript E for "equality"). The rest are satisfied more loosely in the inequality sense (thus the subscript S for "slack").

The theorem indicates that any feasible solution $\mathbf{m}$ is the minimum solution only if the direction in which one would have to perturb $\mathbf{m}$ to decrease the total error E causes the solution to cross some constraint hyperplane and become infeasible. The direction of decreasing error is $-1/2\nabla E=\mathbf{G}^T[\mathbf{d}-\mathbf{Gm}]$. The constraint hyperplanes have normals $+\nabla[\mathbf{Hm}]=\mathbf{H}^T$ which point into the feasible side. As $\mathbf{H}_E\mathbf{m}-\mathbf{h}_E=0$, the solution lies exactly on the bounding hyperplanes of the $\mathbf{H}_E$ constraints. As $\mathbf{H}_S\mathbf{m}-\mathbf{h}_S>0$, it lies within the feasible volume of the $\mathbf{H}_S$ constraints. An infinitesimal perturbation $\delta\mathbf{m}$ of the solution can, therefore, only violate the $\mathbf{H}_E$ constraints. If it is not to violate these constraints, the perturbation must be made in the direction of feasibility so that it must be expressible as a nonnegative combination of hyperplane normals, that is, $\delta\mathbf{m}\cdot\nabla[\mathbf{Hm}]\geq0$. On the other hand, if it is to decrease the total prediction error it must satisfy $\delta\mathbf{m}\cdot\nabla E\leq0$. These two conditions are incompatible with the Kuhn-Tucker theorem, as dotting Eq. (7.57) with $\delta\mathbf{m}$ yields $\delta\mathbf{m}\cdot 1/2\nabla E=\delta\mathbf{m}\cdot\nabla[\mathbf{Hm}]\cdot\mathbf{y}\geq0$ as both $\delta\mathbf{m}\cdot\nabla[\mathbf{Hm}]$ and $\mathbf{y}$ are positive. These solutions are indeed minimum solutions to the constrained problem (Fig. 7.6).

To demonstrate how the Kuhn-Tucker theorem can be used, we consider the simplified problem

$$\text{minimize} \quad E=\|\mathbf{d}-\mathbf{Gm}\|_2^2 \quad \text{subject to} \quad \mathbf{m}\geq\mathbf{0} \tag{7.59}$$

which is called *nonnegative least squares*. We find the solution using an iterative scheme of several steps (Lawson and Hanson, 1974):

Step 1. Start with an initial guess for $\mathbf{m}$. As $\mathbf{H}=\mathbf{I}$ each model parameter is associated with exactly one constraint. These model parameters can be separated into a set $\mathbf{m}_E$ that

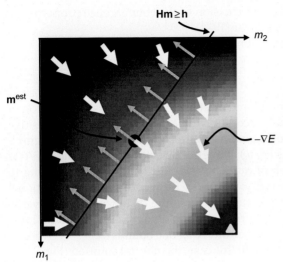

FIG. 7.6 The error $E(\mathbf{m})$ *(colors)* has a single minimum *(yellow triangle)*. The linear equality constraint, $\mathbf{Hm} \geq \mathbf{h}$, divides $S(\mathbf{m})$ into two half-spaces *(black line, with gray arrows* pointing into the feasible half-space). Solution *(circle)* lies on the boundary between the two half-spaces and therefore satisfies the constraint in the equality sense. At this point, the normal of the constraint hyperplane *(gray arrow)* is antiparallel to $-\nabla E$ *(white arrows)*. MatLab script gda07_05.

satisfies the constraints in the equality sense and a set $\mathbf{m}_S$ that satisfies the constraints in the inequality sense. The particular initial guess $\mathbf{m} = 0$ is clearly feasible and has all its elements in $\mathbf{m}_E$.

Step 2. Any model parameter m_i in $\mathbf{m}_E$ that has associated with it a negative gradient $[\nabla E]_i$ can be changed both to decrease the error and to remain feasible. Therefore, if there is no such model parameter in $\mathbf{m}_E$, the Kuhn-Tucker theorem indicates that this $\mathbf{m}$ is the solution to the problem.

Step 3. If some model parameter m_i in $\mathbf{m}_E$ has a corresponding negative gradient, then the solution can be changed to decrease the prediction error. To change the solution, we select the model parameter corresponding to the most negative gradient and move it to the set $\mathbf{m}_S$. All the model parameters in $\mathbf{m}_S$ are now recomputed by solving the system $\mathbf{G}_S \mathbf{m}_S' = \mathbf{d}_S$ in the least squares sense. The subscript S on the matrix indicates that only the columns multiplying the model parameters in $\mathbf{m}_S$ have been included in the calculation. All the $\mathbf{m}_E$s are still zero. If the new model parameters are all feasible, then we set $\mathbf{m} = \mathbf{m}'$ and return to Step 2.

Step 4. If some of the elements of $\mathbf{m}_S'$ are infeasible, however, we cannot use this vector as a new guess for the solution. Instead, we compute the change in the solution $\delta \mathbf{m} = \mathbf{m}_S' - \mathbf{m}_S$ and add as much of this vector as possible to the solution $\mathbf{m}_S$ without causing the solution to become infeasible. We therefore replace $\mathbf{m}_S$ with the new guess $\mathbf{m}_S + \alpha \delta \mathbf{m}$, where $\alpha = \min_i \{ m_{Si} / [m_{Si} - m_{Si}'] \}$ is the largest choice that can be made without some $\mathbf{m}_S$ becoming infeasible. At least one of the m_{Si}s has its constraint satisfied in the equality sense and must be moved back to $\mathbf{m}_E$. The process then returns to Step 3.

This algorithm contains two loops, one nested within the other. The outer loop successively moves model parameters from the group that is constrained to the group that minimizes the prediction error. The inner loop ensures that the addition of a variable to this latter group has not caused any of the constraints to be violated. Discussion of the convergence properties of this algorithm can be found in Lawson and Hanson (1974).

MatLab provides a function that implements nonnegative least squares

```
mest = lsqnonneg(G,dobs);
```

(*MatLab* script gda07_06)

As an example of its use, we analyze the problem of determining the mass distribution of an object from observations of its gravitational force (Fig. 7.7). Mass is an inherently positive quantity, so the positivity constraint constitutes very accurate prior information. The gravitational force d_i is measured at $N = 600$ points (x_i, y_i) above the object. The object (Fig. 7.7A) is subdivided into a grid of 20×20 pixels, each of mass m_j and position (x_j, y_j). According to Newton's inverse-square law, the data kernel is $G_{ij} = \gamma \cos(\theta_{ij})/R_{ij}^2$, where γ is the gravitational constant, θ_{ij} is the angle with respect to the vertical, and R_{ij} is the distance. Singular-value decomposition of **G** indicates that this mixed-determined problem is extremely nonunique, with only about 20 nonzero eigenvalues. Thus, the solution that one obtains depends critically on the type of prior information that one employs. In this case, the natural solution (Fig. 7.7E), which contains 136 negative masses, is completely different from the nonnegative solution (Fig. 7.7F), which contains none. Nevertheless, both satisfy the data almost exactly (Fig. 7.7A), with the prediction error E of the nonnegative solution about 1% larger than that

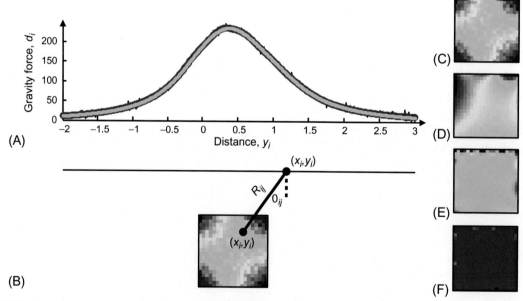

FIG. 7.7   (A) The vertical force of gravity is measured on a horizontal line above a massive cube-shaped object. This cube contains a grid of 20×20 model parameters representing spatially varying density, **m** *(colors)*. The equation **Gm**=**d** embodies Newton's inverse-square law of gravity. (B) The $N=600$ gravitational force observations **d**^obs *(black curve)* and the gravitational force predicted by the natural solution *(red curve, p=15)* and nonnegative least squares *(green curve)*. (C) True model. (D) Natural estimate of model, with $p=4$. (E) Natural estimate of model, with $p=15$. (F) Nonnegative estimate of model. *MatLab* script gda07_06.

of the natural solution. Ironically, neither the nonnegative solution nor the natural solution looks at all like the true solution (Fig. 7.7C), but a heavily damped solution (Fig. 7.7D) does.

The nonnegative least squares algorithm can also be used to solve the problem (Lawson and Hanson, 1974)

$$\text{minimize } E = \|\mathbf{m}\|_2^2 \text{ subject to } \mathbf{Hm} \geq \mathbf{h} \tag{7.60}$$

and, by virtue of the transformation described in Section 7.8.2, the completely general problem. The method consists of forming the $(M+1) \times p$ equation

$$\mathbf{G'm'} = \mathbf{d'} = \begin{bmatrix} \mathbf{H}^T \\ \mathbf{h}^T \end{bmatrix} \mathbf{m'} = \begin{bmatrix} \mathbf{0} \\ 1 \end{bmatrix} \tag{7.61}$$

and finding the $\mathbf{m'}$ that minimizes $\|\mathbf{d'} - \mathbf{G'm'}\|_2^2$ subject to $\mathbf{m'} \geq 0$ by the nonnegative least squares algorithm described earlier. If the prediction error $\mathbf{e'} = \mathbf{d'} - \mathbf{G'm'}$ is identically zero, then the constraints $\mathbf{Hm} \geq \mathbf{h}$ are inconsistent. Otherwise, the solution is $m_i = -e_i'/e_{M+1}'$ (which can also be written as $\mathbf{e'} = [-\mathbf{m}, 1]^T e_{M+1}'$).

In *MatLab*, the solution is computed as

```
Gp = [H, h]';
dp = [zeros(1,length(H(1,:))), 1]';
mp = lsqnonneg(Gp,dp);
ep = dp - Gp*mp;
m = -ep(1:end-1)/ep(end);
```

(*MatLab* script gda07_07)

An example with $M = 2$ model parameters and $N = 3$ constraints is shown in Fig. 7.8. The *MatLab* code for converting a *least squares with inequality constraints* problem into a *nonnegative least squares problem* using the procedure in Section 7.8.2 is:

```
[Up, Lp, Vp] = svd(G,0);
lambda = diag(Lp);
```

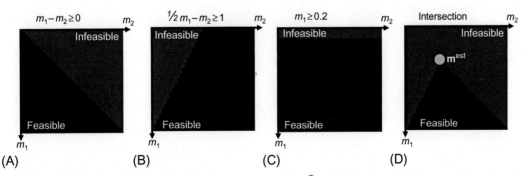

(A)   (B)   (C)   (D)

FIG. 7.8 Exemplary solution of the problem of minimizing $\mathbf{m}^T\mathbf{m}$ with $N=3$ inequality constraints $\mathbf{Hm} \geq \mathbf{h}$. (A–C) Each constraint divides the (m_1, m_2) plane into two half-places, one feasible and the other infeasible. (D) The intersection of the three feasible half-planes is polygonal in shape. The solution $\mathbf{m}^{\text{est}}$ (*green circle*) is the point in feasible area that is closest to the origin. Note that two of the three constrains are satisfied in the equality sense. *MatLab* script gda07_07.

```
rlambda = 1./lambda;
Lpi = diag(rlambda);
% transformation 1
Hp = -H*Vp*Lpi;
hp = h + Hp*Up'*dobs;
% transformation 2
Gp = [Hp, hp]';
dp = [zeros(1,length(Hp(1,:))), 1]';
mpp = lsqnonneg(Gp,dp);
ep = dp - Gp*mpp;
mp = -ep(1:end-1)/ep(end);
% take mp back to m
mest = Vp*Lpi*(Up'*dobs-mp);
dpre = G*mest;
```

(*MatLab* script gda07_07)

Note that svd( ) is called a second argument, set to zero, which causes it to compute $\mathbf{U}_p$ rather than the default, the $\mathbf{U}$. An exemplary problem is illustrated in Fig. 7.9.

We now demonstrate that this method does indeed solve the indicated problem (adapted from Lawson and Hanson, 1974). Step 1 is to show that e'_{M+1} is nonnegative. We first note that the gradient of the error satisfies $1/2 \nabla E' = -\mathbf{G}'^T[\mathbf{d}' - \mathbf{G}'\mathbf{m}'] = -\mathbf{G}'^T\mathbf{e}'$, and that because of the Kuhn-Tucker theorem, $\mathbf{m}'$ and $\nabla E'$ satisfy

$$\mathbf{m}'_E = 0 \qquad [\nabla E']_E < 0$$
$$\mathbf{m}'_S > 0 \qquad [\nabla E']_S = 0 \tag{7.62}$$

Note that these conditions imply $\mathbf{m}'^T \nabla E' = 0$. The length of the error E' is therefore

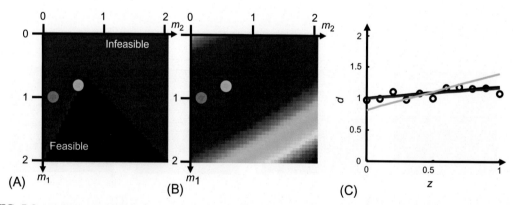

FIG. 7.9   Problem of minimizing the prediction error E subject to inequality constraints, applied to the straight line problem. (A) Feasible region of the model parameters (the intercept m_1 and slope m_2 of the straight line). The unconstrained solution *(orange circle)* is outside the feasible region, but the constrained solution *(green circle)* is on its boundary. (B) The unconstrained solution *(orange circle)* is at the global minimum of the prediction error E, while the constrained solution is not. (C) Plot of the data $d(z)$ showing true data *(black line)*, observed data *(black circles)*, unconstrained prediction *(red line)*, and constrained prediction *(green line)*. *MatLab* script gda07_08.

$$E' = e'^T e' = [d' - G'm']^T e' = d'^T e' - m'^T G'^T e'$$

$$= e'_{M+1} + {}^1/_2 m'^T \nabla E' = e'_{M+1} \tag{7.63}$$

as $d'^T e' = [0,1]e' = e'_{M+1}$. The error E' is necessarily a nonnegative quantity, so if it is not identically zero, then e'_{M+1} must be greater than zero.

Step 2 is to show that the solution satisfies the inequality constraints, $Hm - h \geq 0$. We start with the gradient of the error $\nabla E'$, which must have all nonnegative elements (or else the solution m' could be further minimized without violating the constraints $m' \geq 0$):

$$0 \leq \frac{1}{2} \nabla E' = -G'^T e' = -[H, h][-m, 1]^T e'_{M+1} = [Hm - h]e'_{M+1} \tag{7.64}$$

As $e'_{M+1} > 0$, we have $Hm - h \geq 0$.

Step 3 is to show that the solution minimizes $\|m\|_2^2$. This follows from the Kuhn-Tucker condition that, at a valid solution, the gradient of the error ∇E be represented as a nonnegative combination of the rows of H:

$$\nabla E = \nabla \|m\|_2^2 = 2m = -2 \left[e'_1 \cdots e'_M \right]^T / e'_{M+1} = 2H^T \left[m'_1 \cdots m'_M \right]^T / e'_{M+1} \tag{7.65}$$

Here we have used the fact that $e' = d' - G'm'$, together with the fact that the first M elements of d' are zero. Note that m' and e'_{M+1} are nonnegative, so the Kuhn-Tucker condition is satisfied.

Finally, we can also show that a feasible solution exists only when the error is not identically zero. Consider the contradiction that the error is identically zero but that a feasible solution exists. Then

$$0 = e'^T e' / e'_{M+1} = [-m^T, 1] [d' - G'm'] = 1 + [Hm - h]^T m' \tag{7.66}$$

As $m' \geq 0$, the relationship $Hm < h$ is implied. This contradicts the constraint equations $Hm \geq h$ so that an identically zero error implies that no feasible solution exists and that the constraints are inconsistent.

7.10 PROBLEMS

7.1 This problem builds upon the discussion in Section 6.2. Use *MatLab*'s `svd()` function to compute the null vectors associated with the data kernel $G = [1,1,1,1]$. (A) The null vectors are different from that given in Eq. (6.5). Why? Show that the two sets are equivalent.

7.2 Modify the weighted damped least squares "gap-filling" problem of Fig. 3.10 (*MatLab* script gda03_09) so that the prior information is applied only to the part of the solution in the null space. First, transform the weighted problem $Gm = d$ into an unweighted one $G'm' = d'$ using the transformation given by Eq. (7.23). Then, find the minimum-length solution (or the natural solution) m'^{est}. Finally, transform back to obtain m^{est}. How different is this solution from the one given in the figure? Compare the two prediction errors. In order to insure that $W_e = D^T D$ has an inverse, which is required by the transformation, you should make **D** square by adding two rows, one at the top and the other at the bottom, that constrain the first and last model parameters to known values.

7.3 (A) Compute and plot the null vectors for the data kernel shown in Fig. 7.4A. (B) Suppose that the elements of $\mathbf{m}^{\text{true}}$ are drawn from a uniform distribution between 0 and 1. How large a contribution do the null vectors make to the true solution $\mathbf{m}^{\text{true}}$? (*Hint*: Write the model parameters as a linear combination of all the eigenvectors **V** by writing $\mathbf{m}^{\text{true}} = \mathbf{Va}$, where **a** are coefficients, and then solving for **a**. Then examine the size of the elements of **a**. You may wish to refer to *MatLab* script gda07_03 for the definition of **G**).

7.4 Consider fitting a cubic polynomial $d_i = m_1 + m_2 z_i + m_3 z_i^2 + m_4 z_i^3$ to $N = 20$ data d_i^{obs}, where the zs are evenly spaced on the interval (0,1), where $m_2 = m_3 = m_4 = 1$, and where m_1 is varied from -1 to 1, as described later. (A) Write a *MatLab* script that generates synthetic observed data (including Gaussian-distributed noise) for a particular choice of m_1, estimates the model parameters using both simple least squares and nonnegative least squares, plots the observed and predicted data, and outputs the total error. (B) Run a series of test cases for a range of values of m_1 and comment upon the results.

7.5 (A) Modify the *MatLab* gda07_07 script so that the last constraint is $m_1 \geq 1.2$. How do the feasible region and the solution change? (B) Modify the script to add a constraint $m_2 \geq 0.2$. How do the feasible region and the solution change?

References

Lanczos, C., 1961. Linear Differential Operators. Van Nostrand-Reinhold, Princeton, NJ.

Lawson, C.L., Hanson, D.J., 1974. Solving Least Squares Problems. Prentice-Hall, Englewood Cliffs, NJ.

Menke, W., Menke, J., 2011. Environmental Data Analysis with MatLab. Academic Press, Elsevier Inc., Oxford, UK. 263 pp.

Penrose, R.A., 1955. A generalized inverse for matrices. Proc. Cambridge Philos. Soc. 51, 406–413.

Wiggins, R.A., 1972. The general linear inverse problem: implication of surface waves and free oscillations for Earth structure. Rev. Geophys. Space Phys. 10, 251–285.

Wunsch, C., Minster, J.F., 1982. Methods for box models and ocean circulation tracers: mathematical programming and non-linear inverse theory. J. Geophys. Res. 87, 5647–5662.

Linear Inverse Problems and Non-Gaussian Statistics

8.1 L_1 NORMS AND EXPONENTIAL PROBABILITY DENSITY FUNCTIONS

In Chapter 5, we showed that the method of least squares and the more general use of L_2 norms could be rationalized through the assumption that the data and prior model parameters followed Gaussian statistics. This assumption is not always appropriate; however, some data sets follow other distributions. The *two-sided exponential probability density function* is one simple alternative. Here "two-sided" means that the function is defined on the interval $(-\infty, +\infty)$ and has tails in both directions. When Gaussian and exponential distributions of the same mean $\langle d \rangle$ and variance σ^2 are compared, the exponential distribution is found to be much longer tailed (Fig. 8.1, Table 8.1):

© 2018 Elsevier Inc. All rights reserved.

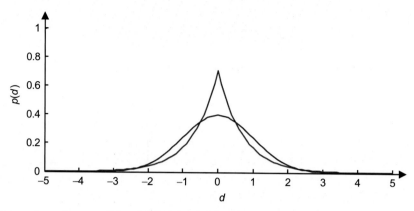

FIG. 8.1 Comparison of exponential probability density distribution *(red)* with a Gaussian probability density distribution *(blue)* of equal variance, $\sigma^2 = 1$. The exponential probability density distribution is longer tailed. *MatLab* script gda08_01.

TABLE 8.1 The Area Beneath $\langle d \rangle \pm n\sigma$ for the Gaussian and Exponential Distributions

n	Gaussian Area (%)	Exponential Area (%)
1	68.2	76
2	95.4	94
3	99.7	98.6
4	99.999 +	99.7

$$\text{Gaussian} \qquad\qquad\qquad\qquad \text{exponential}$$

$$p(d) = \frac{1}{(2\pi)^{1/2}\sigma} \exp\left\{ -\frac{1}{2}\frac{(d - \langle d \rangle)^2}{\sigma^2} \right\} \qquad p(d) = \frac{1}{(2)^{1/2}\sigma} \exp\left\{ -(2)^{1/2}\frac{|d - \langle d \rangle|}{\sigma} \right\} \qquad (8.1)$$

Note that the probability of realizing data far from $\langle d \rangle$ is much higher for the exponential probability density function than for the Gaussian probability density function. A few data, say, 4 standard deviations from the mean, are reasonably probable in a data set of, say, 1000 samples drawn from an exponential probability density function, but very improbable for data drawn from a Gaussian probability density function. We therefore expect that methods based on the exponential probability density function will be able to handle a data set with a few "bad" data (outliers) better than Gaussian methods. Methods that can tolerate a few outliers are said to be *robust* (Claerbout and Muir, 1973).

In *MatLab*, the exponential probability density function with mean dbar and variance sd^2 is calculated as

```
pE = (1/sqrt(2))*(1/sd)*exp(−sqrt(2)*abs((d−dbar))/sd);
```

(*MatLab* script gda08_01)

MatLab's random() function provides only a one-sided version of the exponential probability density function, defined on the interval (0,+∞), but a two-sided version can be created by multiplication of its realizations by a random sign, created here from realizations of the discrete uniform probability density function

```
mu = sd/sqrt(2);
rsign = (2*(random('unid',2,Nr,1)-1)-1);
dr = dbar + rsign .* random('exponential',mu,Nr,1);
```

(*MatLab* script gda08_01)

8.2 MAXIMUM LIKELIHOOD ESTIMATE OF THE MEAN OF AN EXPONENTIAL PROBABILITY DENSITY FUNCTION

Exponential probability density functions bear the same relationship to L_1 norms as Gaussian probability density function bear to L_2 norms. To illustrate this relationship, we consider the joint distribution for N independent data, each with the same mean m_1 and variance σ^2. Since the data are independent, the joint probability density function is just the product of N univariate functions:

$$p(d) = (2)^{-N/2}\sigma^{-N}\exp\left[-\frac{(2)^{1/2}}{\sigma}\sum_{i=1}^{N}|d_i - m_1|\right] \tag{8.2}$$

To maximize the likelihood of $p(d)$, we must maximize the argument of the exponential, which involves minimizing the sum of absolute residuals as

$$\text{minimize } E = \sum_{i=1}^{N}|d_i - m_1| \tag{8.3}$$

This is the L_1 norm of the prediction error in a linear inverse problem of the form $\mathbf{Gm}=\mathbf{d}$, where $M=1$, $\mathbf{G}=[1, 1, \dots, 1]^T$, and $\mathbf{m}=[m_1]$. Applying the principle of maximum likelihood, we obtain

$$\text{maximize } L = \log P = -\frac{N}{2}\log(2) - N\log(\sigma) - \frac{(2)^{1/2}}{\sigma}\sum_{i=1}^{N}|d_i - m_1| \tag{8.4}$$

Setting the derivatives to zero yields

$$\frac{\partial L}{\partial m_1} = 0 = \frac{(2)^{1/2}}{\sigma}\sum_{i=1}^{N}\text{sign}(d_i - m_1)$$

$$\frac{\partial L}{\partial \sigma} = 0 = \frac{N}{\sigma} - \frac{(2)^{1/2}}{\sigma^2}\sum_{i=1}^{N}|d_i - m_1| \tag{8.5}$$

The sign function sign(x) equals +1 if $x>0$, −1 if $x<0$, and 0 if $x=0$. Note that the derivative $d|x|/dx=\text{sign}(x)$, since if x is positive, then it is just equal to $dx/dx=1$ and if it is negative, then to −1. The first equation yields the implicit expression for $m_1=\langle d\rangle^{\text{est}}$ for which $\sum_i\text{sign}(d_i - m_1)=0$. The second equation can then be solved for an estimate of the variance as

$$\sigma^{\text{est}} = \frac{(2)^{1/2}}{N} \sum_{i=1}^{N} |d_i - m_1| \tag{8.6}$$

The equation for m_1^{est} is exactly the sample median; one finds an m_1 such that half the ds are less than m_1 and half are greater than m_1. There are then an equal number of negative and positive signs and the sum of the signs is zero. The median is a robust property of a set of data. Adding one outlier can at worst move the median from one central datum to another nearby central datum. While the maximum likelihood estimate of a Gaussian distribution's true mean is the sample arithmetic mean, the maximum likelihood estimate of an exponential distribution's true mean is the sample median.

The estimate of the variance also differs between the two distributions: in the Gaussian case, it is the square of the sample standard deviation, but in the exponential case, it is not. If there are an odd number of samples, then m_1^{est} equals the middle d_i. If there is an even number of samples, any m_1^{est} between the two middle data maximizes the likelihood. In the odd case, the error E attains a minimum only at the middle sample, but in the even case, it is flat between the two middle samples (Fig. 8.2). We see, therefore, that L_1 problems of

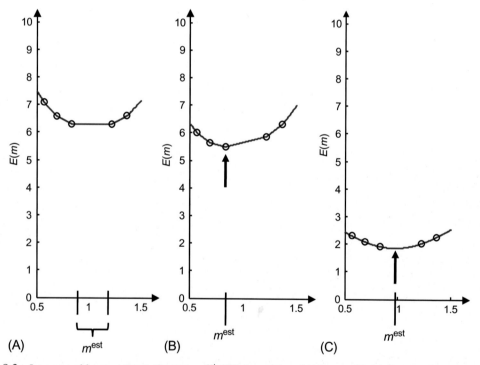

FIG. 8.2   Inverse problem to estimate the mean m^{est} of N observations. (A) L_1 error $E(m)$ (red curve) with even N. The error has a flat minimum, bounded by two observations (circles), and the solution is nonunique. (B) L_1 error with odd N. The error is minimum at one of the observation points, and the solution is unique. (C) The L_2 error, both odd and even N. The error is minimum at a single point, which may not correspond to an observation point, and the solution is unique. MatLab script gda08_02.

minimizing the prediction error of $\mathbf{Gm} = \mathbf{d}$ can possess nonunique solutions that are distinct from the type of nonuniqueness encountered in the L_2 problems. The L_1 problems can still possess nonuniqueness owing to the existence of null solutions since the null solutions cannot change the prediction error under any norm. That kind of nonuniqueness leads to a completely unbounded range of estimates. The new type of nonuniqueness, on the other hand, permits the solution to take on any values between *finite bounds*.

We also note that regardless of whether N is even or odd, we can choose m_1^{est} so that one of the equations $\mathbf{Gm} = \mathbf{d}$ is satisfied exactly (in this case, $m_1^{\text{est}} = d_k$, where k is the index of the "middle" datum). This can be shown to be a general property of L_1 norm problems. Given N equations and M unknowns related by $\mathbf{Gm} = \mathbf{d}$, it is possible to choose $\mathbf{m}$ so that the L_1 prediction error is minimized and so that M of the equations are satisfied exactly.

In *MatLab*, the L_1 estimate of the mean $\langle d \rangle$ and square root of the variance σ are computed as

```
dbarest = median(dr);
sdest = (sqrt(2)/Nr)*sum(abs(dr-dbarest));
```

(*MatLab* script gda08_01)

Here `dr` is a vector of the data of length `Nr`.

8.3 THE GENERAL LINEAR PROBLEM

Consider the linear inverse problem $\mathbf{Gm} = \mathbf{d}$ in which the data and prior model parameters are uncorrelated with known means $\mathbf{d}^{\text{obs}}$ and $\langle \mathbf{m} \rangle$ and known variances σ_d^2 and σ_m^2, respectively. Their joint distribution is then

$$p(\mathbf{d}, \mathbf{m}) = 2^{-(M+N)/2} \prod_{i=1}^{N} \sigma_{d_i}^{-1} \exp\left[-2^{1/2} \frac{|e_i|}{\sigma_{d_i}}\right] \prod_{j=1}^{M} \sigma_{mj}^{-1} \exp\left[-2^{1/2} \frac{|l_j|}{\sigma_{mj}}\right] \tag{8.7}$$

where the prediction error is given by $\mathbf{e} = \mathbf{d} - \mathbf{Gm}$ and the solution length by $\mathbf{l} = \mathbf{m} - \langle \mathbf{m} \rangle$. The maximum likelihood estimate of the model parameters occurs when the exponential is a minimum, that is, when the sum of the weighted L_1 prediction error and the weighted L_1 solution length is minimized:

$$\text{minimize } E + L = \sum_{i=1}^{N} \frac{|e_i|}{\sigma_{d_i}} + \sum_{j=1}^{M} \frac{|l_j|}{\sigma_{mj}} \tag{8.8}$$

In this case, the weighting factors are the reciprocals of the square root of the variance, in contrast to the Gaussian case, in which they are the reciprocals of the variances. Note that linear combinations of exponentially distributed random variables are not themselves exponential (unlike Gaussian variables, which give rise to Gaussian combinations). The covariance matrix of the estimated model parameters is, therefore, difficult both to calculate and to interpret since the manner in which it is related to confidence intervals varies from case to case. We shall focus our discussion on estimating only the model parameters themselves.

8.4 SOLVING L_1 NORM PROBLEMS BY TRANSFORMATION TO A LINEAR PROGRAMMING PROBLEM

We shall show that this problem can be transformed into a "linear programming" problem (see Section 6.6)

$$\text{find } \mathbf{x} \text{ that minimizes } z = \mathbf{f}^T \mathbf{x}$$
$$\text{with the constraints } \mathbf{Ax} \le \mathbf{b} \text{ and } \mathbf{Cx} = \mathbf{d} \text{ and } \mathbf{x}^{(1)} \le \mathbf{x} \le \mathbf{x}^{(u)} \tag{8.9}$$

We first define the L_1 versions of the weighted solution length L and the weighted prediction error E

$$L = \sum_{i=1}^{M} \frac{|m_i - \langle m_i \rangle|}{\sigma_{mi}} \quad \text{and} \quad E = \sum_{i=1}^{N} \frac{|d_i^{\text{obs}} - d_i^{\text{pre}}|}{\sigma_{di}} \quad \text{with} \quad \mathbf{d}^{\text{pre}} = \mathbf{Gm} \tag{8.10}$$

Note that these formulas are weighted by the square root of the variances of the prior model parameters and measurement error (σ_m and σ_d, respectively).

First, we shall consider the completely underdetermined linear problem with prior model parameters, mean $\langle \mathbf{m} \rangle$, and variance σ_m^2. The problem is to minimize the weighted length L subject to the constraint $\mathbf{Gm} = \mathbf{d}$. We first introduce $5M$ new variables m'_i, m''_i, α_i, x_i, and x'_i, where $i = 1, \ldots, M$. The linear programming problem may be stated as follows (Cuer and Bayer, 1980)

$$\text{minimize } z = \sum_{i=1}^{M} \frac{\alpha_i}{\sigma_{mi}} \text{ subject to the constraints}$$
$$\mathbf{G}(\mathbf{m}' - \mathbf{m}'') = \mathbf{d} \text{ and } \mathbf{m}' - \mathbf{m}'' + \mathbf{x} - \boldsymbol{\alpha} = \langle \mathbf{m} \rangle \text{ and } \mathbf{m}' - \mathbf{m}'' - \mathbf{x}' - \boldsymbol{\alpha} = \langle \mathbf{m} \rangle \tag{8.11}$$
$$\text{and}$$
$$\mathbf{m}' \ge 0 \text{ and } \mathbf{m}'' \ge 0 \text{ and } \boldsymbol{\alpha} \ge 0 \text{ and } \mathbf{x} \ge 0 \text{ and } \mathbf{x}' \ge 0$$

This linear programming problem has $5M$ unknowns, $N + 2M$ equality constraints, and $5M$ inequality constraints. If one makes the identification $\mathbf{m} = \mathbf{m}' - \mathbf{m}''$, the first equality constraint is equivalent to $\mathbf{Gm} = \mathbf{d}$. The signs of the elements of $\mathbf{m}$ are not constrained even though those of $\mathbf{m}'$ and $\mathbf{m}''$ are. The remaining equality constraints can be rewritten as

$$\boldsymbol{\alpha} - \mathbf{x} = [\mathbf{m} - \langle \mathbf{m} \rangle] \quad \text{and} \quad \boldsymbol{\alpha} - \mathbf{x}' = -[\mathbf{m} - \langle \mathbf{m} \rangle] \tag{8.12}$$

where the α_i, x_i, and x'_i are nonnegative. Now if $[m_i - \langle m_i \rangle]$ is positive, the first equation requires $\alpha_i \ge [m_i - \langle m_i \rangle]$ since x_i cannot be negative. The second constraint can always be satisfied by choosing some appropriate x'_i. On the other hand, if $[m_i - \langle m_i \rangle]$ is negative, then the first constraint can always be satisfied by choosing some appropriate x_i, but the second constraint requires that $\alpha_i \ge -[m_i - \langle m_i \rangle]$. Taken together, these two constraints imply that $\alpha_i \ge |[m_i - \langle m_i \rangle]|$. Minimizing $\sum \alpha_i / \sigma_{mi}$ is therefore equivalent to minimizing the weighted solution length L. The L_1 minimization problem has been converted to a linear programming problem.

The completely overdetermined problem of minimizing E with no prior information can be converted into a linear programming problem in a similar manner. We introduce $2M + 3N$ new variables, m'_i, m''_i, $i = 1, M$ and α_i, x_i, x'_i, $i = 1, N$, $2N$ equality constraints and $2M + 3N$

inequality constraints. The equivalent linear programming problem is (Cuer and Bayer, 1980):

$$\text{minimize } E = \sum_{i=1}^{N} \frac{\alpha_i}{\sigma_{d_i}} \text{ subject to the constraints}$$

$$\mathbf{G}[\mathbf{m}' - \mathbf{m}''] + \mathbf{x} - \boldsymbol{\alpha} = \mathbf{d} \text{ and } \mathbf{G}[\mathbf{m}' - \mathbf{m}''] - \mathbf{x}' + \boldsymbol{\alpha} = \mathbf{d} \quad (8.13)$$

and

$$\mathbf{m}' \geq 0 \text{ and } \mathbf{m}'' \geq 0 \text{ and } \boldsymbol{\alpha} \geq 0 \text{ and } \mathbf{x} \geq 0 \text{ and } \mathbf{x}' \geq 0$$

In *MatLab*, the solution is computed as

```
% L1 solution to overdetermined problem
% linear programming problem is
% min f*x subject to A x <= b, Aeq x = beq

% variables
% m = mp - mpp
% x = [mp', mpp', alpha', x', xp']'
% with mp, mpp length M and alpha, x, xp, length N
L = 2*M+3*N;
x = zeros(L,1);

f = zeros(L,1);
f(2*M+1:2*M+N) = 1./sd;

% equality constraints
Aeq = zeros(2*N,L);
beq = zeros(2*N,1);

% first equation G(mp-mpp)+x-alpha = d
Aeq(1:N,1:M)              = G;
Aeq(1:N,M+1:2*M)          = -G;
Aeq(1:N,2*M+1:2*M+N)      = -eye(N,N);
Aeq(1:N,2*M+N+1:2*M+2*N) = eye(N,N);
beq(1:N)                  = dobs;

% second equation G(mp-mpp)-xp+alpha=d
Aeq(N+1:2*N,1:M)              = G;
Aeq(N+1:2*N,M+1:2*M)          = -G;
Aeq(N+1:2*N,2*M+1:2*M+N)      = eye(N,N);
Aeq(N+1:2*N,2*M+2*N+1:2*M+3*N) = -eye(N,N);
beq(N+1:2*N)                  = dobs;

% inequality constraints A x <= b
% part 1: everything positive
A = zeros(L+2*M,L);
b = zeros(L+2*M,1);
A(1:L,:) = -eye(L,L);
b(1:L) = zeros(L,1);
```

```
% part 2; mp and mpp have an upper bound.
A(L+1:L+2*M,:) = eye(2*M,L);
mls = (G'*G)\(G'*dobs); % L2 solution — sure easier!
mupperbound=10*max(abs(mls));
b(L+1:L+2*M) = mupperbound;

% solve linear programming problem
[x, fmin] = linprog(f,A,b,Aeq,beq);
fmin=-fmin;
mest = x(1:M) — x(M+1:2*M);
```

(*MatLab* script gda08_03)

Note that the *MatLab* implementation adds upper bounds for $\mathbf{m}'$ and $\mathbf{m}''$, 10 times the largest element of the least-squares solution (an amount that might need to be adjusted for the problem at hand). Without this constraint, the algorithm could possibly return very large values for these variables, leading to a solution $\mathbf{m}=\mathbf{m}'-\mathbf{m}''$ that is inaccurate because of round-off error. An example of the L_1 problem applied to curve fitting is shown in Fig. 8.3.

The mixed-determined problem can be solved by any of several methods. By analogy to the L_2 methods described in Chapters 3 and 7, we could either pick some prior model parameters and minimize $E+L$ or separate the overdetermined model parameters from the under-determined ones and apply prior information to the underdetermined ones only. The first method leads to a linear programming problem similar to the two cases stated earlier but

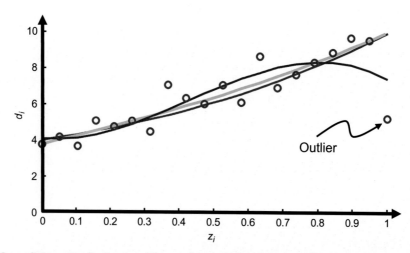

FIG. 8.3 Curve fitting using the L_1 norm. The true data *(red curve)* follow a cubic polynomial in an auxiliary variable, z. Observations *(red circles)* have additive noise with zero mean and variance $\sigma^2=1$ drawn from an exponential probability density function. Note the outlier at $z\approx1$. The L_1 fit *(green curve)* is not as affected by the outlier as a standard L_2 (least-squares) fit *(blue curve)*. *MatLab* script gda08_03.

with even more variables $(5M+3N)$, equality constraints $(2M+2N)$, and inequality constraints $(5M+2N)$:

$$\text{minimize } \sum_{i=1}^{M} \frac{\alpha_i}{\sigma_{m_i}} + \sum_{i=1}^{N} \frac{\alpha_i'}{\sigma_{d_i}} \text{ subject to the constraints}$$

$$[\mathbf{m}' - \mathbf{m}''] + \mathbf{x} - \boldsymbol{\alpha} = \langle\mathbf{m}\rangle \text{ and } [\mathbf{m}' - \mathbf{m}''] - \mathbf{x}' + \boldsymbol{\alpha} = \langle\mathbf{m}\rangle \text{ and}$$

$$\mathbf{G}[\mathbf{m}' - \mathbf{m}''] + \mathbf{x}'' - \boldsymbol{\alpha}' = \mathbf{d} \text{ and } \mathbf{G}[\mathbf{m}' - \mathbf{m}''] - \mathbf{x}''' + \boldsymbol{\alpha}' = \mathbf{d} \text{ and} \qquad (8.14)$$

$$\mathbf{m}' \geq 0 \text{ and } \mathbf{m}'' \geq 0 \text{ and } \boldsymbol{\alpha} \geq 0 \text{ and } \boldsymbol{\alpha}' \geq 0$$

$$\text{and } \mathbf{x} \geq 0 \text{ and } \mathbf{x}' \geq 0 \text{ and } \mathbf{x}'' \geq 0 \text{ and } \mathbf{x}''' \geq 0$$

The second method is more interesting. First, we use the singular-value decomposition to identify the null space of $\mathbf{G}$. The solution then has the form

$$\mathbf{m}^{\text{est}} = \sum_{i=1}^{p} a_i \mathbf{v}_p^{(i)} + \sum_{i=p+1}^{M} b_i \mathbf{v}_0^{(i)} = \mathbf{V}_p \mathbf{a} + \mathbf{V}_0 \mathbf{b} \qquad (8.15)$$

where the $\mathbf{v}$s are eigenvectors and $\mathbf{a}$ and $\mathbf{b}$ are vectors of unknown coefficients. Only the vector $\mathbf{a}$ can affect the prediction error, so one uses the overdetermined algorithm to determine it

$$\text{find } \mathbf{a} \text{ that minimizes } E = \|\mathbf{d} - \mathbf{Gm}\|_1 = \sum_{i=1}^{N} \frac{\left| d_i - [\mathbf{U}_p \boldsymbol{\Lambda}_p \mathbf{a}]_i \right|}{\sigma_{d_i}} \qquad (8.16)$$

Next, one uses the underdetermined algorithm to determine $\mathbf{b}$:

$$\text{find } \mathbf{b} \text{ that minimizes } E = \|\mathbf{m} - \langle\mathbf{m}\rangle\|_1 = \sum_{i=1}^{N} \frac{\left| [\mathbf{V}_0 \mathbf{b}]_i - (\langle m_i\rangle - [\mathbf{V}_p \mathbf{a}]_i) \right|}{\sigma_{m_i}} \qquad (8.17)$$

It is also possible to implement the basic underdetermined and overdetermined L_1 algorithms in such a manner that the many extra variables are never explicitly calculated (Cuer and Bayer, 1980). This procedure vastly decreases the storage and computation time required, making these algorithms practical for solving moderately large $(M=1000)$ inverse problems.

8.5 SOLVING L_1 NORM PROBLEMS BY REWEIGHTED L_2 MINIMIZATION

In the previous section, we solved the L_1 minimization problem by transforming it to an equivalent linear programming problem. The transformation is exact, but the formulation is cumbersome because many new variables are introduced (which, among other things, increases computer memory requirements). In this section, we show that the problem can also be transformed into an equivalent L_2 problem and solved with the standard least-squares methods. The transformation is only approximate and the method requires iteration to achieve an accurate solution, but is usually much faster than the linear programming transformation introduced in the previous section. (We will return to the issue of choosing between methods at the end of this section.)

We begin by considering the L_n norm; we will later focus on the the L_1 norm by setting $n=1$. Our goal is to manipulate $\|\mathbf{v}\|_n^n$, where $\mathbf{v}$ is an arbitrary vector, so that it looks like the square of a *weighted L_2 norm* $\|\mathbf{W}^{\frac{1}{2}}\mathbf{v}\|_2^2$. The former is defined as:

$$\|\mathbf{v}\|_n^n = \sum_k |v_k|^n \tag{8.18}$$

and the latter is defined as:

$$\left\|\mathbf{W}^{\frac{1}{2}}\mathbf{v}\right\|_2^2 = \mathbf{v}^\mathsf{T}\mathbf{W}\mathbf{v} = \sum_k w_k v_k^2 \tag{8.19}$$

where $\mathbf{W} = \mathrm{diag}(\mathbf{w})$ is a diagonal weight matrix. Consider the choice:

$$w_k = \left(v_k^\gamma + \delta^\gamma\right)^{(n-2)/\gamma} \text{ with } 0 < \delta \ll 1 \text{ and } 1 \le \gamma \le 2 \tag{8.20}$$

Here δ is a small positive number that prevents w_k from being singular should v_k^γ ever be zero and the factor γ is added for numerical stability. Frommlet and Nuel (2016) suggest values of $\delta = 10^{-5}$ and $\gamma = 2$. Inserting Eq. (8.20) into Eq. (8.19) yields:

$$\left\|\mathbf{W}^{\frac{1}{2}}\mathbf{v}\right\|_2^2 = \mathbf{v}^\mathsf{T}\mathbf{W}\mathbf{v} = \sum_k \frac{v_k^n}{\left(v_k^\gamma + \delta^\gamma\right)^{-(n-2)/\gamma}} \approx \sum_k \frac{|v_k|^2}{|v_k|^{(2-n)}} = \sum_k |v_k|^n = \|\mathbf{v}\|_n^n \tag{8.21}$$

Thus, this choice of weighting makes the two norms equal. Frommlet and Nuel (2016) point out that significant round-off error can occur when Eq. (8.20) is used to evaluate the weights, and recommend instead the numerically more stable (and algebraically equivalent) formula:

$$w_k = \begin{cases} \delta^{n-2} \exp\left[\dfrac{n-2}{\gamma} \log 1\mathrm{p}\left(\left|\dfrac{v_k}{\delta}\right|^\gamma\right)\right] & \text{if } |v_k| \le \delta \\[3mm] |v_k|^{n-2} \exp\left[\dfrac{n-2}{\gamma} \log 1\mathrm{p}\left(\left|\dfrac{\delta}{v_k}\right|^\gamma\right)\right] & \text{if } |v_k| > \delta \end{cases} \tag{8.22}$$

Here $\log 1\mathrm{p}(x) \equiv \log(1+x)$; MATLAB's version is also called `log1p()`.

We now consider the special case of the L_1 norm; that is, $n=1$. While the procedure has made $\|\mathbf{v}\|_1$ look like $\|\mathbf{v}\|_2^2$, it has introduced the undesirable complication of a weight matrix $\mathbf{W}$ that is a function of $\mathbf{v}$. This behavior is problematical for a minimization problem where one seeks the model parameters $\mathbf{m}^{\mathrm{est}}$ that minimize $\|\mathbf{W}^{\frac{1}{2}}\mathbf{v}\|_2^2$, for now both $\mathbf{W}(\mathbf{m})$ and $\mathbf{v}(\mathbf{m})$ are functions of $\mathbf{m}$, and the standard least squares formula, which requires $\mathbf{W}$ to be constant, do not apply. This problem can be handled by successive approximation. Starting with a weight matrix approximated as $\mathbf{W}^{(0)} = \mathbf{I}$, the standard least-squares formula is used to determine an initial approximation $\mathbf{m}^{(0)}$ for the model parameters. This solution is then iteratively improved, by first updating the weight matrix $\mathbf{W}^{(j)} = \mathbf{W}(\mathbf{m}^{(j-1)})$ and then using the least-squares formula to update the solution $\mathbf{m}^{(j)}$. This *reweighting* process is terminated when the solution no longer changes between successive iterations, whence $\mathbf{m}^{\mathrm{est}} = \mathbf{m}^{(j)}$. Candes et al. (2008) analyze the properties of this algorithm (but only for the $n=1$ case) and conclude that it will always converge to the correct solution.

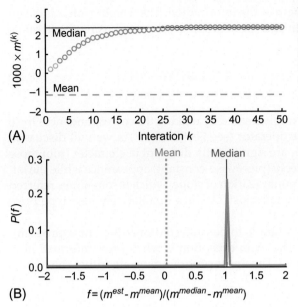

(A)

(B)

FIG. 8.4 (A) When applied to the first exemplary problem, the iterative algorithm rapidly converges to the median of the data (*red line*, the correct value). (B) An empirical probability density function $p(f)$ for the solution quality factor f is strongly peaked around $f=1$ (*red line*, the correct value).

As an example, consider the problem of finding the single model parameter m_1 that minimizes the error $\|\mathbf{e}\|_1$ where $\mathbf{e}=\mathbf{d}^{obs}-\mathbf{Gm}$. Here $\mathbf{d}^{obs}=\mathbf{Gm}^{true}+\mathbf{n}$ are the N observed data, $\mathbf{n}$ is exponentially distributed noise, and $\mathbf{G}=[1 \cdots 1]^T$; that is, each datum equals the model parameter plus noise. We have previously shown that the solution of this L_1 problem is $m_1^{est}=m_1^{median}=$ median($\mathbf{d}^{obs}$). The L_1 problem is transformed into the equivalent L_2 problem of minimizing $\|\mathbf{W}^{\frac{1}{2}}\mathbf{e}\|_2^2$ by making the correspondence $\mathbf{v}=\mathbf{e}$. The iterative form of the standard least-squares formula is:

$$\mathbf{m}^{(j)} = \left[\mathbf{G}^T\mathbf{W}^{(j-1)}\mathbf{G}\right]^{-1}\mathbf{G}^T\mathbf{W}^{(j-1)}\mathbf{d}^{obs} \qquad (8.23)$$

(see Eq. 3.47).

In the example, the solution is initialized to $m_1^{(0)}=m_1^{mean}=$ mean($\mathbf{d}^{obs}$). The iterations quickly converge—typically win about 25 iterations—to the expected solution of $m_1^{est}=m_1^{median}$ (Fig. 8.4A). The quality of the solution can be assessed using the function:

$$f = \frac{(m^{est}-m^{mean})}{(m^{median}-m^{mean})} \qquad (8.24)$$

since it takes on a value of unity when the solution is correct. An empirical p.d.f., created by solving the problem 1000 times for different realizations of the noise $\mathbf{n}$, is very strongly peaked at $f=1$ (Fig. 8.4B), which verifies that the method very reliably converges to the correct answer.

The iterative algorithm can also be used to solve problems with prior information of the form $\mathbf{Hm}-\mathbf{h}$. Consider, for instance the problem of finding the $\mathbf{m}^{est}$ that minimizes:

$$\|\mathbf{e}\|_2^2 + \mu\|\mathbf{h}-\mathbf{Hm}\|_1 \qquad (8.25)$$

where the parameter μ quantifies the strength of the information. The weight matrix is calculated using the correspondence $\mathbf{v} = \mathbf{h} - \mathbf{Hm}$ and the solution is calculated using the iterative formula:

$$\mathbf{m}^{(j)} = \left[\mathbf{G}^T\mathbf{G} + \mu\mathbf{H}^T\mathbf{W}^{(j-1)}\mathbf{H}\right]^{-1}\left[\mathbf{G}^T\mathbf{d}^{obs} + \mu\mathbf{H}^T\mathbf{W}^{(j-1)}\mathbf{Hh}\right] \tag{8.26}$$

(see Eq. 3.50).

The prior information of flatness can be implemented by making the correspondences $\mathbf{h} = \mathbf{0}$ and $\mathbf{H} = \mathbf{D}$, where $\mathbf{D}$ is the first difference operator (see Section 3.9). As we will discuss in Section 8.7, solutions that minimize $\|\mathbf{Dm}\|_1$ are significantly different in character from those that minimize $\|\mathbf{Dm}\|_2^2$; the former have a *blocky* (piecewise-constant) appearance while the latter tend to be smooth as well as flat. The L_1 minimization of slope, which is sometimes referred to as *Total Variation Regularization* (or sometimes just *TV*), is a qualitatively new type prior information that has wide application.

In most instances, the reweighting algorithm is to be preferred over the linear programming algorithm presented in Section 8.4. The main exception is when prior information in the form of inequality constraints is available. The linear programming algorithm can handle such constraints very naturally.

8.6 THE L_∞ NORM

While the L_1 norm weights "bad" data less than the L_2 norm, the L_∞ norm weights it more:

$$\text{minimize } L + E = \|\mathbf{e}\|_\infty + \|\mathbf{l}\|_\infty = \max_i \frac{|e_i|}{\sigma_{d_i}} + \max_i \frac{|l_i|}{\sigma_{m_i}} \tag{8.27}$$

The prediction error and solution length are weighted by the reciprocal of their some prior standard deviations. Normally, one does not want to emphasize outliers, so the L_∞ form is useful mainly in that it can provide a "worst-case" estimate of the model parameters for comparison with estimates derived on the basis of other norms. If the estimates are in fact close to one another, one can be confident that the data are highly consistent. Since the L_∞ estimate is controlled only by the worst error or length, it is usually nonunique.

The general linear equation $\mathbf{Gm} = \mathbf{d}$ can be solved in the L_∞ sense by transformation into a linear programming problem using a variant of the method used to solve the L_1 problem. In the underdetermined problem, we again introduce new variables, m'_i, m''_i, x_i, and x'_i, each of length M, and a single parameter α ($4M + 1$ variables, total). The linear programming problem is

$$\text{minimize } \alpha \text{ subject to the constraints}$$
$$\mathbf{G}[\mathbf{m}' - \mathbf{m}''] = \mathbf{d} \text{ and}$$
$$[m'_i - m''_i] + x_i - \alpha\sigma_{m_i} = \langle m_i \rangle \text{ and } [m'_i - m''_i] - x'_i + \alpha\sigma_{m_i} = \langle m_i \rangle \tag{8.28}$$
$$\text{and } \mathbf{m}' \geq 0 \text{ and } \mathbf{m}'' \geq 0 \text{ and } \alpha \geq 0 \text{ and } \mathbf{x} \geq 0 \text{ and } \mathbf{x}' \geq 0$$

where $\mathbf{m} = \mathbf{m}' - \mathbf{m}''$. We note that the new constraints can be written as

$$\alpha - \frac{x_i}{\sigma_{m_i}} = \frac{[m_i - \langle m_i \rangle]}{\sigma_{m_i}} \text{ and } \alpha - \frac{x'_i}{\sigma_{m_i}} = -\frac{[m_i - \langle m_i \rangle]}{\sigma_{m_i}} \tag{8.29}$$

where α, x_i, and x'_i are nonnegative. Using the same argument as was applied in the L_1 case, we conclude that these constraints require that $\alpha \geq |[m_i - \langle m_i \rangle]/\sigma_{mi}|$ for all i. Since this problem has but a single parameter α, it must therefore satisfy

$$\alpha \geq \max_i \left| \frac{[m_i - \langle m_i \rangle]}{\sigma_{m_i}} \right| \tag{8.30}$$

Minimizing α yields the L_∞ solution.

The linear programming problem for the overdetermined case is

minimize α subject to the constraints

$$\sum_{j=1}^{M} G_{ij} \left[m'_j - m''_j \right] + x_i - \alpha \sigma_{d_i} = d_i \quad \text{and} \quad \sum_{j=1}^{M} G_{ij} \left[m'_j - m''_j \right] - x'_i + \alpha \sigma_{d_i} = d_i \tag{8.31}$$

and $\mathbf{m}' \geq 0$ and $\mathbf{m}'' \geq 0$ and $\alpha \geq 0$ and $\mathbf{x} \geq 0$ and $\mathbf{x}' \geq 0$

Here $\mathbf{m} = \mathbf{m}' - \mathbf{m}''$, where $\mathbf{m}'$ and $\mathbf{m}''$ are two unknown vectors of length M, and $\mathbf{x}$ and $\mathbf{x}'$ are two unknown vectors of length N. The *MatLab* implementation is

```
% Linf solution to overdetermined problem
% linear programming problem is
% min f*x subject to A x <= b, Aeq x = beq

% variables
% m = mp − mpp
% x = [mp', mpp', alpha', x', xp']'
% with mp, mpp length M; alpha length 1, x, xp, length N
L = 2*M+1+2*N;
x = zeros(L,1);

% f is length L
% minimize alpha
f = zeros(L,1);
f(2*M+1:2*M+1)=1;

% equality constraints
Aeq = zeros(2*N,L);
beq = zeros(2*N,1);

% first equation G(mp−mpp)+x−alpha=d
Aeq(1:N,1:M)               = G;
Aeq(1:N,M+1:2*M)           = −G;
Aeq(1:N,2*M+1:2*M+1)       = −1./sd;
Aeq(1:N,2*M+1+1:2*M+1+N) = eye(N,N);
beq(1:N)                   = dobs;

% second equation G(mp−mpp)−xp+alpha=d
Aeq(N+1:2*N,1:M)               = G;
Aeq(N+1:2*N,M+1:2*M)           = −G;
Aeq(N+1:2*N,2*M+1:2*M+1)       = 1./sd;
```

```
Aeq(N+1:2*N,2*M+1+N+1:2*M+1+2*N) = -eye(N,N);
beq(N+1:2*N) = dobs;

% inequality constraints A x <= b
% part 1: everything positive
A = zeros(L+2*M,L);
b = zeros(L+2*M,1);
A(1:L,:) = -eye(L,L);
b(1:L) = zeros(L,1);

% part 2; mp and mpp have an upper bound
A(L+1:L+2*M,:) = eye(2*M,L);
mls = (G'*G)\(G'*dobs); % L2 solution - sure easier!
mupperbound=10*max(abs(mls));
b(L+1:L+2*M) = mupperbound;

% solve linear programming problem
[x, fmin] = linprog(f,A,b,Aeq,beq);
fmin = -fmin;
mest = x(1:M) - x(M+1:2*M);
```

(*MatLab* script gda08_05)

As in the L_1 case, this implementation adds upper bounds on the variables **m'** and **m''**. An example of the L_1 problem applied to curve fitting is shown in Fig. 8.5.

The mixed-determined problem can be solved by applying these algorithms and either of the two methods described for the L_1 problem.

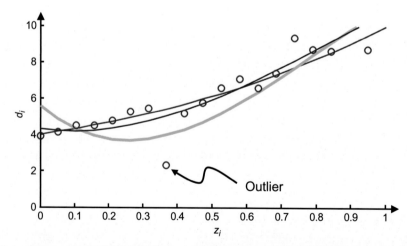

FIG. 8.5   Curve fitting using the L_∞ norm. The true data *(red curve)* follow a cubic polynomial in an auxiliary variable z. Observations *(red circles)* have additive noise with zero mean and variance $\sigma^2 = 1$ drawn from an exponential probability density function. Note the outlier at $z \approx 0.37$. The L_∞ fit *(green curve)* is more affected by the outlier than is a standard L_2 (least-squares) fit *(blue curve)*. *MatLab* script gda08_05.

8.7 THE L_0 NORM AND SPARSITY

As discussed previously in Section 3.2, the pseudo-norm:

$$\|\mathbf{m}\|_0^0 \equiv \lim_{n \to 0} \|\mathbf{m}\|_n^n = \lim_{n \to 0} \sum_k |m_k|^n = \text{number of nonzero elements in } \mathbf{m} \qquad (8.32)$$

(where $\lim_{n \to 0} 0^n$ is taken to be zero) is a measure of the *sparsity* of $\mathbf{m}$; that is, the number of its nonzero elements.

Sparsity is a form of prior information that has many applications. Consider, for instance, the problem of fitting spectral peaks to data that was discussed in Section 1.3.5. One problem with implementing this inversion is that, initially, one does not know the number q of spectral peaks. One strategy, based on already-presented techniques, is to solve a sequence of problems, with $q = 1, 2, 3, \cdots$ peaks and to select the smallest q for which the prediction error $E(q)$ is acceptably small. An alternative strategy is to choose q to be a large number, but then to impose sparsity as prior information; that is, to minimize a weighted sum of the prediction error $\|\mathbf{e}\|_2^2$ and sparsity $\|\mathbf{m}\|_0^0$. This strategy suppresses the amplitude of any peak that is not needed to fit the data.

In another example, consider the CT tomography problem discussed in Section 1.3.4, where measurements of X-ray attenuation are used to determine an image of the opacity of body tissue $c(x, y)$. A useful type of prior information is that different tissues each tend to have its own distinct opacity, so that, to first approximation, the CT image is composed of patches (organs) of more-or-less uniform opacity, separated by sharp boundaries. Mathematically, this idea implies that the first differences of model parameters, measured in the x and y direction, are sparse and suggests the strategy of minimizing:

$$\|\mathbf{e}\|_2^2 + \mu \|\mathbf{D}_x \mathbf{m}\|_0^0 + \mu \|\mathbf{D}_y \mathbf{m}\|_0^0 \qquad (8.33)$$

Here, μ is a positive constant that quantifies the strength of the prior information and $\mathbf{D}_x$ and $\mathbf{D}_y$ are matrices of first-differences in the x and y directions, respectively. This strategy selects a *blocky* model.

While the reweighting algorithm presented in Section 8.4 cannot be used to solve a L_0 problem, it usually works for a fractional n that is close to zero (say $n = 0.1$) and a solution obtained for that value can be considered an approximation for the L_0 solution. The word "usually' is used above, because the convergence properties of the reweighting algorithm are not well-understood for $n < 1$ and may depend upon the matrices involved and the initial guess of the solution. That being said, the algorithm is one of the few available for solving L_0 problems. It should be used, but cautiously. The solution $\mathbf{m}^{\text{est}}$ should be compared to the corresponding L_1 solution to verify that it at least as sparse and its prediction error should be examined to verify that it is acceptably small.

When applied to the minimum length problem:

$$\text{minimize} \|\mathbf{e}\|_2^2 + \mu \|\mathbf{m}\|_0^0 \qquad (8.34)$$

the reweighting algorithm is known to suffer from a solution *shrinkage* problem, meaning that the solution is always a bit smaller than if the zero-valued model parameters were to be

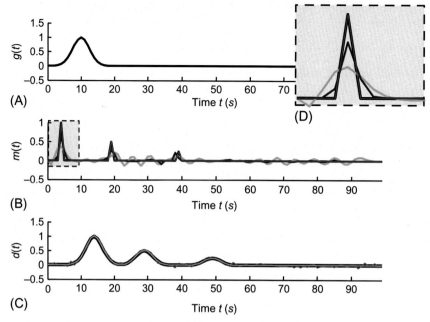

FIG. 8.6   The filter problem $d = g * m$ with a spiky filter $\mathbf{m}$. (A) The time series g is a smooth pulse. (B) The true filter $\mathbf{m}^{true}$ *(black)* is sparse, with three widely separated spikes. The estimated L_2 *(green)*, L_1 *(blue)*, and L_0 *(red)* filters have increasing spikiness. (C) All three filters *(green)* fit data *(black)*. (D) Enlargement of the shaded region in C. *MatLab* gda08_06.

removed and the problem resolved. This behavior suggests that value of μ in the iterative formula (call it μ'):

$$\mathbf{m}^{(j)} = \left[\mathbf{G}^T\mathbf{G} + \mu'\mathbf{W}^{(j-1)} \right]^{-1} \mathbf{G}^T\mathbf{d}^{obs} \tag{8.35}$$

should be smaller than the μ in Eq. (8.34). An analysis by Frommlet and Nuel (2016) indicates that $\mu' \approx \mu/4$. However, the issue is moot in many practical applications, where the value of μ is chosen by trial and error, as contrasted to being guided by a quantitative estimate of the certainty of the prior information.

The following example illustrates the process of finding a sparse solution. We seek the filter $\mathbf{m}^{est}$ that minimizes a weighted sum of prediction error and solution length, $\|\mathbf{e}\|_2^2 + \mu\|\mathbf{m}\|_n^n$. Here the prediction error is $\mathbf{e} = \mathbf{d}^{obs} - \mathbf{Gm}$, the N observed data are $\mathbf{d}^{obs} = \mathbf{Gm}^{true} + \mathbf{n}$, and the Normally distributed noise is $\mathbf{n}$. The data kernel $\mathbf{G}$ is a Toeplitz matrix that implements convolution by a filter $\mathbf{g}$ (see Section 1.3.7). We choose $\mathbf{g}$ to consist of a single smooth pulse (Fig. 8.6A) and $\mathbf{m}^{true}$ to be sparse and to consist of three widely separated spikes (Fig. 8.6B). We compare the L_0, L_1, L_2 solutions, where the L_0 and L_1 solutions are computed using the reweighting algorithm with the correspondence $\mathbf{v} = \mathbf{m}$ and with the least squares formula:

$$\mathbf{m}^{(j)} = \left[\mathbf{G}^T\mathbf{G} + \mu\mathbf{W}^{(j-1)} \right]^{-1} \mathbf{G}^T\mathbf{d}^{obs} \tag{8.36}$$

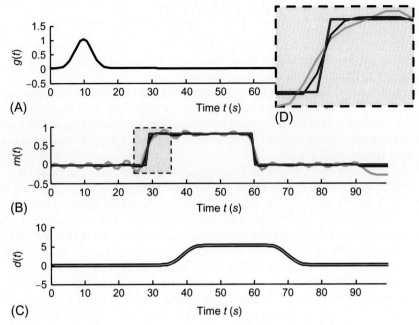

FIG. 8.7 The filter problem $\mathbf{d} = \mathbf{g} * \mathbf{m}$ with a piecewise constant filter $\mathbf{m}$. (A) The time series g is a smooth pulse. (B) The true filter $\mathbf{m}^{\text{true}}$ *(black)* consists of a single boxcar function. The estimated L_2 *(green)*, L_1 *(blue)*, and L_0 *(red)* filters have increasingly sharp steps. (C) All three filters *(green)* fit data *(black)*. (D) Enlargement of the shaded region in C. *MatLab gda08_07.*

All three solutions (Fig. 8.6B) fit the data (Fig. 8.6C) to within similar error. The L_0 solution matches the true solution extremely well, recovering the three spikes almost exactly. The L_1 solution is also fairly spiky, although not nearly as sparse as the L_0 solution. These two solutions can be contrasted to the L_2 solution, which is extremely ringy.

If the true model had been a boxcar function (Fig. 8.7B), as contrasted to a spike, then a solution $\mathbf{m}^{\text{est}}$ that minimizes a weighted sum of prediction error and solution slope, $\|\mathbf{e}\|_2^2 + \mu\|\mathbf{Dm}\|_0^0$ would have been preferable. In this case, the L_0 and L_1 solutions are computed using the reweighting algorithm with the correspondence $\mathbf{v} = \mathbf{Dm}$ and with the least squares formula:

$$\mathbf{m}^{(j)} = \left[\mathbf{G}^{\text{T}}\mathbf{G} + \mu\mathbf{D}^{\text{T}}\mathbf{W}^{(j-1)}\mathbf{D}\right]^{-1}\mathbf{G}^{\text{T}}\mathbf{d}^{\text{obs}} \tag{8.37}$$

As before, all three solutions (Fig. 8.7B) fit the data (Fig. 8.7C) to within similar error. The L_0 solution matches the true solution extremely well, recovering the boxcar function almost exactly. The L_1 solution also does fairly well, except that it bevels the step slightly. Once again, the L_2 solution is extremely ringy.

8.8 PROBLEMS

8.1 Solve the best-fitting plane problem of Fig. 3.6 under the L_1 norm and compare the estimated model parameters with those determined under the L_2 norm. How much does the estimated *strike and dip* of the plane change?

8.2 Solve the constrained best-fitting line problem of Fig. 3.11 under the L_1 norm, by these two methods: (A) by considering the point (z',d') as normal data with very small variance and (B) by explicitly including the constraint as a linear equality constraint within the linear programming problem. Compare the estimated model parameters with those determined under the L_2 norm.

8.3 This problem builds upon Problem 7.4. Consider fitting a cubic polynomial $d_i = m_1 + m_2 z_i + m_3 z_i^2 + m_4 z_i^3$ to $N=20$ data d_i^{obs}, where the zs are evenly spaced on the interval $(0, 1)$, where $m_2 = m_3 = m_4 = 1$, and where m_1 is varied from -1 to 1, as described below. (A) Write a *MatLab* script that generates synthetic observed data (including exponential-distributed noise) for a particular choice of m_1, estimates the model parameters under the L_1 norm, with the extra inequality constraint that $\mathbf{m} \geq 0$. (B) Run a series of test cases for a range of values of m_1 and comment upon the results.

8.4 Consider the problem of finding $M=100$ model parameters $\mathbf{m}$ that minimize $\|\mathbf{e}\|_2^2 + \mu \|\mathbf{Hm}\|_n^n$, where μ is a small positive number, $\mathbf{e} = \mathbf{d}^{obs} - \mathbf{Gm}$ is a vector of errors, $\mathbf{d}^{obs} = \mathbf{Gm}^{true} + \mathbf{n}$ are the observed data, $\mathbf{n}$ is Normally distributed noise, and:

$$\mathbf{m}^{true} = \left[[1 \ \cdots \ 1]_{M/2} \ [2 \ \cdots \ 2]_{M/2} \right]^T$$

$$\mathbf{G} = \begin{bmatrix} [1 \ \cdots \ 1]_{M/2} & [0 \ \cdots \ 0]_{M/2} \\ [0 \ \cdots \ 0]_{M/2} & [1 \ \cdots \ 1]_{M/2} \end{bmatrix}$$

$$\mathbf{H} = \mathrm{diag}(\mathbf{a}) \ \text{with} \ a_k = 1 + 0.001k \ \text{and} \ k = 1 \cdots M$$

The true model parameters $\mathbf{m}^{true}$ are divided into two groups of size $M/2$, with equal values within each group. The data kernel $\mathbf{G}$ sums each group separately, to yield $N=2$ data. The problem is very underdetermined, since $N \ll M$. The matrix $\mathbf{H}$ is almost the identity matrix, but weights the model parameters near the top of $\mathbf{m}$ slightly less than those at the bottom, so that, all other things being equal, the minimization of $\|\mathbf{Hm}\|_n^n$ selects a model that is *squeezed* toward the top. (A) Show that the true solution $\mathbf{m}^{true}$ is also the L_2 minimum length solution $\mathbf{m}^{ML} = \mathbf{G}[\mathbf{GG}^T]^{-1}\mathbf{d}^{obs}$. (B) Compute the solution for $n=0,1,2$ and plot and interpret the result. (C) Repeat the solution in B but for $a_k = 1 + 0.001(M-k+1)$. In what respect is the solution different, and why?

8.5 Solve the previous problem, but for $\mathbf{H} = \mathbf{D}$, where $\mathbf{D}$ is the $M-2 \times M$ first difference operator:

$$\mathbf{D} = \begin{bmatrix} 1 & -1 & 0 & 0 & 0 & 0 \\ 0 & 1 & -1 & 0 & 0 & 0 \\ \ddots & \ddots & \ddots & \ddots & \ddots & \ddots \\ 0 & 0 & 0 & 0 & 1 & -1 \end{bmatrix}$$

Show that the model parameter vector becomes *piecewise constant* as $p \to 0$.

References

Candes, E.J., Wakin, M.B., Boyd, S.P., 2008. Enhancing sparsity by reweighted $\ell 1$ minimization, J. Fourier Anal. Appl. 14, 877–905.

Claerbout, J.F., Muir, F., 1973. Robust modelling with erratic data. Geophysics 38, 826–844.

Cuer, M., Bayer, R., 1980. FORTRAN routines for linear inverse problems. Geophysics 45, 1706–1719.

Frommlet, F., Nuel, G., 2016. An adaptive ridge procedure for L0 regularization. PLoS ONE 11, e0148620.

9

Nonlinear Inverse Problems

9.1 PARAMETERIZATIONS

Variables representing the data and model parameters—the *parameterization*—must be selected at the start of any inverse problem. In many instances, this selection is rather *ad hoc*; there might be no strong reasons for selecting one parameterization over another. This can become a substantial issue in nonlinear inverse problems because the answer obtained by solving it is dependent on the parameterization. In other words, the solution is not invariant under nonlinear transformations of the variables. This is in contrast to the linear inverse problem with Gaussian statistics, in which solutions are invariant for any linear reparameterization of the data and model parameters.

As an illustration of this difficulty, consider the problem of fitting a straight line to the data pairs $(1, 1)$, $(2, 2)$, $(3, 3)$, $(4, 5)$. Suppose that we regard these data as (z, d) pairs, where z is an auxiliary variable. The least-squares fit is $d = -0.500 + 1.300z$. On the other hand, we might

Geophysical Data Analysis
https://doi.org/10.1016/B978-0-12-813555-6.00009-5

© 2018 Elsevier Inc. All rights reserved.

regard them as (d', z') pairs, where z' is the auxiliary variable. Least squares then gives $d' = 0.457 + 0.743z'$, which can be rearranged as $z' = -0.615 + 1.346d'$. These two straight lines have intercepts that differ by 20% and slopes that differ by 4% (see *MatLab* script gda09_01).

This discrepancy arises from two sources. The first is an inconsistent application of probability theory. In the earlier example, we alternately assumed that z was exactly known but d contained Gaussian noise and that d was exactly known but z contained Gaussian noise. These are two radically different assumptions about the distribution of errors, so it is no wonder that the solutions are different.

This first source of discrepancy can in theory be avoided completely by recognizing and taking into account the fact that reparameterizing a problem changes the associated probability density functions. For instance, consider an inverse problem in which there is a model parameter m that is known to possess a uniform probability density function $p(m)$ on the interval $[0, 1]$ (Fig. 9.1A). If the inverse problem is reparameterized in terms of a new model parameter $m' = m^2$, then the transformed probability density function $p(m')$ can be calculated as

$$p(m)dm = p[m(m')]\left|\frac{dm}{dm'}\right|dm' = p(m')dm' \quad \text{so} \quad p(m') = \frac{1}{2}m'^{-1/2} \tag{9.1}$$

The distribution of m' is not uniform (Fig. 9.1B), and any inverse method developed to solve the problem under this parameterization must account for this fact.

The second source of discrepancy is more serious. Suppose that we could use some inverse theory to calculate the distribution of the model parameters under a particular parameterization. We could then use Eq. (9.1) to find their distribution under any arbitrary parameterization. Insofar as the distribution of the model parameters is the *answer* to the inverse problem, we would have the correct answer in the new parameterization. Probability distributions are invariant under changes of parameterization. However, a distribution is not always the answer for which we are looking. More typically, we need an estimate (a single number) based on a probability distribution (for instance, its maximum likelihood point or mean value).

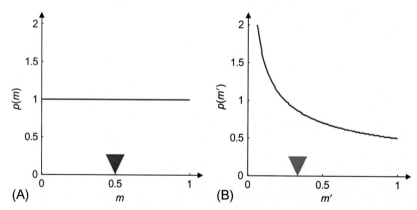

FIG. 9.1   (A) A probability density function, $p(m)$, that is uniform on the interval $0 < m < 1$. (B) The corresponding probability distribution, $p(m')$, for the transformation $m' = m^2$. The mean (expectation) of each probability density function is indicated by a triangle. *MatLab* script gda09_02.

Estimates are *not* invariant under changes in the parameterization. For example, suppose $p(m)$ has a uniform distribution as above. Then, if $m' = m^2$, $p(m') = \frac{1}{2}m'^{-1/2}$. The distribution in m has no maximum likelihood point, whereas the distribution in m' has one at $m' = m = 0$. The distributions also have different means (expectations):

$$\langle m \rangle = \int_0^1 m\, p(m)\mathrm{d}m = \int_0^1 m\, \mathrm{d}m = 1/2$$

$$\langle m' \rangle = \int_0^1 m'\, p(m')\mathrm{d}m' = \int_0^1 m'^{1/2}\mathrm{d}m' = 1/3$$

(9.2)

Even though m' equals the square of the model parameter m, the mean (expectation) of m' is not equal to the square of the expectation of m, that is, $(1/3) \neq (1/2)^2$.

There is some advantage, therefore, in working explicitly with probability distributions as long as possible, forming estimates only at the last step. If **m** and **m'** are two different parameterizations of model parameters, we want to avoid as much as possible sequences like

$$\text{p.d.f. for } \mathbf{m} \rightarrow \text{estimate of } \mathbf{m} \rightarrow \text{estimate of } \mathbf{m'}$$

(9.3)

in favor of sequences like

$$\text{p.d.f. for } \mathbf{m} \rightarrow \text{p.d.f. for } \mathbf{m'} \rightarrow \text{estimate of } \mathbf{m'}$$

(9.4)

Note, however, that the mathematics for this second sequence is typically much more difficult than that for the first.

There are objective criteria for the "goodness" of a particular estimate of a model parameter. Suppose that we are interested in the value of a model parameter **m**. Suppose further that this parameter is either deterministic with a true value or (if it is a random variable) has a well-defined distribution from which the true expectation could be calculated if the distribution were known. Of course, we cannot know the true value; we can only perform experiments and then apply inverse theory to derive an estimate of the model parameter. As any one experiment contains noise, the estimate we derive will not coincide with the true value of the model parameter. But we can at least expect that if we perform the experiment enough times, the estimated values will scatter about the true value. If they do, then the method of estimating is said to be *unbiased*. Estimating model parameters by taking nonlinear combinations of estimates of other model parameters almost always leads to bias.

9.2 LINEARIZING TRANSFORMATIONS

One of the reasons for changing parameterizations is that an inverse problem can sometimes be transformed into a simpler form that can be solved by a known method. The problems that most commonly benefit from such transformations involve fitting exponential and power functions to data. Consider a set of (z, d) data pairs that are thought to obey the model $d_i = m_1 \exp(m_2 z_i)$. By making the transformation

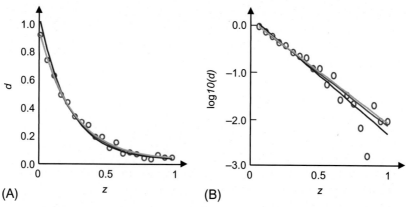

FIG. 9.2   (A) Least-squares fit to the exponential function, $d_i = m_1 \exp(m_2 z_i)$. *(Red curve)* The true function, $d(z)$, for $(m_1, m_2) = (0.7, -4)$. *(Red circles)* Data that include Normally distributed random noise with zero mean and variance, $\sigma_d^2 = (0.2)^2$. *(Green curve)* Nonlinear least-squares solution using Newton's method. *(Blue curve)* Least-squares solution using the linearizing transformation, $\log d_i = \log m_1 + m_2 z_i$. Note that the two solutions are different. (B) Log-linear version of the graph in (A). Note that the scatter of the data increases with z, and that the solution based on the linearizing transformation is strongly affected by outliers associated with it. *MatLab* script gda09_03.

$$m_1' = \log(m_1) \quad \text{and} \quad m_2' = m_2 \quad \text{and} \quad d_i' = \log(d_i) \tag{9.5}$$

we can write the model as the linear equation $d_i' = m_1' + m_2' z_i$, which can be solved by simple least-squares techniques. However, we must assume that the d_i' are independent random variables with a Gaussian probability density function with uniform variance in order to justify rigorously the application of least-squares techniques to this problem. The distribution of the data in their original parameterization must therefore be non-Gaussian. (It must have a *log-normal* probability density function.)

Notice that, if the exponential decays with increasing z for all $m_2 < 0$, the process of taking a logarithm amplifies the scattering of the near-zero points that occurs at large z. The assumption that the d_i' have uniform variance, therefore, implies that the data **d** were measured with an accuracy that increases with z (Fig. 9.2). This assumption may well be inconsistent with the facts of the experiment. Linearizing transformations must be used with some caution.

9.3 ERROR AND LIKELIHOOD IN NONLINEAR INVERSE PROBLEMS

Suppose that the data **d** in an inverse problem has a possibly non-Gaussian probability density function $p(\mathbf{d}; \langle \mathbf{d} \rangle)$, where $\langle \mathbf{d} \rangle$ is the mean and the semicolon is used to indicate that $\langle \mathbf{d} \rangle$ is just a parameter in the probability density function for **d**. The principle of maximum likelihood—that the observed data are the most probable data—holds regardless of the form of $p(\mathbf{d}; \langle \mathbf{d} \rangle)$. As long as the theory is explicit, we can assume that the theory predicts the mean of the data, that is, $\langle \mathbf{d} \rangle = \mathbf{g}(\mathbf{m})$ and write the likelihood $L(\mathbf{m})$ as

$$L(\mathbf{m}) = \log p\left(\mathbf{d}^{\mathrm{obs}}; \mathbf{m}\right) = c - \frac{1}{2} E(\mathbf{m}) \tag{9.6}$$

Here c is some constant, $E(\mathbf{m})$ is some function, and the factor of $1/2$ has been introduced so that $E(\mathbf{m})$ equals the weighted prediction error in the Gaussian case; that is, $E(\mathbf{m}) = [\mathbf{d}^{\mathrm{obs}} - \mathbf{g}(\mathbf{m})]^{\mathrm{T}} [\mathrm{cov}\,\mathbf{d}]^{-1} [\mathbf{d}^{\mathrm{obs}} - \mathbf{g}(\mathbf{m})]$. Thus, it is always possible to define an "error" $E(\mathbf{m})$ such that minimizing $E(\mathbf{m})$ is equivalent to maximizing the likelihood $L(\mathbf{m})$. However, although $E(\mathbf{m})$ is the total prediction error in the Gaussian case, it does not necessarily have all the properties of the total prediction error in other cases. For instance, it may not be zero when $\mathbf{d}^{\mathrm{obs}} = \mathbf{g}(\mathbf{m})$.

There is no simple means for deciding whether a nonlinear inverse problem has a unique solution. Consider the very simple nonlinear model $d_i = m_1^2 + m_1 m_2 z_i$ with Gaussian data. This problem can be linearized by the transformation of variables $m_1' = m_1^2$, $m_2' = m_1 m_2$ and can therefore be solved by the least-squares method if $N \geq 2$. Nevertheless, even if the primed parameters are unique, the unprimed ones are not: if $\mathbf{m}^{\mathrm{est}}$ is a solution that minimizes the prediction error, then $-\mathbf{m}^{\mathrm{est}}$ is also a solution with the same error. In this instance, the error $E(\mathbf{m})$ has two minima of equal depth.

We must examine the global properties of the prediction error $E(\mathbf{m})$ in order to determine whether the inverse problem is unique. If the function has but a single minimum point $\mathbf{m}^{\mathrm{est}}$, then the solution is unique. If it has more than one minimum point, the solution is nonunique, and prior information must be added to resolve the indeterminacy. The error surface of a linear problem is always a paraboloid (Fig. 9.3), which can have only a simple range of shapes. An arbitrarily complex nonlinear inverse problem can have an arbitrarily complicated error. If $M = 2$ or 3, it may be possible to investigate the shape of the surface by graphical techniques (Fig. 9.4). For most realistic problems this is infeasible.

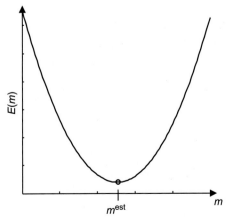

FIG. 9.3 L_2 prediction error $E(m)$, as a function of model parameter m, for a typical linear inverse problem. The solution m^{est} minimizes the error. In the linear case, $E(m)$ is always a paraboloid. *MatLab* script gda09_04.

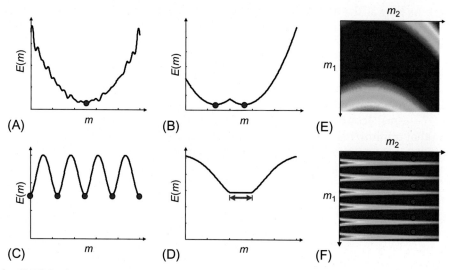

FIG. 9.4 (A–D) Prediction error, E, as a function of a single model parameter, m. (A) A single minimum *(red dot)* corresponds to an inverse problem with a unique solution. (B) Two solutions. (C) Many well-separated solutions. (D) Finite range of solutions *(red arrow)*. (E and F) Error *(colors)* as a function of two model parameters, m_1 and m_2. (E) A single solution, with the minimum occurring within a nearly flat valley. (F) Many well-separated solutions. *MatLab* scripts gda09_05 and gda09_06.

9.4 THE GRID SEARCH

One strategy for solving a nonlinear inverse problem is to exhaustively consider "every possible" solution and pick the one with the smallest error $E(\mathbf{m})$. Of course, it is impossible to examine "every possible" solution; but it is possible to examine a large set of trial solutions. When the trial solutions are drawn from a regular grid in model space, this procedure is called a *grid search*. Grid searches are most practical when

(1) The total number of model parameters is small, say $M < 7$. The grid is M-dimensional, so the number of trial solutions is proportional to L^M, where L is the number of trial solutions along each dimension of the grid.
(2) The solution is known to lie within a specific range of values, which can be used to define the limits of the grid.
(3) The forward problem $\mathbf{d} = \mathbf{g}(\mathbf{m})$ can be computed rapidly enough that the time needed to compute L^M of them is not prohibitive.
(4) The error function $E(\mathbf{m})$ is smooth over the scale of the grid spacing, Δm, so that the minimum is not missed through the grid spacing being too coarse.

As an example, we use *MatLab* to solve the nonlinear problem $d(x_i) = \sin(\omega_0 m_1 x_i) + m_1 m_2$. The data are assumed to be Gaussian and uncorrelated with uniform variance so that the error is $E(m_1, m_2) = \|\mathbf{d}^{\text{obs}} - \mathbf{d}^{\text{pre}}(\mathbf{m})\|_2$. The script below has three sections: the first section defines the grid of trial $\mathbf{m}$ values; the second evaluates $E(\mathbf{m})$ for every trial $\mathbf{m}$ and stores the results

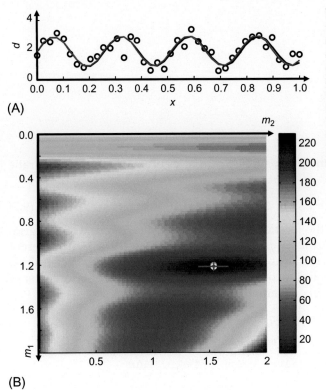

(A)

(B)

FIG. 9.5 A grid search is used to solve the nonlinear curve fitting problem, $d(x_i) = \sin(\omega_0 m_1 x_i + m_1 m_2)$. (A) The true data *(black curve)* are for $m_1 = 1.21$, $m_2 = 1.54$. The observed data *(black circles)* have additive noise with variance $\sigma_d^2 = (0.4)^2$. The predicted data *(red curve)* are based on results of the grid search. (B) Error surface (colors), showing true solution *(green circle)*, estimated solution *(white circle)* and refined estimated solution with 95% confidence intervals *(red bars)*. *MatLab* script gda09_07.

in a matrix E; and the third searches the matrix for the smallest value of E and calculates $\mathbf{m}^{est}$ on the basis of its row and column indices. A useful by-product of the grid search is a tabulation of E on the grid, which can be turned into an informative plot (Fig. 9.5).

```
% 2D grid of m's
L = 101;
Dm = 0.02;
m1min=0;
m2min=0;
m1a = m1min+Dm*[0:L−1]';
m2a = m2min+Dm*[0:L−1]';
m1max = m1a(L);
m2max = m2a(L);
% grid search, compute error, E
E = zeros(L,L);
for j = [1:L]
for k = [1:L]
  dpre = sin(w0*m1a(j)*x) + m1a(j)*m2a(k);
  E(j,k) = (dobs−dpre)'*(dobs−dpre);
```

```
end
end
% find the minimum value of E
[Erowmins, rowindices] = min(E);
[Emin, colindex] = min(Erowmins);
rowindex = rowindices(colindex);
m1est = m1min+Dm*(rowindex-1);
m2est = m2min+Dm*(colindex-1);
```

(*MatLab* script gda09_07)

In the vicinity of its minimum, the error surface can be approximated as a quadratic function of the model parameters:

$$E(m_1, m_2) \approx E_0 + \mathbf{b}^{\mathbf{T}}\Delta\mathbf{m} + \tfrac{1}{2}(\Delta m)^T \mathbf{B}\Delta\mathbf{m} \tag{9.7}$$

Here $\Delta\mathbf{m} = \mathbf{m} - \mathbf{m}_A$, where $\mathbf{m}_A$ is the location of the smallest value of error on the grid, which occurs, say, at at row i_A and column j_A. The parameters $\mathbf{b}$ and $\mathbf{B}$, which represent the first and second derivative of the error near its minimum, respectively, can be determined by using least-squares to fit the parabolic function to a 3×3 portion of the error grid, centered on the minimum value. The 9×6 matrix equation for the unknowns is:

$$
\begin{bmatrix}
E_{i_A-1, j_A-1} \\
E_{i_A, j_A-1} \\
E_{i_A+1, j_A-1} \\
E_{i_A-1, j_A} \\
E_{i_A, j_A} \\
E_{i_A+1, j_A} \\
E_{i_A-1, j_A+1} \\
E_{i_A, j_A+1} \\
E_{i_A+1, j_A+1}
\end{bmatrix}
=
\begin{bmatrix}
1 & \Delta m_{i_A-1} & \Delta m_{j_A-1} & \tfrac{1}{2}\Delta m_{i_A-1}^2 & \Delta m_{i_A-1}\Delta m_{j_A-1} & \tfrac{1}{2}\Delta m_{j_A-1}^2 \\
1 & \Delta m_{i_A} & \Delta m_{j_A-1} & \tfrac{1}{2}\Delta m_{i_A}^2 & \Delta m_{i_A}\Delta m_{j_A-1} & \tfrac{1}{2}\Delta m_{j_A-1}^2 \\
1 & \Delta m_{i_A+1} & \Delta m_{j_A-1} & \tfrac{1}{2}\Delta m_{i_A+1}^2 & \Delta m_{i_A+1}\Delta m_{j_A-1} & \tfrac{1}{2}\Delta m_{j_A-1}^2 \\
1 & \Delta m_{i_A-1} & \Delta m_{j_A} & \tfrac{1}{2}\Delta m_{i_A-1}^2 & \Delta m_{i_A-1}\Delta m_{j_A} & \tfrac{1}{2}\Delta m_{j_A}^2 \\
1 & \Delta m_{i_A} & \Delta m_{j_A} & \tfrac{1}{2}\Delta m_{i_A}^2 & \Delta m_{i_A}\Delta m_{j_A} & \tfrac{1}{2}\Delta m_{j_A}^2 \\
1 & \Delta m_{i_A+1} & \Delta m_{j_A} & \tfrac{1}{2}\Delta m_{i_A+1}^2 & \Delta m_{i_A+1}\Delta m_{j_A} & \tfrac{1}{2}\Delta m_{j_A}^2 \\
1 & \Delta m_{i_A-1} & \Delta m_{j_A+1} & \tfrac{1}{2}\Delta m_{i_A-1}^2 & \Delta m_{i_A-1}\Delta m_{j_A+1} & \tfrac{1}{2}\Delta m_{j_A+1}^2 \\
1 & \Delta m_{i_A} & \Delta m_{j_A+1} & \tfrac{1}{2}\Delta m_{i_A}^2 & \Delta m_{i_A}\Delta m_{j_A+1} & \tfrac{1}{2}\Delta m_{j_A+1}^2 \\
1 & \Delta m_{i_A+1} & \Delta m_{j_A+1} & \tfrac{1}{2}\Delta m_{i_A+1}^2 & \Delta m_{i_A+1}\Delta m_{j_A+1} & \tfrac{1}{2}\Delta m_{j_A+1}^2
\end{bmatrix}
\begin{bmatrix}
E_0 \\
b_1 \\
b_2 \\
B_{11} \\
B_{12} \\
B_{22}
\end{bmatrix}
\tag{9.8}
$$

(with $B_{21} = B_{21}$). The location of the minimum can be improved upon using the principle that the first derivative of the error is zero there:

$$0 = \frac{\partial E}{\partial \mathbf{m}}\bigg|_{\mathbf{m}^{\text{est}}} \approx \mathbf{b} + \mathbf{B}\left[\mathbf{m}^{\text{est}} - \mathbf{m}_A\right] \text{ so } \left[\mathbf{m}^{\text{est}} - \mathbf{m}_A\right] = -\mathbf{B}^{-1}\mathbf{b} \tag{9.9}$$

Finally, since the second derivative is:

$$\frac{\partial^2 E}{\partial \mathbf{m}^2}\bigg|_{\mathbf{m}^{\text{est}}} \approx \mathbf{B} \tag{9.10}$$

Eq. (3.74) can be used to estimate the covariance of the model parameters. We provide a *MatLab* function gda_Esurface() that calculates the refined solution (see *MatLab* script gda09_07 for an example of its use).

9.5 THE MONTE CARLO SEARCH

The Monte Carlo search is a modification of the grid search in which the trial solutions are randomly generated, in contrast to being drawn from a regular grid. In its pure form, where each trial solution is generated independently of previous ones, it has only minor advantages over the grid search. In a later section, we will show that it can be improved by the introduction of correlation between successive trial solutions.

The accompanying *MatLab* script, which implements the algorithm, has two sections: the first section generates an initial trial solution and its corresponding error; the second is a loop that randomly generates a trial solution, calculates its corresponding error, and accepts it if the error is less than the previously accepted solution.

```
% initial guess and corresponding error
mg=[1,1]';
dg = sin(w0*mg(1)*x) + mg(1)*mg(2);
Eg = (dobs-dg)'*(dobs-dg);
% randomly generate pairs of model parameters and check
% if they further minimize the error
ma = zeros(2,1);
for k = [1:Niter]
  % randomly generate a solution
  ma(1) = random('unif',m1min,m1max);
  ma(2) = random('unif',m2min,m2max);
  % compute its error
  da = sin(w0*ma(1)*x) + ma(1)*ma(2);
  Ea = (dobs-da)'*(dobs-da);
  % adopt it if it is better
  if( Ea < Eg )
  mg=ma;
  Eg=Ea;
  end
end
```

(*MatLab* script gda09_08)
An example is shown in Fig. 9.6.

9.6 NEWTON'S METHOD

A completely different approach to solving the nonlinear inverse problem is to use information about the shape of the error $E(\mathbf{m})$ in the vicinity of a trial solution $\mathbf{m}^{(p)}$ to devise a better solution $\mathbf{m}^{(p+1)}$. One source of shape information is the derivatives of $E(\mathbf{m})$ at a trial solution $\mathbf{m}^{(p)}$, as Taylor's theorem indicates that the entire function can be built up from them. Expanding $E(\mathbf{m})$ in a Taylor series about the trial solution and keeping the first three terms, we obtain the parabolic approximation

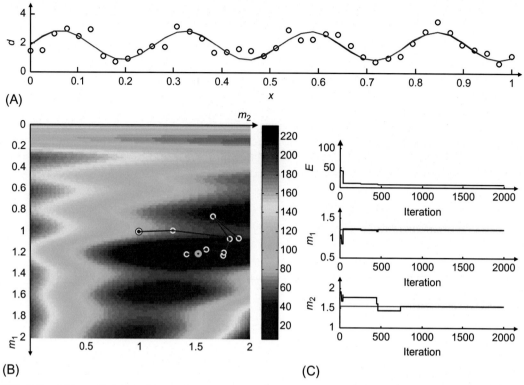

FIG. 9.6   A Monte Carlo search is used to solve the same nonlinear curve-fitting problem as in Fig. 9.5. (A) The observed data *(black circles)* are computed from the true data *(black curve)* by adding random noise. The predicted data *(red curve)* are based on the results of the method. (B) Error surface *(colors)*, showing true solution *(green circle)*, and a series of improved solutions *(white circles* connected by *red lines)* determined by the method. (C) Plot of error E and model parameters m_1 and m_2 as a function of iteration number. *MatLab* script gda09_08.

$$E(\mathbf{m}) \approx E\left(\mathbf{m}^{(p)}\right) + \sum_{i=1}^{M} b_i \left(m_i - m_i^{(p)}\right) + \frac{1}{2} \sum_{i=1}^{M} \sum_{j=1}^{M} B_{ij} \left(m_i - m_i^{(p)}\right) \left(m_j - m_j^{(p)}\right)$$

$$(9.11)$$

$$\text{with } b_i = \frac{\partial E}{\partial m_i}\bigg|_{\mathbf{m}^{(p)}} \text{ and } B_{ij} = \frac{\partial^2 E}{\partial m_i \partial m_j}\bigg|_{\mathbf{m}^{(p)}}$$

Here **b** is a vector of first derivatives and **B** is a matrix of second derivatives of $E(\mathbf{m})$, evaluated at the trial solution $\mathbf{m}^{(p)}$. These derivatives can be calculated either by analytically differentiating $E(\mathbf{m})$, if its functional form is known, or by approximating it with finite differences

$$\frac{\partial E}{\partial m_i}\Big|_{\mathbf{m}^{(p)}} \approx \frac{1}{\Delta m}\left\{E\left(\mathbf{m}+\Delta\mathbf{m}^{(i)}\right)-E(\mathbf{m})\right\}$$

$$\frac{\partial^2 E}{\partial m_i \partial m_j}\Big|_{\mathbf{m}^{(p)}} \approx \begin{cases} \frac{1}{(\Delta m^2)}\left\{E\left(\mathbf{m}+\Delta\mathbf{m}^{(i)}\right)-2E(\mathbf{m})+E\left(\mathbf{m}-\Delta\mathbf{m}^{(i)}\right)\right\} & (i=j) \\[2mm] \frac{1}{4(\Delta m)^2}\left\{E\left(\mathbf{m}+\Delta\mathbf{m}^{(i)}+\Delta\mathbf{m}^{(j)}\right)-E\left(\mathbf{m}+\Delta\mathbf{m}^{(i)}-\Delta\mathbf{m}^{(j)}\right)\right. & (i\neq j) \\[2mm] \left.-E\left(\mathbf{m}-\Delta\mathbf{m}^{(i)}+\Delta\mathbf{m}^{(j)}\right)+E\left(\mathbf{m}-\Delta\mathbf{m}^{(i)}-\Delta\mathbf{m}^{(j)}\right)\right\} & \end{cases} \tag{9.12}$$

Here $\Delta\mathbf{m}^{(i)}$ is a small increment Δm in the ith direction; that is, $\Delta\mathbf{m}^{(i)}=\Delta m\,[0,\dots,0,1,0,\dots,0]^{\mathrm{T}}$, where the ith element is unity and the rest are zero. Note that these approximations are computationally expensive, in the sense that E must be evaluated many times for each instance of $\mathbf{b}$ and $\mathbf{G}$.

We can now find the minimum by differentiating this approximate form of $E(\mathbf{m})$ with respect to m_q and setting the result to zero:

$$\frac{\partial E(\mathbf{m})}{\partial m_q}=0=b_q+\sum_{j=1}^{M}B_{qj}\left(m_j-m_j^{(p)}\right) \quad \text{or} \quad \mathbf{m}-\mathbf{m}^{(p)}=-\mathbf{B}^{-1}\mathbf{b} \tag{9.13}$$

In the case of uncorrelated Gaussian data with uniform variance and the *linear* theory $\mathbf{d}=\mathbf{Gm}$, the error is $E(\mathbf{m})=[\mathbf{d}-\mathbf{Gm}]^{\mathrm{T}}[\mathbf{d}-\mathbf{Gm}]$, from whence we find that $\mathbf{b}=-2\mathbf{G}^{\mathrm{T}}(\mathbf{d}-\mathbf{Gm}^{(p)})$, $\mathbf{B}=2\mathbf{G}^{\mathrm{T}}\mathbf{G}$, and $\mathbf{m}=[\mathbf{G}^{\mathrm{T}}\mathbf{G}]^{-1}\mathbf{G}^{\mathrm{T}}\mathbf{d}$, which is the familiar least-squares solution. In this case, the result is independent of the trial solution and is exact.

In the case of uncorrelated Gaussian data with uniform variance and the nonlinear theory $\mathbf{d}-\mathbf{g}(\mathbf{m})=0$, the error is $E(\mathbf{m})=[\mathbf{d}-\mathbf{g}(\mathbf{m})]^{\mathrm{T}}[\mathbf{d}-\mathbf{g}(\mathbf{m})]$, from which we conclude

$$\mathbf{b}=-2\mathbf{G}^{(p)\mathrm{T}}\left[\mathbf{d}-\mathbf{g}(\mathbf{m}^{(p)})\right] \quad \text{and} \quad \mathbf{B}\approx 2\left[\mathbf{G}^{(p)\mathrm{T}}\mathbf{G}(p)\right] \quad \text{with} \quad G_{ij}^{(p)}=\frac{\partial g_i}{\partial m_j}\Big|_{\mathbf{m}^{(p)}}$$

$$\text{so} \quad \mathbf{m}-\mathbf{m}^{(p)}\approx \left[\mathbf{G}^{(p)\mathrm{T}}\mathbf{G}^{(p)}\right]^{-1}\mathbf{G}^{(p)\mathrm{T}}\left[\mathbf{d}-\mathbf{g}(\mathbf{m}^{(p)})\right] \tag{9.14}$$

Note that we have omitted the term involving the gradient of $\mathbf{G}$ in the formula for $\mathbf{B}$. In a linear theory, $\mathbf{G}$ is constant and its gradient is zero. We assume that the theory is sufficiently close to linear that here too it is insignificant.

The form of Eq. (9.14) is very similar to simple least squares. The matrix $\mathbf{G}^{(p)}$, which is the gradient of the model $\mathbf{g}(\mathbf{m})$ at the trial solution, acts as a data kernel. It can be calculated analytically, if its functional form is known, or approximated by finite differences (see Eq. 9.12). The generalized inverse $\mathbf{G}^{-g}=[\mathbf{G}^{(p)\mathrm{T}}\mathbf{G}^{(p)}]^{-1}\mathbf{G}^{(p)\mathrm{T}}$ relates the deviation of the data $\Delta\mathbf{d}=[\mathbf{d}-\mathbf{g}(\mathbf{m}^{(p)})]$ from what is predicted by the trial solution to the deviation of the solution $\Delta\mathbf{m}=\mathbf{m}-\mathbf{m}^{(p)}$ from the trial solution, that is, $\Delta\mathbf{m}=\mathbf{G}^{-g}\Delta\mathbf{d}$. However, because it is based on a truncated Taylor Series, the solution is only approximate; it yields a solution that is improved over the trial solution, but not the exact solution. However, it can be iterated to yield a succession of improvements

$$\mathbf{m}^{(p+1)}=\mathbf{m}^{(p)}+\left[\mathbf{G}^{(p)\mathrm{T}}\mathbf{G}^{(p)}\right]^{-1}\mathbf{G}^{(p)\mathrm{T}}\left[\mathbf{d}-\mathbf{g}\left(\mathbf{m}^{(p)}\right)\right] \tag{9.15}$$

until the error declines to an acceptably low level (Fig. 9.7). Unfortunately, while convergence of $\mathbf{m}^{(p)}$ to the value that globally minimizes $E(\mathbf{m})$ is often very rapid, it is not guaranteed. One common problem is the solution converging to a local minimum instead of the global minimum (Fig. 9.8). A critical part of the algorithm is picking the initial trial solution $\mathbf{m}^{(1)}$. The method may not find the global minimum if the trial solution is too far from it.

FIG. 9.7 The iterative method locates the global minimum m^{GM} of the error $E(m)$ (black curve) by determining the paraboloid (red curve) that is tangent to E at the trial solution m_n^{est}. The improved solution m_{n+1}^{est} is at the minimum (red circle) of this paraboloid and, under favorable conditions, can be closer to the solution corresponding to the global minimum than is the trial solution. MatLab script gda09_09.

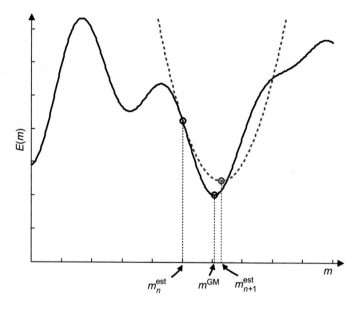

FIG. 9.8 If the trial solution, m_n^{est}, is too far from the global minimum m^{GM}, the method may converge to a local minimum. MatLab script gda09_10.

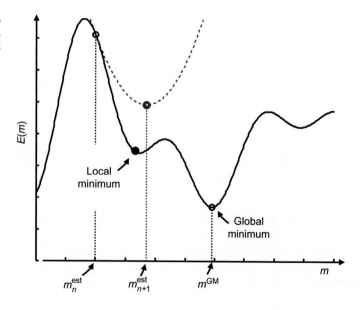

As an example, we consider the problem from Section 9.4, in which $g_i(\mathbf{m}) = \sin(\omega_0 m_1 x_i) + m_1 m_2$. The matrix of partial derivatives is

$$
\mathbf{G}^{(p)} = \begin{bmatrix}
\omega_0 x_1 \cos\left(\omega_0 x_1 m_1^{(p)}\right) + m_2^{(p)} & m_1^{(p)} \\
\omega_0 x_2 \cos\left(\omega_0 x_2 m_1^{(p)}\right) + m_2^{(p)} & m_1^{(p)} \\
\vdots & \vdots \\
\omega_0 x_N \cos\left(\omega_0 x_N m_1^{(p)}\right) + m_2^{(p)} & m_1^{(p)}
\end{bmatrix}
\tag{9.16}
$$

In *MatLab*, the main part of the script is a loop that iteratively updates the trial solution. First, the deviation $\Delta\mathbf{d}$ (dd in the script) and the data kernel $\mathbf{G}$ are calculated, using the trial solution $\mathbf{m}^{(p)}$. Then the deviation $\Delta\mathbf{m}$ is calculated using least squares. Finally, the trial solution is updated as $\mathbf{m}^{(p+1)} = \mathbf{m}^{(p)} + \Delta\mathbf{m}$. In *MatLab*

```
% initial guess and corresponding error
mg=[1,1]';
dg = sin(w0*mg(1)*x) + mg(1)*mg(2);
Eg = (dobs-dg)'*(dobs-dg);
% iterate to improve initial guess
Niter = 20;
G = zeros(N,M);
for k = [1:Niter]
  dg = sin(w0*mg(1)*x) + mg(1)*mg(2);
  dd = dobs-dg;
  Eg=dd'*dd;
  G = zeros(N,2);
  G(:,1) = w0 * x .* cos( w0 * mg(1) * x ) + mg(2);
  G(:,2) = mg(2)*ones(N,1);
  % least squares solution
  dm = (G'*G)\(G'*dd);
  % update
  mg = mg+dm;
end
```

(*MatLab* script gda09_11)

The results of this script are shown in Fig. 9.9. This exemplary script iterates a fixed number of times, as specified by the variable Niter. A more elegant approach would be to perform a test that terminates the iterations when the error declines to an acceptable level or when the solution no longer changes significantly from iteration to iteration.

9.7 THE IMPLICIT NONLINEAR INVERSE PROBLEM WITH GAUSSIAN DATA

We now generalize the iterative method of the previous section to the general case of the implicit theory $\mathbf{f}(\mathbf{d}, \mathbf{m}) = 0$, where $\mathbf{f}$ is of length $L \leq M + N$. We assume that the data $\mathbf{d}$ and prior

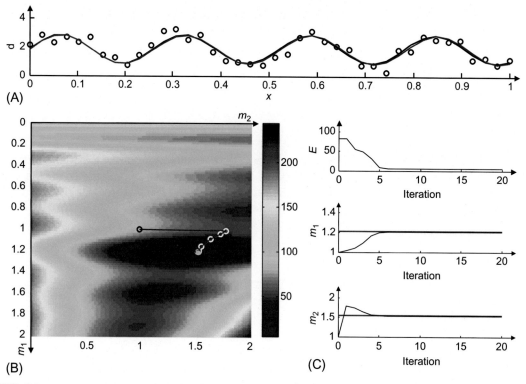

FIG. 9.9   Newton's method (linearized least squares) is used to solve the same nonlinear curve-fitting problem as in Fig. 9.5. (A) The observed data *(black circles)* are computed from the true data *(black curve)* by adding random noise. The predicted data *(red curve)* are based on the results of the method. (B) Error surface *(colors)*, showing true solution *(green dot)*, and a series of improved solutions *(white circles* connected by *red lines)* determined by the method. (C) Plot of error E and model parameters, m_1 and m_2 as a function of iteration number. *MatLab* script gda09_11.

model parameters $\langle \mathbf{m} \rangle$ have Gaussian distributions with covariance [cov **d**] and [cov **m**]$_A$, respectively. If we let $\mathbf{x} = [\mathbf{d}^T, \mathbf{m}^T]^T$, we can think of the prior distribution of the data and model as a cloud in the space $S(\mathbf{x})$ centered about the observed data and mean prior model parameters, with a shape determined by the covariance matrix [cov **x**] (Fig. 9.10). The matrix [cov **x**] contains [cov **d**] and [cov **m**] on diagonal blocks, as in Eq. (5.27). In principle, the off-diagonal blocks could be made nonzero, indicating correlation between observed data and prior model parameters. However, specifying prior constraints is typically an *ad hoc* procedure for which one can seldom find motivation for introducing such a correlation. The prior distribution is therefore

$$p_A(\mathbf{x}) \propto \exp\left\{ -\frac{1}{2}[\mathbf{x} - \langle \mathbf{x} \rangle]^T[\text{cov } \mathbf{x}]^{-1}[\mathbf{x} - \langle \mathbf{x} \rangle] \right\} \tag{9.17}$$

where $\langle \mathbf{x} \rangle = [(\mathbf{d}^{\text{obs}})^T, \langle \mathbf{m} \rangle^T]^T$ is a vector containing the observed data and prior model parameters.

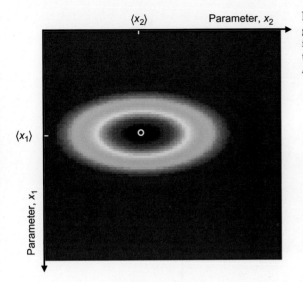

FIG. 9.10 The data and model parameters are grouped together in a vector, **x**. The prior information for **x** is then represented as a probability density function *(colors)* in the $(M+N)$-dimensional space, $S(\mathbf{x})$. *MatLab* script gda09_12.

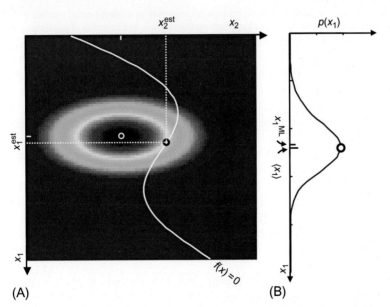

FIG. 9.11 (A) The estimated solution $\mathbf{x}^{est}$ *(black circle)* is at the point on the surface $\mathbf{f}(\mathbf{x})=0$ *(white curve)* where the prior probability density function *(colors)* attains its largest value. (B) The probability density function $p(x_1)$ evaluated along the surface, as a function of position x_1. As the function is nonnormal, its mean $\langle x_1 \rangle$ may be distinct from its mode x_1^{ML} (although in this case they are similar). *MatLab* script gda09_13.

The theory $\mathbf{f}(\mathbf{x})=0$ defines a surface in $S(\mathbf{x})$ on which the predicted data and estimated model parameters $\mathbf{x}^{est}=[\mathbf{d}^{pre\ T}, \mathbf{m}^{est\ T}]^T$ must lie. The probability distribution for $\mathbf{x}^{est}$ is, therefore, $p_A(\mathbf{x})$, evaluated on this surface (Fig. 9.11). If the surface is plane, this is just the linear case

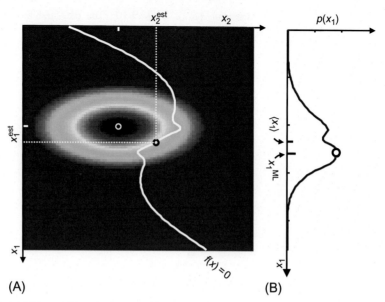

FIG. 9.12   (A) A highly nonlinear inverse problem corresponds to a complicated surface $\mathbf{f}(\mathbf{x})=0$ (white curve). (B) The probability density function $p(x_1)$ evaluated along the surface, as a function of position x_1. It may have several peaks, and as it is nonnormal, its mean $\langle x_1 \rangle$ may be distinct from its mode x_1^{ML}. *MatLab* script gda09_14.

described in Chapter 5 and the final distribution is Gaussian. On the other hand, if the surface is very "bumpy," the distribution on the surface will be very non-Gaussian and may even possess several maxima (Fig. 9.12).

One approach to estimating the solution is to find the maximum likelihood point of $p_A(\mathbf{x})$ on the surface $\mathbf{f}(\mathbf{x})=0$ (Fig. 9.11). This point can be found without explicitly determining the distribution on the surface. One just maximizes $p_A(\mathbf{x})$ with the constraint that $\mathbf{f}(\mathbf{x})=0$. One should keep in mind, however, that the maximum likelihood point of a non-Gaussian distribution may not be the most sensible estimate that can be made from that distribution. Gaussian distributions are symmetric, so their maximum likelihood point always coincides with their mean value. In contrast, the maximum likelihood point can be arbitrarily far from the mean of a non-Gaussian distribution (Fig. 9.12). Computing the mean, however, requires one to compute explicitly the distribution on the surface and then take its expectation (a much more difficult procedure).

These caveats aside, we proceed with the calculation of the maximum likelihood point by minimizing the argument of the exponential in $p_A(\mathbf{x})$ with the constraint that $\mathbf{f}(\mathbf{x})=0$ (adapted from Tarantola and Valette, 1982):

$$\text{minimize } \boldsymbol{\Phi} = [\mathbf{x} - \langle \mathbf{x} \rangle]^{\mathrm{T}} [\text{cov } \mathbf{x}]^{-1} [\mathbf{x} - \langle \mathbf{x} \rangle] \text{ subject to } \mathbf{f}(\mathbf{x}) = 0 \tag{9.18}$$

The Lagrange multiplier equations are

$$\frac{\partial \boldsymbol{\Phi}}{\partial x_i} - \sum_{j=1}^{L} 2\lambda_j \frac{\partial f_j}{\partial x_i} = 0 \text{ or } [\mathbf{x} - \langle \mathbf{x} \rangle]^{\mathrm{T}} [\text{cov } \mathbf{x}]^{-1} = \mathbf{F}^{\mathrm{T}} \boldsymbol{\lambda} \tag{9.19}$$

where λ is a vector of Lagrange multipliers and F is the matrix of derivatives $F_{ij} = \frac{\partial f_i}{\partial x_j}$. The Lagrange multipliers can be determined by premultiplying the transformed equation by F as

$$F[x - \langle x \rangle] = F[\text{cov } x]F^T \lambda \tag{9.20}$$

and then premultiplying $\{F[\text{cov } x]F^T\}^{-1}$ as

$$\lambda = \{F[\text{cov } x]F^T\}^{-1} F[x - \langle x \rangle] \tag{9.21}$$

Substitution into the transpose of the original equation yields

$$[x - \langle x \rangle] = [\text{cov } x]F^T \{F[\text{cov } x]F^T\}^{-1} F[x - \langle x \rangle] \tag{9.22}$$

which must be solved simultaneously with the constraint equation $f(x) = 0$. These two equations are equivalent to the single equation

$$[x - \langle x \rangle] = [\text{cov } x]F^T \{F[\text{cov } x]F^T\}^{-1} \{F[x - \langle x \rangle] - f(x)\} \tag{9.23}$$

as the original two equations can be recovered by premultiplying this equation by F. The form of this equation is very similar to the linear solution of Eq. (5.37); in fact, it reduces to it in the case of the exact linear theory $f(x) = Fx$. As the unknown x appears on both sides of the equation and f and F are functions of x, this equation may be difficult to solve explicitly. We now examine an iterative method of solving it. This method consists of starting with some initial trial solution, say, $x^{(p)}$, where $p = 1$, and then generating successive approximations as

$$x^{(p+1)} = \langle x \rangle + [\text{cov } x]F^{(p)T} \left\{F^{(p)}[\text{cov } x]F^{(p)T}\right\}^{-1} \left\{F^{(p)}\left[x^{(p)} - \langle x \rangle\right] - f\left(x^{(p)}\right)\right\} \tag{9.24}$$

The superscript on $F^{(p)}$ implies that it is evaluated at $x^{(p)}$. If the initial guess is close enough to the maximum likelihood point, the successive approximations will converge to the true solution x^{est}; else it may converge to a local minimum.

If the theory is explicit (i.e., if $f(x) = d - g(m) = 0$) and if the data and prior model parameters are uncorrelated, the iterative formula can be rewritten as

$$m^{(p+1)} = \langle m \rangle + G_{(p)}^{-g} \left\{d - g\left(m^{(p)}\right) + G^{(p)}\left[m^{(p)} - \langle m \rangle\right]\right\} \tag{9.25}$$

$$
\begin{aligned}
G_{(p)}^{-g} &= [\text{cov } m]_A G^{(p)T} \left\{G^{(p)}[\text{cov } m]_A G^{(p)T} + [\text{cov } d]\right\}^{-1} \\
&= \left\{G^{(p)T}[\text{cov } d]^{-1} G^{(p)} + [\text{cov } m]_A^{-1}\right\}^{-1} G^{(p)T}[\text{cov } d]^{-1}
\end{aligned}
\tag{9.26}
$$

or solved using simple least squares

$$
\begin{bmatrix} [\text{cov } d]^{-1/2} \ G^{(p)} \\ [\text{cov } m]_A^{-1/2} \ I \end{bmatrix} m^{(p+1)} = \begin{bmatrix} [\text{cov } d]^{-1/2}\left\{d - g(m^{(p)}) + G^{(p)}m^{(p)}\right\} \\ [\text{cov } m]_A^{-1/2}\langle m \rangle \end{bmatrix}
\tag{9.27}
$$

Here $[G^{(p)}]_{ij} = \partial g_i / \partial m_j$ is evaluated at $m^{(p)}$ and the generalized inverse notation has been used for convenience. The two versions of the generalized inverse in Eq. (9.26)—one with the form of the minimum length solution and the other with the form of the least-squares solution—are equivalent, as was shown in Eq. (5.39). Eqs. (9.25)–(9.27) are the nonlinear, iterative analog to the linear, noniterative formulas stated for the linear inverse problem in Eq. (5.37), except that in this case the theory has been assumed to be exact. One obtains formulas for an inexact theory with covariance $[\text{cov}\,\mathbf{g}]$ with the substitution $[\text{cov}\,\mathbf{d}] \to [\text{cov}\,\mathbf{d}] + [\text{cov}\,\mathbf{g}]$ and for prior information of the form $\mathbf{Hm}=\mathbf{h}$ with the substitutions $\mathbf{I} \to \mathbf{H}$, $\langle\mathbf{m}\rangle \to \mathbf{h}$, and $[\text{cov}\,\mathbf{m}]_A \to [\text{cov}\,\mathbf{h}]_A$.

Provided the problem is overdetermined, Eqs. (9.25)–(9.27) reduce to Newton's method in the limit where $\langle\mathbf{m}\rangle \to 0$, $[\text{cov}\,\mathbf{m}]_A^{-1} \to 0$, and $[\text{cov}\,\mathbf{d}]^{-1} \to \mathbf{I}$

$$\Delta\mathbf{m} = \mathbf{G}^{-g}\Delta\mathbf{d} + \left(\mathbf{I} - \mathbf{R}^{(p)}\right)\mathbf{m}^{(p)} \to \mathbf{G}^{-g}\Delta\mathbf{d} \tag{9.28}$$

as the resolution matrix $\mathbf{R}^{(p)} = \mathbf{G}_{(p)}^{-g}\mathbf{G}^{(p)} \to \mathbf{I}$ in the overdetermined case. However, Eqs. (9.25)–(9.27) indicate that Newton's method cannot be *patched* for underdetermined problems merely by using a damped least-squares version of $\mathbf{G}_{(p)}^{-g}$. The problem is that damped least squares drives $\Delta\mathbf{m}$ toward zero, rather than (as is more sensible) driving $\mathbf{m}^{(p+1)}$ toward zero. Instead, *MatLab* scripts should solve Eq. (9.27), using the `bicg()` solver together with the `weightedleastsquaresfcn()` function.

In a linear problem, the error $E(\mathbf{m})$ is a paraboloid and the estimated model parameters $\mathbf{m}^{\text{est}}$ are at its minimum. While a nonlinear problem has an error with a more complicated shape, it may still be approximately paraboloid in the vicinity of its minimum. This is the region of the space of model parameters where the inverse problem behaves linearly and the probability density function $p(\mathbf{m})$ is approximately Gaussian in shape. If the patch is big enough to encompass a large percentage of the total probability, then one might use the linear formula

$$\left[\text{cov}\,\mathbf{m}^{\text{est}}\right] = \mathbf{G}_{(p)}^{-g}[\text{cov}\,\mathbf{d}]\mathbf{G}_{(p)}^{-g\text{T}} + \left[\mathbf{I} - \mathbf{R}^{(p)}\right][\text{cov}\,\mathbf{m}]_A\left[\mathbf{I} - \mathbf{R}^{(p)}\right]^{\text{T}} \tag{9.29}$$

where p is the final iteration, to calculate approximate variances of the model parameters. Whether 95% confidence intervals inferred from these variances are correct will depend upon the size of the patch because only the central part of the probability density function, and not its tails, is approximately Gaussian. For this reason, estimates of confidence intervals based on the Bootstrap method (Section 9.11) are usually preferred.

The same caveat applies to interpretations of the resolution matrices $\mathbf{N}^{(p)}$ and $\mathbf{R}^{(p)}$. As the problem is nonlinear, they do not describe the true resolution of the problem. On the other hand, they give the resolution of a linear problem that is in some sense close to the nonlinear one.

9.8 GRADIENT METHOD

Occasionally, one encounters an inverse problem in which the error $E(\mathbf{m})$ and its gradient $[\nabla E]_i = \partial E/\partial m_i$ are especially easy to calculate. It is possible to solve the inverse problem using this information alone, as the unit vector

$$v = -\frac{\nabla E}{|\nabla E|} \tag{9.30}$$

points in the direction in which the error E is minimized. Thus a trial solution $\mathbf{m}^{(j)}$ can be improved to $\mathbf{m}^{(j+1)} = \mathbf{m}^{(j)} + \alpha \boldsymbol{\nu}$, where α is a positive number. The only problem is that one does not immediately know how large α should be. Too large and the minimum may be skipped over; too small and the convergence will be very slow. *Armijo's rule* provides an acceptance criterion for α:

$$E\left(\mathbf{m}^{(k+1)}\right) \leq E\left(\mathbf{m}^{(k)}\right) + c\alpha \boldsymbol{\nu}^{\mathrm{T}} \nabla E|_{m^{(k)}} \tag{9.31}$$

Here c is an empirical constant in the range $(0, 1)$ that is usually chosen to be about 10^{-4}. One strategy is to start the iteration with a "largish" value of α and to use it as long as it passes the test, but to decrease it whenever it fails, say using the rule $\alpha \to \alpha/2$. An example is shown in Fig. 9.13.

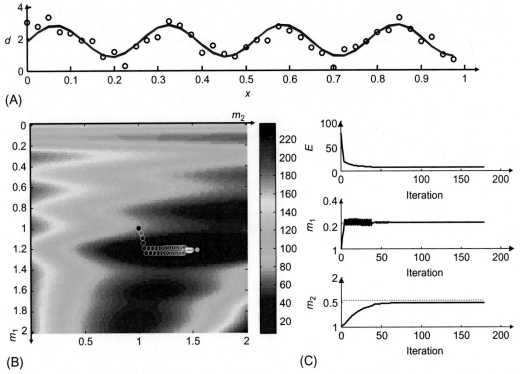

(A)

(B)

(C)

FIG. 9.13 The gradient method is used to solve the same nonlinear curve-fitting problem as in Fig. 9.5. (A) The observed data *(black circles)* are computed from the true data *(black curve)* by adding random noise. The predicted data *(red curve)* are based on the results of the method. (B) Error surface *(colors)*, showing true solution *(green dot)*, and a series of improved solutions *(white circles)* determined by the method. (C) Plot of error E and model parameters m_1 and m_2 as a function of iteration number. *MatLab* script gda09_15.

9.9 SIMULATED ANNEALING

The Monte Carlo method (Section 9.5) is completely *undirected*. The whole model space is sampled randomly so that the global minimum of the error *eventually* is found, provided that enough trial solutions are examined. Unfortunately, the number of trial solutions that need to be computed may be very large—perhaps millions.

In contrast, Newton's method (Section 9.6) is completely directed. Local properties of the error—its slope and curvature—are used to determine the optimal improvement to a trial solution. Convergence to the global minimmum of the error, when it occurs, is rapid and just a few trial solutions need be computed—perhaps just 10. Unfortunately, the method may only find a local minimum of the error that corresponds to an incorrect estimate of the model parameters.

The *simulated annealing* method combines the best features of these two methods. Like the Monte Carlo method, it samples the whole model space and so can avoid getting stuck in local minima. And like Newton's method, it uses local information to direct the sequence of trial solutions. Its development was inspired by the physical annealing of metals (Kirkpatrick et al., 1983), where an orderly minimum-energy crystal structure develops within the metal as it slowly cooled from a red hot state. Initially, when the metal is hot, atomic motions are completely dominated by random thermal fluctuations, but as the temperature is slowly lowered, interatomic forces become more and more important. In the end, the atoms become a crystal lattice that represents a minimum-energy configuration.

In the simulated annealing algorithm, a parameter T is the analog to temperature and the error E is the analog to energy. Large values of T cause the algorithm to behave like a Monte Carlo search; small values cause it to act in a more directed way. As in physical annealing, one starts with a large T and then slowly decreases it as more and more trial solutions are examined. Initially, a very large volume of model space is randomly sampled, but the search becomes increasingly directed as it progresses.

The simulated annealing algorithm starts with a trial solution $\mathbf{m}^{(p)}$ with corresponding error $E(\mathbf{m}^{(p)})$. A test solution $\mathbf{m}^*$ with corresponding error $E(\mathbf{m}^*)$ is then generated that is in the neighborhood of $\mathbf{m}^{(p)}$, say by adding to $\mathbf{m}^{(p)}$ an increment $\Delta\mathbf{m}$ drawn from a Gaussian distribution. The test solution is always accepted as the new trial solution $\mathbf{m}^{(p+1)}$ when $E(\mathbf{m}^*) \leq E(\mathbf{m}^{(p)})$, but it is also sometimes accepted even when $E(\mathbf{m}^*) > E(\mathbf{m}^{(p)})$. To decide the latter case, a test parameter

$$t = \frac{\exp\{-E(\mathbf{m}^*)/T\}}{\exp\{-E(\mathbf{m}^{(p)})/T\}} = \exp\left\{-\frac{[E(\mathbf{m}^*) - E(\mathbf{m}^{(p)})]}{T}\right\} \tag{9.32}$$

is computed. Then, a random number r that is uniformly distributed on the interval $[0, 1]$ is generated and the solution $\mathbf{m}^*$ is accepted if $t > r$. When T is large, the parameter t is close to unity and $\mathbf{m}^*$ is almost always accepted, regardless of the value of the error. This corresponds to the "thermal motion" case where the space of model parameters is explored in an undirected way. When T is small, the parameter t is close to zero and $\mathbf{m}^*$ is almost never accepted. This corresponds to the directed search case, as then the only solutions that decrease the error are accepted. An astute reader will recognize that this is just the Metropolis-Hastings algorithm (Section 2.8) applied to the *Boltzmann* probability density function

$$p(\mathbf{m}) \propto \exp\left\{\frac{-E(\mathbf{m})}{T}\right\} \tag{9.33}$$

The maximum likelihood point of this probability density function corresponds to the point of minimum error, regardless of the value of the parameter T. However, T controls the width of the probability density function, with a larger T corresponding to a wider function. At first, T is high and the distribution is wide. The sequence of realizations samples a broad region of model space that includes the global minimum and, possibly, local minima as well. As T is decreased, the probability density function becomes increasingly peaked at the global minimum. In the limit of $T=0$, all the realizations are at the maximum likelihood point; that is, the sought-after point that minimizes the error.

Note that when $T=2$, we recover the probability density function $p(\mathbf{d}^{obs}; \mathbf{m})$, as defined in Section 9.3. An alternate strategy for "solving" the inverse problem is to stop the cooling at this temperature and then to produce a large set of realizations of this distribution (see Section 1.4.4). Either the entire set of solutions, itself, or a single parameter derived from it, such as the mean, can be considered the "solution" of the inverse problem.

In *MatLab*, the main part of the Metropolis-Hastings algorithm is a loop that computes the current value of T, randomly computes a $\Delta\mathbf{m}$ to produce the solution $\mathbf{m}^*$ and a corresponding error $E(\mathbf{m}^*)$, and applies the Metropolis rules to either accept or reject $\mathbf{m}^*$.

```
Dm = 0.2;
Niter=400;
for k = [1:Niter]
  % temperature falls off with iteration number
  T = 0.1 * Eg0 * ((Niter−k+1)/Niter)^2;
  % randomly pick model parameters and evaluate error
  ma(1) = random('Normal',mg(1),Dm);
  ma(2) = random('Normal',mg(2),Dm);
  da = sin(w0*ma(1)*x) + ma(1)*ma(2);
  Ea = (dobs−da)'*(dobs−da);
  % accept according to Metropolis rules
  if( Ea < Eg )
  mg=ma;
  Eg=Ea;
  p1his(k+1)=1;
  else
  p1 = exp( −(Ea−Eg)/T );
  p2 = random('unif',0,1);
  if( p1 > p2 )
  mg=ma;
  Eg=Ea;
  end
  end
end
```

(*MatLab* script gda09_16)
An example is shown in Fig. 9.14.

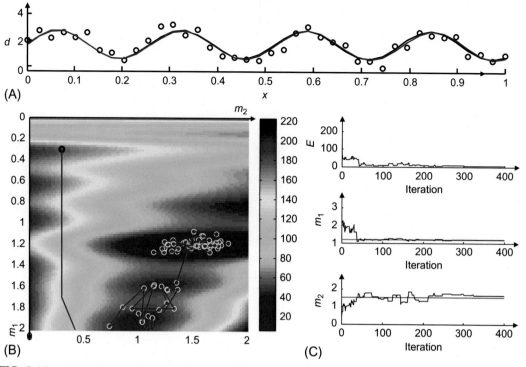

FIG. 9.14   Simulated annealing is used to solve the same nonlinear curve-fitting problem as in Fig. 9.5. (A) The observed data *(black circles)* are computed from the true data *(black curve)* by adding random noise. The predicted data *(red curve)* are based on the results of the method. (B) Error surface *(colors)*, showing true solution *(green circle)*, and a series of solutions *(white circles* connected by *red lines)* determined by the method. (C) Plot of error E and model parameters m_1 and m_2 as a function of iteration number. *MatLab* script gda09_16.

9.10 THE GENETIC ALGORITHM

Evolution is a natural process that drives a population of living organisms to the optimal condition of biological fitness, in the same sense that physical annealing, discussed in the previous section, drives a crystalline metal to the optimal condition of minimum energy. Just as mathematical analogs of annealing can be used solve inverse problems, so can mathematical analogs of biological evolution. Called *genetic algorithms*, they evolve a population of *candidate* solutions toward a best-estimate solution.

Biological evolution operates on a *population* of organisms that is composed of a large number of individuals and that reproduces for a large number of generations. The inheritable attributes of an individual are specified by its *genome*, which is composed of sequence of *genes*; most typically, a given attribute is influenced by just a few genes. The genome of a child differs from that of its parent (or parents), because of *mutation*, which randomly modifies the genome as it is copied from parent to child, and sexual reproduction, which causes a child to obtain some of its genes from each of two parents. The process of *natural selection* leads to the preferential survival of individuals who, by virtue of their particular genes, are better fitted to their environment. Selection leads to the average fitness of a population of organisms increasing from generation to generation.

In the genetic algorithm, the analog to the genome of an individual k is a candidate model vector $\mathbf{m}^{(k)}$, the analog to a population of size K is a set of the model vectors $\{k=1\ldots K\}$ and the analog to fitness is a function $F(E_k)$ that declines with error $E_k = E(\mathbf{m}^{(k)})$. The genomes of individuals within the population are initialized to a set of trial values and the population is allowed to evolve for a large number L of generations. The best estimate of the solution is the one, in the final generation, with the highest fitness (and lowest error).

A key issue is how to encode the model vector $\mathbf{m}^{(k)}$ into a genome. One possibility is to concatenate the binary representation of each element of the vector to form a single bit string $\mathbf{s}^{(k)}$ (where each element of $\mathbf{s}^{(k)}$ is 0 or 1). The binary floating point representation of an element m_i could be used in the genome, but then mutations of the exponent part of the representation would lead to very large changes in value, which is undesirable. Usually better is representing m_i by a nonnegative integer p_i with a fixed number of bits (e.g., 4 bits, which can represent the numbers 0–15; or 12 bits, which can represent the numbers 0–4095; or 16 bits, which can represent the numbers 0–65,535; etc.) and then linearly scaling it to the expected range of the model parameter; that is, $m_i = a_i + b_i p_i$, where a_i and b_i are prescribed constants. Then, model parameters always vary within a fixed range.

In the following example, the 16-bit genome represents four 4-bit integer model parameters, each in the range 0–15:

	$\mathbf{m}^{(k)}$	$\mathbf{s}^{(k)}$
Individual k	$[2,6,5,4]^\mathrm{T}$	0010011001010100

A mutation is accomplished by randomly choosing a bit—say bit 10, highlighted in bold in the example, below—and flipping its value as the genome is copied from parent to child:

	$\mathbf{m}^{(k)}$	$\mathbf{s}^{(k)}$
Parent k	$[2,6,5,4]^\mathrm{T}$	0010011001010100
Child k	$[2,6,1,4]^\mathrm{T}$	0010011000010100

Sexual reproduction, or rather the simplified version of it called *recombination* (or *crossover*), is accomplished in the following steps: (1) two parents, called i and j in the example, below, are randomly selected; (2) their genomes are cut into two pieces at the same randomly chosen point (between bits 6 and 7 in the example below, with one piece highlighted in bold); and (3) the pieces are swapped as the genomes are copied from the two parents to their two children:

	$\mathbf{m}^{(k)}$	$\mathbf{s}^{(i)}$	$\mathbf{m}^{(k)}$		$\mathbf{s}^{(j)}$
Parent i	$[2,6,5,4]^\mathrm{T}$	**001001**1001010100	$[1,3,0,15]^\mathrm{T}$	Parent j	0001001**100001111**
Child i	$[2,7,0,15]^\mathrm{T}$	**001001**1100001111	$[1,2,5,4]^\mathrm{T}$	Child j	0001001001010100

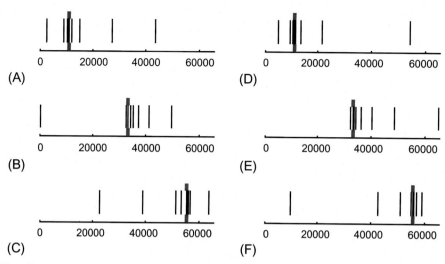

FIG. 9.15   Effect on one-bit mutations of the 16-bit standard binary code for the numbers (A) 11,111, (B) 33,333, and (C) 55,555. (C)–(E) Same as (A)–(C), except for the 16-bit Gray code. In each case, the original number is shown in *red* and the mutations in *black*. *MatLab* script gda09_17.

Note that the point at which the genomes were cut is not a multiple of 4 bits. Consequently, one pair of integer values is altered (an effect similar to mutation).

Mutation and cross-over have somewhat different effects on a genome. Most single-bit mutations change an integer by only a small amount (Fig. 9.15). They allow the population to undergo, from generation to generation, a directed random walk toward greater fitness—a process akin to the one used in simulated annealing. Such a process will rapidly hone in on the nearest minima in error (though it may be only a local minima). Some single-bit mutations change an integer by a large amount. They allow individuals to explore wildly different regions of the space of model parameters and hence provide a way—when those individuals are especially fit—for the population to jump out of a local minimum into the global one. Except in very homogeneous populations, recombination can also lead to a large change in the genome and hence provides another mechanism for the population to jump out of a local minimum. Additionally, in cases where the inverse problem is approximately *separable* into two or more approximately independent subproblems, they provide an effective means of bringing together fit subsolutions into a single, extremely fit overall solution (discussed further below).

One idiosyncrasy of the standard binary representation of integers (Table 9.1) is that some adjacent integers are separated by only one mutation, but others are separated by many mutations. The integers 2 and 3 (0010 and 0011) differ by only one bit so that a single mutation can change a 2 to a 3, and vice versa. On the other hand, 7 and 8 (0111 and 1000) differ by four bits, so four mutations are required to change a 7 to an 8, and vice versa. Thus, individuals of one generation may be very close to the exact solution, and yet the probability may be low that an individual in the next generation will mutate to it—an undesirable characteristic called a *Hamming wall*. It can be avoided by using *Gray coding* (Table 9.1), a nonstandard binary representation in which all adjacent integers differ by precisely 1 bit. In the earlier example,

TABLE 9.1 The Standard Binary and Gray Codes for Integers in the 0–15 Range

Decimal	Binary	Gray Code	Decimal	Binary	Gray Code
0	0000	0000	8	1000	1100
1	0001	0001	9	1001	1101
2	0010	0011	10	1010	1111
3	0011	0010	11	1011	1110
4	0100	0110	12	1100	1010
5	0101	0111	13	1101	1011
6	0110	0101	14	1110	1001
7	0111	0100	15	1111	1000

the Gray codes for 2 and 3 are 0011 and 0010, respectively, and the codes for 7 and 8 are 1000 and 1001, respectively. Both pairs differ by exactly 1 bit.

Choices encountered when implementing a genetic algorithm include: the size of the population and whether or not its size varies between generations; the values to which the set of genomes; are initialized; the rates of mutation and conjugation; the formula used to assess fitness; and the criteria used to decide when an acceptable best-estimate solution has been reached. Unfortunately, little theoretical guidance is available for making these choices; they usually need to be made by trial and error.

In our exemplary implementation of the nonlinear curve-fitting problem (as in Fig. 9.5), we use 16-bit Gray codes for each of the $M=2$ model parameters, and scale their $0-65,535$ range onto the $(0,2)$ interval. We implement two functions, $s=\text{int2gray}(i)$ and $i=\text{gray2int}(s)$ that translate between integers and character strings. The strings are of length-16 and represent the Gray code as a sequence of '0' and '1' characters. The use of character strings, as contrasted to bit strings, leads to some computational inefficiency, but greatly simplifies the manipulation and display of the codes. The functions are table-based and need to be initialized by placing the code:

```
clear all;
global gray1 gray1index pow2;
load('gray1table.mat');
```

at the beginning of the main script.

A population of $K=100$ candidate solutions (black dots in Fig. 9.16A) is initialized so that individuals scatter around $\mathbf{m}=[0.25,0.30]^T$, a point that is far from the true solution $\mathbf{m}^{true}=[1.21,1.54]$ (green circle in the figure). This population is then evolved for 40 generations.

New generations, all of size K, are formed by duplicating each individual of the previous generation to create a population of size $2K$, performing mutations and recombinations, determining each individual's fitness and the accepting only the K most fit individuals.

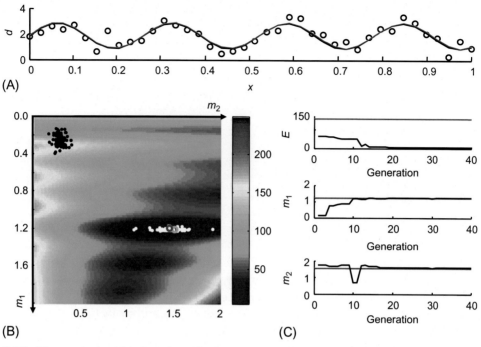

FIG. 9.16   The genetic algorithm is used to solve the same nonlinear curve-fitting problem as in Fig. 9.5. (A) The observed data *(black circles)* are computed from the true data *(black curve)* by adding random noise. The predicted data *(red curve)* are based on the results of the method. (B) Error surface *(colors)*, showing true solution *(green circle)*, the initial population of solutions *(black dots)*, the population of solutions after 40 generations and the best estimate solution *(red circle)*. (C) Plot of error E and model parameters m_1 and m_2 as a function of generation. *MatLab* script gda09_18.

```
Kold = K;
genes = [genes;genes];
K = 2*K;
```

Single-bit mutations are performed at randomly chosen bit locations in half of individuals in each generation:

```
k = unidrnd(16*M,1); % one random mutation
if( genes(i,k)=='0' )
    genes(i,k)='1';
else
    genes(i,k)='0';
end
```

Here `genes` is a $K \times 16M$ character array, whose rows represent the genome of different individuals.

A recombination is performed on one randomly selected pair of individuals in each generation:

```
j1 = unidrnd(K,1); % random individual
j2 = unidrnd(K,1); % another random individual
k = unidrnd(M*16-2,1)+1; % genome cut a random 2<=k<=16M-1
g1(1,1:16*M) = genes(j1,:);
g1(1,k+1:16*M) = genes(j2,k+1:16*M);
g2(1,1:16*M) = genes(j2,:);
g2(1,k+1:16*M) = genes(j1,k+1:16*M);
genes(j1,:) = g1;
genes(j2,:) = g2;
```

The fitness F_i of individual i is assessed as:

$$F_i = r_i \exp\left(-cE_i/E_{max}\right) \text{ with } c = 5 \tag{9.34}$$

Here E_{max} is the maximum error of any individual in that generation and r_i is a random number drawn from a uniform distribution on the $(0, 1)$ interval. The exponential implies that, on average, fitness declines rapidly with error. The multiplicative random number implies that some less-fit organisms survive—a strategy that prevents the population from becoming too homogeneous (and hence getting stuck in a local minimum).

The K individuals with the largest fitness survive into the next generation. They are indentified by sorting the fitness vector:

```
[F_sorted, k] = sort(F);
```

The vector `k(K+1:2*K)` contains the indices of individuals who survive.

```
k = k(Kold+1:2*Kold,1);
genes = genes(k,1:16*M);
K = Kold;
```

About 20 generations are needed for the population to evolve to a state that is close to the true solution (Fig. 9.16C). Little variation in the lowest-error individual is observed thereafter. Individuals of the final generation (white dots in the figure) scatter about the elliptical region of low error, with the most fit individual (red circle in the figure) being very close to the true solution (green circle).

In biology, different parts of an organism's genome code for different cellular components, such as proteins, and the function of these components is to some degree independent; that is, they operate separately to good approximation. Hypothetically, when present individually, a mutation-improved protein involved in the detection of food may lead to a modest gain in an individual's overall fitness, as might also a mutation-improved protein involved in digestion. The gain might be much more substantial, however, when both mutated genes are present in the same individual (since the organism can then digest the extra food that it finds). Recombination provides an efficient mechanism for bringing together two favorable versions of genes.

Many inverse problems share this trait. Their model parameters can be separated into several groups, each of which affect one group of data and leave the rest approximately

FIG. 9.17 Genetic algorithm applied to the approximately separable localized average problem discussed in the text. The minimum error E_i^{min} in a population of 100, declines with generation i. The rate of decline is faster for a combination of evolutionary parameters that includes recombination (*red curves*, 12.5% mutation rate, 5% recombination rate) than for one that omits it (*black curves*, 50% mutation, 0% recombination). Histories of five populations are shown for each parameter set. *MatLab* script gda09_19.

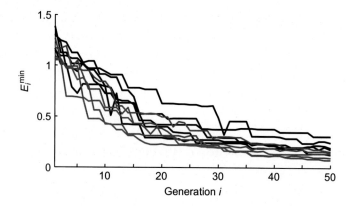

unaffected. Recombination can be especially effective in such cases, since it brings together several groups of model parameters, each of which fit their subset of data well, into a solution with low overall error. However, the model parameters must be ordered in such a way that the members of a group are contiguous on the genome; else the recombination will not tend to swap groups as a whole. (Alternatively, the recombination algorithm can be redesigned to swap groups even though they are not contiguous).

As an example, consider an inverse problem in which each of $N=M=20$ data is the arithmetic mean of three adjacent model parameters:

$$d_i = \frac{m_i + m_{i-1} + m_{i-2}}{3} \text{ with } m_i \equiv 0 \text{ if } i < 1 \tag{9.35}$$

This problem is separable, in the sense that the error for a group of data at the left hand side of the $(1, N)$ interval depends mostly on the values of the model parameters at the left hand side of the interval, and the data on the right hand side depend mostly on the model parameters on the right hand side. Fig. 9.17 shows the minimum error as a function of generation, in each of two sets of five runs of the genetic algorithm, all with a population of population of $K=100$ candidate solutions. The first set (black curves) is for a genetic algorithm that implements a 50% mutation rate per generation but omits recombination. The second set (red curves) is for a genetic algorithm that implements both mutation and recombination. It has a reduced mutation rate, compared with the mutation-only set, of 12.5%, and a 10% recombination rate. The tendency of recombination to create random variation by combining partial genes has been suppressed by cutting the genome only at 16-bit gene boundaries. The algorithm that includes recombination converges, on average, faster, even though it has a much lower mutation rate.

9.11 CHOOSING THE NULL DISTRIBUTION FOR INEXACT NON-GAUSSIAN NONLINEAR THEORIES

As we found in Section 5.2.4, the starting place for analyzing inexact theories is the rule for combining probability density functions to yield the total probability density function p_T. It is built up by combining three probability density functions p_A (prior), p_f or p_g (theory), and p_N (null)

$$p_T(\mathbf{d}, \mathbf{m}) = \frac{p_A(\mathbf{d})p_A(\mathbf{m})p_g(\mathbf{d}, \mathbf{m})}{p_N(\mathbf{d})p_N(\mathbf{m})} \quad \text{or} \quad p_T(\mathbf{x}) = \frac{p_A(\mathbf{x})p_f(\mathbf{x})}{p_N(\mathbf{x})} \quad (9.36)$$

In the Gaussian case, we have been assuming $p_N \propto$ constant. However, this definition is sometimes inadequate. For instance, if $\mathbf{x}$ is the Cartesian coordinate of an object in three-dimensional space, then $P_N \propto$ constant means that the object could be anywhere with equal probability. This is an adequate definition of the null distribution. On the other hand, if the position of the object is specified by the spherical coordinates $\mathbf{x} = [r, \theta, \varphi]^T$, then the statement $P_N \propto$ constant actually implies that the object is near the origin. The statement that the object could be anywhere is $P_N(\mathbf{x}) \propto r^2 \sin(\theta)$. The null distribution must be chosen with the physical significance of the vector $\mathbf{x}$ in mind.

Unfortunately, it is sometimes difficult to find a guiding principle with which to choose the null distribution. Consider the case of an acoustics problem in which a model parameter is the acoustic velocity v. At first sight it may seem that a reasonable choice for the null distribution is $p_N(v) \propto$ constant. Acousticians, however, often work with the acoustic slowness $s = 1/v$, and the distribution $p_N(v) \propto$ constant implies $p_N(s) \propto s^2$. This is somewhat unsatisfactory, as one could, with equal plausibility, argue that $p_N(s) \propto$ constant, in which case $p_N(v) \propto v^2$. One possible solution to this dilemma is to choose a null solution whose *form* is invariant under the reparameterization. The distribution that works in this case is $p_N(v) \propto 1/v$, as this leads to $p_N(s) \propto 1/s$.

Thus, while a non-Gaussian inexact implicit theory can be handled with the same machinery as was applied to the Gaussian case of Section 9.7, more care must be taken when choosing the parameterization and defining the null distribution.

9.12 BOOTSTRAP CONFIDENCE INTERVALS

The probability density function of a nonlinear problem is non-Gaussian, even when the data have Gaussian-distributed error. The simple formulas that we developed for error propagation are not accurate in such cases, except perhaps when the nonlinearity is very weak. We describe here an alternative method of computing confidence intervals for the model parameters that performs the error propagation in an alternative way.

If many *repeat* data sets were available, the problem of estimating confidence intervals could be approached empirically. If we had repeated the experiment 1000 times, each time with the exact same experimental conditions, we would have a group of 1000 data sets, all similar to one another, but each containing a different pattern of observational noise. We could then solve the inverse problem 1000 times, once for each repeat data set, make histograms of the resulting estimates of the model parameters and infer confidence intervals from them.

Repeat data sets are rarely available. However, it is possible to construct an approximate repeat data set by the *random resampling with duplication* of a single data set. The idea is to treat a set of N data as a pool of hypothetical observations and randomly draw N *realized* observations, which together constitute one repeat data set, from them. Even though the pool and the repeat data set are both of length N, they are not the same because the latter contains duplications. It can be shown that this process leads to a group of data sets that approximately have the probability density function of the original data.

As an example, suppose that a data set consists of N pairs of (x_i, d_i) observations, where x_i is an auxiliary variable. In *MatLab*, the resampling is performed as

```
rowindex = unidrnd(N,N,1);
xresampled = x( rowindex );
dresampled = dobs( rowindex );
```

(*MatLab* script gda09_20)

Here, the function `unidrnd(N,N,1)` returns a vector of N uniformly distributed integers in the range 1 through N. The inverse problem is then solved for many such repeat data sets, and the resulting estimates of the model parameters are saved. Confidence intervals are estimated as

```
Nbins=50;
m1hmin=min(m1save);
m1hmax=max(m1save);
Dm1bins = (m1hmax−m1hmin)/(Nbins−1);
m1bins=m1hmin+Dm1bins*[0:Nbins−1]';
m1hist = hist(m1save,m1bins);
pm1 = m1hist/(Dm1bins*sum(m1hist));
Pm1 = Dm1bins*cumsum(pm1);
m1low=m1bins(find(Pm1>0.025,1));
m1high=m1bins(find(Pm1>0.975,1));
```

(*MatLab* script gda09_20)

Here, estimates of the model parameter m_1 for all the repeat data sets have been saved in the vector `m1save`. A histogram `m1hist` is computed from them using the `hist()` function, after determining a reasonable range of m_1 values for its bins. The histogram is converted to an empirical probability density function `pm1`, by scaling it so that its integral is unity, and the cumulative sum function `cumsum()` is used to integrate it to a cumulative probability distribution `Pm1`. The `find()` function is then used to determine the values of m_1 that enclose 95% of the area, defining 95% confidence limits for m_1. An example is shown in Fig. 9.18.

9.13 PROBLEMS

9.1 Use the principle of maximum likelihood to estimate the mean μ of N uncorrelated data, each of which is drawn from the same one-sided exponential probability density function $p(d_i) = \mu^{-1} \exp(-d_i/\mu)$ on the interval $(0, \infty)$. Suppose all the N data are equal to unity. What is the variance?

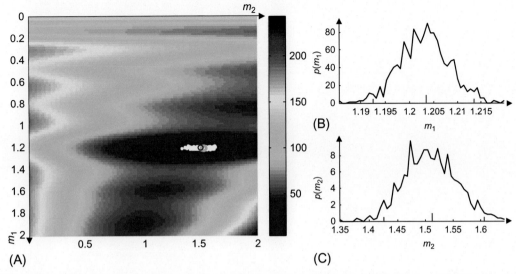

FIG. 9.18 Bootstrap confidence intervals for model parameters estimated using Newton's method for the same problem as in Fig. 9.5. (A) Error surface *(colors)*, showing true solution *(red circle)*, estimated solution *(green circle)*, and bootstrap solutions *(white dots)*. (B) Empirically derived probability density function $p(m_1)$, with m_1^{est} *(large red tick)* and 95% confidence limits *(small red ticks)*. (C) Same as (B), but for $p(m_2)$. *MatLab* script gda09_20.

9.2 Experiment with the Newton's method example of Fig. 9.9, by changing the initial guess mg of the solution. Map out the region of the (m_1, m_2) plane for which the solution converges to the global minimum.

9.3 Solve the nonlinear inverse problem $d_i = m_1 z_i^{m_2}$, with z_i an auxiliary variable on the interval $(1, 2)$ and with $\mathbf{m}^{true} = [3, 2]^T$, using both a linearizing transformation based on taking the logarithm of the equation and by Newton's method and compare the result. Use noisy synthetic data to test your scripts.

9.4 Suppose that the zs in the straight line problem $d_i = m_1 + m_2 z_i$ are considered data, not auxiliary variables. (A) How should the inverse problem be classified? (B) Write a *MatLab* script that solves a test case using an appropriate method.

9.5 A vector $\mathbf{d}$ is constructed by adding together scaled and shifted versions of vectors $\mathbf{a}$, $\mathbf{b}$, and $\mathbf{c}$. The vector $\mathbf{d}$, of length N, is observed. The vectors, $\mathbf{a}$, $\mathbf{b}$, and $\mathbf{c}$, also of length N, are auxiliary variables. They are related by $d_i = A a_{i+p} + B b_{i+q} + C c_{i+r}$, where A, B, and C are unknown constants and p, q, and r are unknown positive integers (assume $a_{i+p} = 0$ if $i+p > N$ and similarly for $\mathbf{b}$ and $\mathbf{c}$). This problem can be solved using a three-dimensional grid search over p, q, and r, only, as A, B, and C can be found using least squares once p, q, and r are specified. Write a *MatLab* script that solves a test case for $N = 50$. Assume that the error is $E = \mathbf{e}^T \mathbf{e}$ with $e_i = d_i - (A a_{i+p} + B b_{i+q} + C c_{i+r})$.

References

Kirkpatrick, S., Gelatt, C.D., Vecchi, M.P., 1983. Optimization by simulated annealing. Science 220 (4598), 671–680.
Tarantola, A., Valette, B., 1982. Generalized non-linear inverse problems solved using the least squares criterion. Rev. Geophys. Space Phys. 20, 219–232.

Factor Analysis

10.1 THE FACTOR ANALYSIS PROBLEM

Consider an ocean whose sediment is derived from the simple mixing of continental source rocks A and B (Fig. 10.1). Suppose that the concentrations of three elements are determined for many samples of sediment and then plotted on a graph whose axes are percentages of those elements. Since all the sediments are derived from only two source rocks, the sample compositions lie on the triangular portion of a plane bounded by the compositions of A and B (Fig. 10.2).

The factor analysis problem is to deduce the number of the source rocks (called *factors*) and their composition from observations of the composition of the sediments (called *samples*). It is therefore a problem in inverse theory. We shall discuss it separately, since it provides an interesting example of the use of some of the vector space analysis techniques developed in Chapter 7.

The model proposes that the samples are simple mixtures (linear combinations) of the factors. If there are N samples containing M elements and if there are p factors, we can state this model algebraically with the equation

$$\mathbf{S} = \mathbf{CF} \tag{10.1}$$

© 2018 Elsevier Inc. All rights reserved.

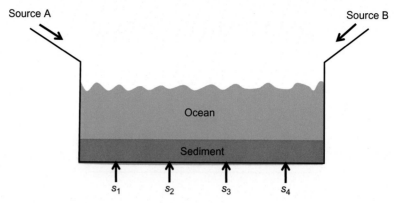

FIG. 10.1 Material from sources A and B is eroded into the ocean and deposited to form sediment. Samples s_i of the sediment are collected and their chemical composition is determined. The data are used to infer the composition of the sources.

FIG. 10.2 The composition of the samples s_i *(black arrows)* lies on a triangular sector of a plane bounded by the composition of the sources A and B *(red arrows)*. *MatLab* script gda10_01.

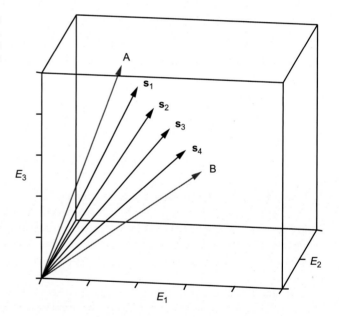

where S_{ij} is the fraction of element j in sample i:

$$\mathbf{S} = \begin{bmatrix} \text{element 1 in sample 1} & \cdots & \text{element } M \text{ in sample 1} \\ \text{element 1 in sample 2} & \cdots & \text{element } M \text{ in sample 2} \\ \vdots & \ddots & \vdots \\ \text{element 1 in sample } N & \cdots & \text{element } M \text{ in sample } N \end{bmatrix} \quad (10.2)$$

(The word "element" is used in the generic sense, since most factor analysis problems will not involve *chemical* elements.) We will refer to individual samples as $\mathbf{s}^{(i)}$, with $\mathbf{s}^{(i)T}$ a row of $\mathbf{S}$. Similarly, F_{ij} is the fraction of element j in factor i:

$$\mathbf{F} = \begin{bmatrix} \text{element 1 in factor 1} & \cdots & \text{element } M \text{ in factor 1} \\ \text{element 1 in factor 2} & \cdots & \text{element } M \text{ in factor 2} \\ \vdots & \ddots & \vdots \\ \text{element 1 in factor } p & \cdots & \text{element } M \text{ in factor } p \end{bmatrix} \tag{10.3}$$

We will refer to individual factors as $\mathbf{f}^{(i)}$, with $\mathbf{f}^{(i)T}$ a row of $\mathbf{F}$. C_{ij} is the fraction of factor i in sample j:

$$\mathbf{C} = \begin{bmatrix} \text{factor 1 in sample 1} & \cdots & \text{factor } p \text{ in sample 1} \\ \text{factor 1 in sample 2} & \cdots & \text{factor } p \text{ in sample 2} \\ \vdots & \ddots & \vdots \\ \text{factor 1 in sample } N & \cdots & \text{factor } p \text{ in sample } N \end{bmatrix} \tag{10.4}$$

The elements of the matrix $\mathbf{C}$ are referred to as the *factor loadings* (or sometimes just the *loadings*).

The inverse problem is to factor the matrix $\mathbf{S}$ into $\mathbf{C}$ and $\mathbf{F}$. Each sample (each row of $\mathbf{S}$) is represented as a linear combination of factors (rows of $\mathbf{F}$), with the elements of $\mathbf{C}$ giving the coefficients of the combination. As long as we pick an $\mathbf{F}$ whose rows span the space spanned by the rows of $\mathbf{S}$, we can perform the factorization. For $p \geq M$, any linearly independent set of factors will do, so in this sense the factor analysis problem is completely nonunique. It is much more interesting to ask what the minimum number of factors is that can be used to represent the samples. Then the factor analysis problem is equivalent to examining the space spanned by $\mathbf{S}$ and determining its dimension. This problem can be easily solved by representing the sample matrix with its singular-value decomposition as

$$\mathbf{S} = \mathbf{U}_p \mathbf{\Lambda}_p \mathbf{V}_p^T = (\mathbf{U}_p \mathbf{\Lambda}_p)\left(\mathbf{V}_p^T\right) = \mathbf{CF} \tag{10.5}$$

Only the eigenvectors with nonzero singular values appear in the decomposition. The number of factors is given by the number of nonzero singular values. One possible set of factors is the p eigenvectors. This set of factors is not unique (Fig. 10.3). Any set of factors that spans the p space will do. Mathematically, we transform the factors with any matrix $\mathbf{T}$ that possesses an inverse, in which case $\mathbf{S} = \mathbf{CF} = \mathbf{CIF} = (\mathbf{CT}^{-1})(\mathbf{TF}) = \mathbf{C'F'}$ with the transformed factors being $\mathbf{F'} = (\mathbf{TF})$ and the new loadings being $\mathbf{C'} = \mathbf{CT}^{-1}$.

If we write out Eq. (10.5), we find that the composition of the ith sample $\mathbf{s}^{(i)}$ is related to the eigenvectors $\mathbf{v}^{(i)}$ and singular values λ_i by

$$\mathbf{s}^{(1)} = \left[\mathbf{U}_p\right]_{11}\lambda_1\mathbf{v}^{(1)} + \left[\mathbf{U}_p\right]_{12}\lambda_2\mathbf{v}^{(2)} + \cdots + \left[\mathbf{U}_p\right]_{1p}\lambda_p\mathbf{v}^{(p)}$$
$$\mathbf{s}^{(N)} = \left[\mathbf{U}_p\right]_{N1}\lambda_1\mathbf{v}^{(1)} + \left[\mathbf{U}_p\right]_{N2}\lambda_2\mathbf{v}^{(2)} + \cdots + \left[\mathbf{U}_p\right]_{Np}\lambda_p\mathbf{v}^{(p)} \tag{10.6}$$

If the singular values are arranged in descending order, then most of each sample is composed of factor 1, with a smaller contribution from factor 2, etc. Because $\mathbf{U}_p$ and $\mathbf{v}^{(p)}$ are composed of unit vectors, on average their elements are of equal size. We have identified the most "important" factors (Fig. 10.4). Even if $p = M$, it might be possible to neglect some of the

FIG. 10.3 Eigenvectors $\mathbf{v}_1$ and $\mathbf{v}_2$ lie in the plane of the samples ($\mathbf{v}_1$ is closest to the mean sample). Eigenvector $\mathbf{v}_3$ is normal to the plane. *MatLab* script and gda10_02.

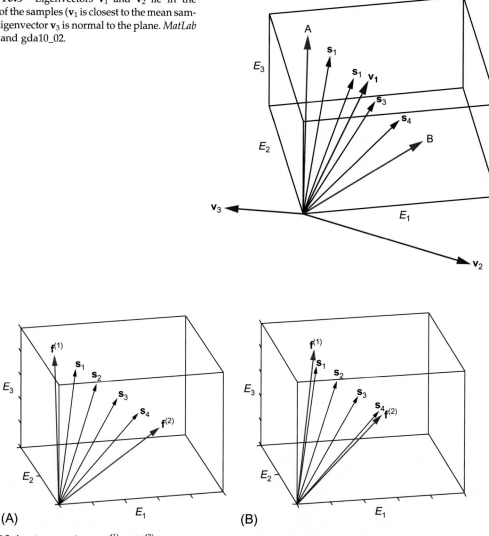

FIG. 10.4 Any two factors $\mathbf{f}^{(1)}$ and $\mathbf{f}^{(2)}$ *(red arrows)* that lie in the plane of the samples and that bound the range of sample compositions *(black arrows)* are acceptable, such as those shown in (A) and (B). *MatLab* script gda10_03.

smaller singular values and still achieve a reasonably good prediction of the sample compositions; that is, $\mathbf{S} \approx \mathbf{CF} = (\mathbf{U}_q \mathbf{\Lambda}_q)(\mathbf{V}_q^T)$ with $q < p$.

The eigenvector with the largest singular value is near the mean of the sample vectors. It is easy to show that the sample mean $\langle \mathbf{s} \rangle$ maximizes the sum of dot products with the data $\sum_i [\mathbf{s}^{(i)} \cdot \langle \mathbf{s} \rangle]$, while the eigenvector $\mathbf{v}$ with largest singular value maximizes the sum of squared dot products $\sum_i [\mathbf{s}^{(i)}{}_i \cdot \mathbf{v}]^2$. (To show this, maximize the given functions using Lagrange multipliers, with the constraint that $\langle \mathbf{s} \rangle$ and $\mathbf{v}$ are unit vectors.) As long as most of the samples are in the same quadrant, these two functions have roughly the same maximum.

The *MatLab* code for computing the singular-value decomposition is

```
[U, LAMBDA, V] = svd(S,0);
lambda = diag(LAMBDA);
F = V';
C = U*LAMBDA;
```

(*MatLab* gda10_04)

Here we use the "economy" version of svd(), which has a second argument of zero. In the $N > M$ case, it returns **U** as $N \times M$ and Λ as $M \times M$ (since the bottom $N - M$ rows of Λ are zero and hence the right $N - M$ columns of **U** do not contribute to **G**). The svd() function does not throw out any of the zero (or near-zero) eigenvalues; this is left to the user. The diagonal of LAMBDA has been copied into the column-vector, lambda, for convenience.

We apply factor analysis to rock chemistry data taken from a petrologic database (PetDB at www.petdb.org). This database contains chemical information on igneous and metamorphic rocks collected from the floor of all the world's oceans, but we analyze here $N = 6356$ samples from the Atlantic Ocean that have the following chemical species: SiO_2, TiO_2, Al_2O_3, FeO_{total}, MgO, CaO, Na_2O, and K_2O (units of weight percent).

A plot of the singular values of the Atlantic Rock data set (Fig. 10.5) reveals that the first value is by far the largest, values 2 through 5 are intermediate in size, and values 6 through 8 are near-zero. The fact that the first singular value λ_1 is much larger than all the other reflects the composition of the rock samples having only a small range of variability. Thus, all rock samples contain a large amount of the first factor, $f^{(1)}$—the typical sample. Only four additional factors, $f^{(2)}$, $f^{(3)}$, $f^{(4)}$, and $f^{(5)}$, out of a total of eight are needed to describe the variability about the typical sample:

Element	$f^{(1)}$	$f^{(2)}$	$f^{(3)}$	$f^{(4)}$	$f^{(5)}$
SiO_2	+0.908	+0.007	−0.161	+0.209	+0.309
TiO_2	+0.024	−0.037	−0.126	+0.151	−0.100
Al_2O_3	+0.275	−0.301	+0.567	+0.176	−0.670
FeO-total	+0.177	−0.018	−0.659	−0.427	−0.585
MgO	+0.141	+0.923	+0.255	−0.118	−0.195
CaO	+0.209	−0.226	+0.365	−0.780	+0.207
Na_2O	+0.044	−0.058	−0.0417	+0.302	−0.145
K_2O	+0.003	−0.007	−0.006	+0.073	+0.015

Each role of each of the factors can be understood by examining its elements. Factor 2, for instance, increases the amount of MgO while decreasing mostly Al_2O_3 and CaO, with respect to the typical sample.

The *factor analysis* has reduced the dimensions of variability of the rock data set from eight elements to four factors, improving the effectiveness of scatter plots. *MatLab*'s three-dimensional plotting capabilities are useful in this case, since any three of the four factors can be used as axes and the resulting three-dimensional scatter plot viewed from a variety

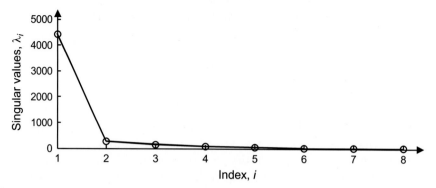

FIG. 10.5 Singular values λ_i of the Atlantic Ocean Rock dataset. *MatLab* script gda10_04.

FIG. 10.6 Three-dimensional perspective view of the coefficients C_i of factors 2, 3, and 4 in each of the rock samples *(dots)* of the Atlantic Ocean Rock data set. *MatLab* script gda10_04.

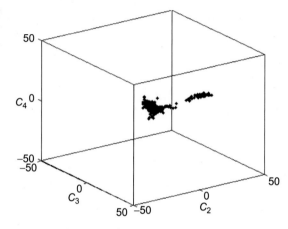

of perspectives. The following *MatLab* command plots the coefficients of factors 2 through 4 for each sample:

```
plot3( C(:,2), C(:,3), C(:,4), 'k.' );
```

(*MatLab* script gda10_04)

The plot can then be viewed from different perspectives by using the rotation controls of the Figure Window (Fig. 10.6). Note that the samples appear to form two populations, one in which the variability is due to $\mathbf{f}^{(2)}$ and another due to $\mathbf{f}^{(3)}$.

10.2 NORMALIZATION AND PHYSICALITY CONSTRAINTS

In many instances, an element can be important even though it occurs only in trace quantities. In such cases, one cannot neglect factors simply because they have small singular values. They may contain an important amount of the trace elements. It is therefore appropriate to normalize the matrix $\mathbf{S}$ so that there is a direct correspondence between singular-value size and importance. This is usually done by defining a diagonal matrix of weights $\mathbf{W}$ (usually proportional to the reciprocal of the standard deviations of measurement of each of the elements) and then forming a new weighted sample matrix $\mathbf{S}' = \mathbf{SW}$.

The singular-value decomposition enables one to determine a set of factors that span, or approximately span, the space of samples. These factors, however, are not unique in the sense that one can form linear combinations of factors that also span the space. This transformation is typically a useful thing to do since, ordinarily, the singular-value decomposition eigenvectors violate prior constraints on what "good" factors should be like. One such constraint is that the factors should have a unit L_1 norm, that is, their elements should sum to one. If the components of a factor represent fractions of chemical elements, for example, it is reasonable that the elements should sum to 100%. Another constraint is that the elements of both the factors and the factor loadings should be nonnegative. Ordinarily a material is composed of a positive combination of components. Given an initial representation of the samples $\mathbf{S} = \mathbf{CFW}^{-1}$, we could imagine finding a new representation consisting of linear combinations of the old factors, defined by $\mathbf{F}' = \mathbf{TF}$, where $\mathbf{T}$ is an arbitrary $p \times p$ transformation matrix. The problem can then be stated.

Find $\mathbf{T}$ subject to the following constraints:

$$\sum_{j=1}^{M} \left[\mathbf{F}'\mathbf{W}^{-1}\right]_{ij} = 1 \text{ for all } i$$

$$\left[\mathbf{CT}^{-1}\right]_{ij} \geq 0 \text{ and } \left[\mathbf{F}'\mathbf{W}^{-1}\right]_{ij} \geq 0 \text{ for all } i \text{ and } j \tag{10.7}$$

These conditions do not uniquely determine $\mathbf{T}$, as can be seen from Fig. 10.4. Note that the second constraint is nonlinear in the elements of $\mathbf{T}$. This is a very difficult constraint to implement and in practice is often ignored.

To find a unique solution, one must add some prior information. One possibility is to find a set of factors that maximize some measure of simplicity. One such measure is *spikiness*: the notion that a factor should have only a few large elements, with the other elements being near-zero. Minerals, for example, obey this principle. While a rock can contain upward of 20 chemical elements, typically it will be composed of minerals such as forsterite (Mg_2SiO_4), anorthite ($CaAl_2Si_2O_8$), rutile (TiO_2), etc., each of which contains just a few elements. Spikiness is more or less equivalent to the idea that the elements of the factors should have *high variance*. The usual formula for estimated variance, σ_d^2, of a data set, $\mathbf{d}$, is

$$\sigma_d^2 = \frac{1}{N}\left(\sum_{i=1}^{N}(d_i - \bar{d})^2\right) = \frac{1}{N^2}\left(N\sum_{i=1}^{N}d_i^2 - \left(\sum_{i=1}^{N}d_i\right)^2\right) \tag{10.8}$$

Its generalization to a factor, f_i, is

$$\sigma_f^2 = \frac{1}{M^2}\left(M\sum_{i=1}^{M}f_i^4 - \left(\sum_{i=1}^{M}f_i^2\right)^2\right) \tag{10.9}$$

FIG. 10.7   Two mutually perpendicular factors f_1 and f_2 are rotated in their plane by an angle, θ, creating two new mutually orthogonal vectors, f'_1 and f'_2.

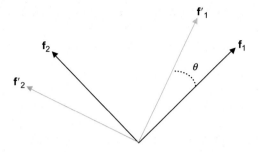

Note that this is the variance of the *squares* of the elements of the factors. Thus, a factor, **f**, has a large variance, σ_f^2, if the absolute values of its elements have high variation. The signs of the elements are irrelevant.

The *varimax* procedure is a way of constructing a matrix, **T**, that increases the variance of the factors while preserving their orthogonality (Kaiser, 1958). It is an iterative procedure, with each iteration operating on only one pair of factors, with other pairs being operated upon in subsequent iterations. The idea is to view the factors as vectors and to rotate them in their plane (Fig. 10.7) by an angle, θ, chosen to maximize the sum of their variances. The rotation changes only the two factors, leaving the other $p-2$ factors unchanged, as in the following example:

$$
\begin{bmatrix}
\mathbf{f}^{(1)\mathrm{T}} \\
\mathbf{f}^{(2)\mathrm{T}} \\
\cos{(\theta)}\mathbf{f}^{(3)\mathrm{T}} + \sin{(\theta)}\mathbf{f}^{(5)\mathrm{T}} \\
\mathbf{f}^{(4)\mathrm{T}} \\
-\sin{(\theta)}\mathbf{f}^{(3)\mathrm{T}} + \cos{(\theta)}\mathbf{f}^{(3)\mathrm{T}} \\
\mathbf{f}^{(6)\mathrm{T}}
\end{bmatrix}
=
\begin{bmatrix}
1 & 0 & 0 & 0 & 0 & 0 \\
0 & 1 & 0 & 0 & 0 & 0 \\
0 & 0 & \cos{(\theta)} & 0 & \sin{(\theta)} & 0 \\
0 & 0 & 0 & 1 & 0 & 0 \\
0 & 0 & -\sin{(\theta)} & 0 & \cos{(\theta)} & 0 \\
0 & 0 & 0 & 0 & 0 & 1
\end{bmatrix}
\begin{bmatrix}
\mathbf{f}^{(1)\mathrm{T}} \\
\mathbf{f}^{(2)\mathrm{T}} \\
\mathbf{f}^{(3)\mathrm{T}} \\
\mathbf{f}^{(4)\mathrm{T}} \\
\mathbf{f}^{(5)\mathrm{T}} \\
\mathbf{f}^{(6)\mathrm{T}}
\end{bmatrix}
\qquad (10.10)
$$

or $\mathbf{F}' = \mathbf{TF}$. Here, only the pair, $\mathbf{f}^{(3)}$ and $\mathbf{f}^{(5)}$, is changed.

In Eq. (10.10), the matrix, **T**, represents a rotation of *one pair* of vectors. The rotation matrix for many such rotations is just the product of a series of pair-wise rotations. Note that the matrix, **T**, obeys the rule, $\mathbf{T}^{-1} = \mathbf{T}^{\mathrm{T}}$ (i.e., **T** is a *unary* matrix). For a given pair of factors, $\mathbf{f}^A$ and $\mathbf{f}^B$, the rotation angle θ is determined by minimizing $\Phi(\theta) = M^2\left(\sigma_{fA}^2 + \sigma_{fB}^2\right)$ with respect to θ (i.e., by solving $d\Phi/d\theta = 0$).

The minimization requires a substantial amount of algebraic and trigonometric manipulation, so we omit it here. The result is (Kaiser, 1958)

$$
\theta = \frac{1}{4}\tan^{-1}\frac{2M\sum_i u_i v_i - \sum_i u_i \sum_i v_i}{M\sum_i\left(u_i^2 - v_i^2\right) - \left(\left(\sum_i u_i\right)^2 - \left(\sum_i v_i\right)^2\right)} \quad \text{with}
\qquad (10.11)
$$

$$
u_i = \left(f_i^A\right)^2 - \left(f_i^B\right)^2 \quad \text{and} \quad v_i = 2f_i^A f_i^B
$$

By way of example, we note that the two vectors

$$\mathbf{f}^A = \frac{1}{2}[1 \ 1 \ 1 \ 1]^T \text{ and } \mathbf{f}^B = \frac{1}{2}[1 \ -1 \ 1 \ -1]^T \tag{10.12}$$

are extreme examples of two *nonspiky* orthogonal vectors because all their elements have the same absolute value. When applied to them, the varimax procedure returns

$$\mathbf{f}^{A'} = \frac{1}{\sqrt{2}}[1 \ 0 \ 1 \ 0]^T \text{ and } \mathbf{f}^{B'} = \frac{1}{\sqrt{2}}[0 \ -1 \ 0 \ -1]^T \tag{10.13}$$

which are significantly spikier than the originals. The *MatLab* code is

```
u = fA.^2 - fB.^2;
v = 2* fA .* fB;
A = 2*M*u'*v;
B = sum(u)*sum(v);
top = A - B;
C = M*(u'*u-v'*v);
D = (sum(u)^2)-(sum(v)^2);
bot = C - D;
q = 0.25 * atan2(top,bot);
cq = cos(q);
sq = sin(q);
fAp = cq*fA + sq*fB;
fBp = -sq*fA + cq*fB;
```

(*MatLab* gda10_05)

Here, the original pair of factors are fA and fB, and the rotated pair are fAp and fBp.

We now apply this procedure to factors $\mathbf{f}_2$ through $\mathbf{f}_5$ of the Atlantic Rock data set (i.e., the factors related to deviations about the typical rock). The varimax procedure is applied to all pairs of these factors and achieves convergence after several such iterations. The *MatLab* code for the loops are

```
FP=F;
% spike these factors using the varimax procedure
k = [2, 3, 4, 5]';
Nk = length(k);
for iter = [1:3]
for ii = [1:Nk]
for jj = [ii+1:Nk]
% spike factors i and j
i=k(ii);
j=k(jj);
% copy factors from matrix to vectors
fA = FP(i,:)';
fB = FP(j,:)';
% standard varimax procedure to determine rotation angle q
```

FIG. 10.8 (A) Factors $\mathbf{f}^{(2)}$ through $\mathbf{f}^{(5)}$ of the Atlantic Rock data set, as calculated by singular-value decomposition. (B) Factors $\mathbf{f}'^{(2)}$ through $\mathbf{f}'^{(5)}$, after application of the varimax procedure. *MatLab* script gda10_05.

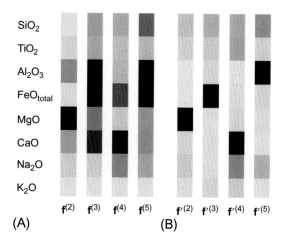

(A)   (B)

(*MatLab* gda10_05)

```
- - -
% copy rotated factors back to matrix
FP(i,:) = fAp';
FP(j,:) = fBp';
end
end
end
```

(*MatLab* gda10_05)

Here the rotated matrix of factors FP is initialized to the original matrix of factors F and then modified by the varimax procedure (omitted and replaced with a "- - -"), with each pass through the inner loop rotating one pair of factors. The procedure converges very rapidly, with three iterations of the outside loop being sufficient. The resulting factors (Fig. 10.8) are much spikier than the original ones. Each now involves mainly variations in one chemical element. For example, $\mathbf{f}'_2$ mostly represents variations in MgO, and $\mathbf{f}'_5$ mostly represents variations in Al_2O_3.

Another possible way of adding prior information is to find factors that are in some sense close to a set of prior factors. If closeness is measured by the L_1 or L_2 norm and if the constraint on the positivity of the factor loadings is omitted, then this problem can be solved using the techniques of Chapters 7 and 12. One advantage of this latter approach is that it permits one to test whether a particular set of prior factors can be factors of the problem (i.e., whether or not the distance between prior factors and actual factors can be reduced to an insignificant amount).

10.3 Q-MODE AND R-MODE FACTOR ANALYSIS

The eigenvectors $\mathbf{U}$ and $\mathbf{V}$ play completely symmetric roles in the singular-value decomposition of the sample matrix $\mathbf{S} = \mathbf{U}\mathbf{\Lambda}\mathbf{V}^T$. We introduced an asymmetry when we grouped them as $\mathbf{S} = (\mathbf{U}\mathbf{\Lambda})(\mathbf{V}^T) = \mathbf{CF}$ to define the loadings $\mathbf{C}$ and factors $\mathbf{F}$.

This grouping is associated with the term *R-mode factor analysis*. It is appropriate when the focus is on patterns among the elements, which is to say, reducing a large number of elements to a smaller number of factors. Thus, for example, we might note that the elements in the Atlantic Rock data set contain a pattern, associated with factor $\mathbf{f}^{(2)}$, in which Al_2O_3 and MgO are strongly and negatively correlated and another pattern, associated with factor $\mathbf{f}^{(3)}$, in which Al_2O_3 and FeO_{total} are strongly and negatively correlated. The effect of these correlations is to reduce the effective number of elements, that is, to allow us to substitute a small number of factors for a large number of elements.

Alternately, we could have grouped the singular-value decomposition as $\mathbf{S} = (\mathbf{U})(\mathbf{\Lambda V}^T)$, an approach associated with the term *Q-mode factor analysis*. The equivalent transposed form $\mathbf{S}^T = (\mathbf{V\Lambda})(\mathbf{U}^T)$ is more frequently encountered in the literature and is also more easily understood, since it can be interpreted as "normal factor analysis" applied to the matrix $\mathbf{S}^T$. The transposition has reversed the sense of samples and elements, so the factor matrix $\mathbf{U}^T$ quantifies patterns of variability among samples, in the same way that $\mathbf{V}^T$ quantifies patterns of variability among elements. To pursue this comparison further, consider an R-mode problem in which there is only one factor, $\mathbf{v}^{(1)} = [1, 1, \ldots]^T$. This factor implies that all of the samples in the data table contain a 1:1 ratio of elements 1 and 2. Similarly, a Q-mode problem in which there is only one factor, $\mathbf{u}^{(1)} = [1, 1, \ldots]^T$ implies that all of the elements in the transposed data table have a 1:1 ratio of samples 1 and 2. This approach is especially useful in detecting clustering among the samples; indeed, Q-mode factor analysis is often referred to as a form of *cluster analysis*.

10.4 EMPIRICAL ORTHOGONAL FUNCTION ANALYSIS

Factor analysis need not be limited to data that contain actual mixtures of components. Given any set of vectors $\mathbf{s}^{(i)}$, one can perform the singular-value decomposition and represent $\mathbf{s}^{(i)}$ as a linear combination of a set of orthogonal factors. Even when the factors have no obvious physical interpretation, the decomposition can be useful as a tool for quantifying the similarities between the $\mathbf{s}^{(i)}$ vectors. This kind of factor analysis is often called *empirical orthogonal function (EOF)* analysis.

As an example of this application of factor analysis, consider the set of $N = 14$ shapes shown in Fig. 10.9. These shapes might represent profiles of mountains or other subjects of interest. The problem we shall consider is how these profiles might be ordered to bring out the similarities and differences between the shapes. A geomorphologist might desire such an ordering because, when combined with other kinds of geological information, it might reveal the kinds of erosional processes that cause the shape of mountains to evolve with time.

We begin by discretizing each profile and representing it as a unit vector (in this case of length $M = 11$). These unit vectors make up the matrix $\mathbf{S}$, on which we perform factor analysis. Since the factors do not represent any particular physical object, there is no need to impose any positivity constraints on them, and we use the untransformed singular-value decomposition factors. The three most important EOFs (factors) (i.e., the ones with the three largest singular values) are shown in Fig. 10.10. The first EOF, as expected, appears to be simply an "average" mountain; the second seems to control the skewness, or degree of asymmetry,

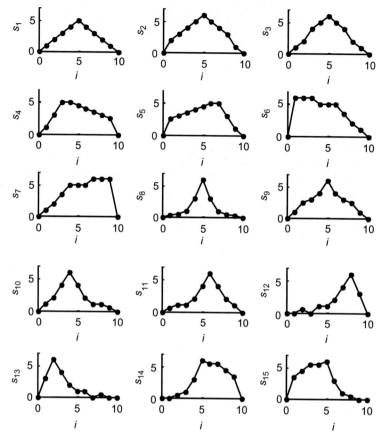

FIG. 10.9 A set of hypothetical mountain profiles. The variability of shape will be determined using factor analysis. *MatLab* script gda10_06.

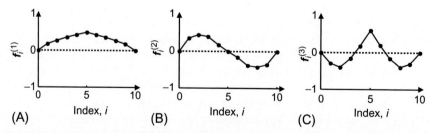

FIG. 10.10 The three largest factors $\mathbf{f}^{(1)}$, $\mathbf{f}^{(2)}$, and $\mathbf{f}^{(3)}$ in the representation of the mountain profiles in Fig. 10.9. (A) The factor with the largest singular value $\lambda_1 = 38.4$ has the shape of the average profile. (B) The factor with the second largest singular value $\lambda_2 = 12.7$ quantifies the asymmetry of the profiles. (C) The factor with the third largest singular value $\lambda_3 = 7.4$ quantifies the sharpness of the mountain summits. *MatLab* script gda10_06.

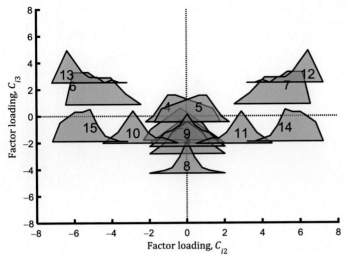

FIG. 10.11 The mountain profiles of Fig. 10.9, arranged according to the relative amounts of factors 2 and 3 contained in each profile's orthogonal decomposition; that is, by the size of the factor loadings C_{i2} and C_{i3}. *MatLab* script gda10_06.

of the mountain; and the third, the sharpness of the mountain's summit. We emphasize, however, that this interpretation was made after the EOF analysis and was not based on any prior notions of how mountains might differ. We can then use the loadings as a measure of the similarities between the mountains. Since the amount of the first factor does not vary much between mountains, we use a two-dimensional ordering based on the relative amounts of the second and third factors in each of the mountain profiles (Fig. 10.11).

EOF analysis is especially useful when the data have an ordering in space, time, or some other sequential variable. For instance, suppose that profiles in Fig. 10.9 are measured at sequential times. Then we can understand the model equation $\mathbf{S} = \mathbf{CF}$ to mean

$$S\left(t_i, x_j\right) = \sum_{k=1}^{p} C_k(t_i) F_k\left(x_j\right) \tag{10.14}$$

Here the quantity $S(t_i, x_j)$, which varies with both time and space, has been broken up into the sum of two sets of function, the EOFs $F_k(x_j)$, which vary only with space, and the corresponding loadings $C_k(t_i)$, which vary only with time. A plot of the kth loading $C_k(t_i)$ as a function of time t_i reveals how the importance of the kth EOF varies with time.

This analysis can be extended to functions of two or more spatial dimensions by writing Eq. (10.14) as

$$S\left(t_i, \mathbf{x}^{(j)}\right) = \sum_{k=1}^{P} C_k(t_i) F_k\left(\mathbf{x}^{(j)}\right) \tag{10.15}$$

Here, the spatial observation points $\mathbf{x}$ are multidimensional, but they have been given a linear ordering through the index k. As an example, suppose that the data are a sequence of two-dimensional images, where each image represents a physical parameter observed on the (x, y) plane. This image can be *unfolded* (reordered) into a vector (Fig. 10.12), which then

FIG. 10.12 A matrix $\mathbf{A}$ representing a discrete version of a two-dimensional function $A_{ij}=a(x_i,\,y_j)$ is unfolded row-wise into a vector $\mathbf{a}$ using the rule $a_k=A_{ij}$ with $k=(i-1)M+j$.

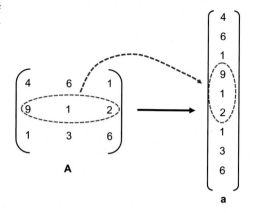

becomes a row of the sample matrix $\mathbf{S}$. The resulting EOFs have this same ordering and must be folded back into two-dimensional images before being interpreted.

As an example, we consider a sequence of $N=25$ images, each of which contains a grid of $20 \times 20 = 400$ pixels (Fig. 10.13). Each image represents the spatial variation of a physical parameter such as pressure or temperature at a fixed time, with the overall sequence being time-sequential. These data are synthetic and are constructed by summing three spatial patterns (EOFs) with coefficients (loadings) that vary systematically with time, and then adding random noise. As expected, only $p=3$ singular values are found to be significant (Fig. 10.14). The corresponding EOFs and loadings are shown in Fig. 10.15. Had this analysis been based upon actual data, the time variation of each of the loadings, which have different periodicities, would be of special interest and might possibly provide insight into the physical processes associated with each of the EOFs. A similar example that uses actual ocean temperature data to examine the El Nino-Southern Oscillation climate instability is given by Menke and Menke (2011, their Section 8.5).

10.5 PROBLEMS

10.1 Suppose that a set of $N>M$ samples are represented as $\mathbf{S}\approx\mathbf{C}_p\mathbf{F}_p$ where the matrix $\mathbf{F}_p$ contains $p<M$ factors whose values are prescribed (i.e., prior information). (A) How can the loadings $\mathbf{C}_p$ be determined? (B) Write a *MatLab* script that implements your procedure for the case $M=3$, $N=10$, $p=2$. (C) Make a three-dimensional plot of your results.

10.2 Write a *MatLab* script that verifies the varimax result given in Eq. (10.12). Use the following steps. (A) Compute the factors $\mathbf{f}'^A$ and $\mathbf{f}'^B$ for a complete suite of angles θ using the rotation

$$\mathbf{f}'^A = \cos(\theta)\mathbf{f}^A + \sin(\theta)\mathbf{f}^B$$
$$\mathbf{f}'^B = -\sin(\theta)\mathbf{f}^A + \cos(\theta)\mathbf{f}^B$$

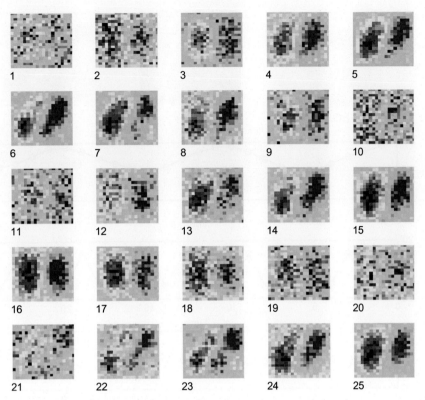

FIG. 10.13 Time sequence of $N = 25$ images. Each image represents a parameter, such as pressure or temperature, that varies spatially in the (x, y) plane. *MatLab* script gda10_07.

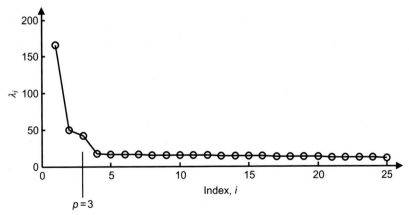

FIG. 10.14 Singular values λ_i of the image sequence shown in Fig. 10.12. Only $p = 3$ singular values have significant amplitudes. *MatLab* script gda10_07.

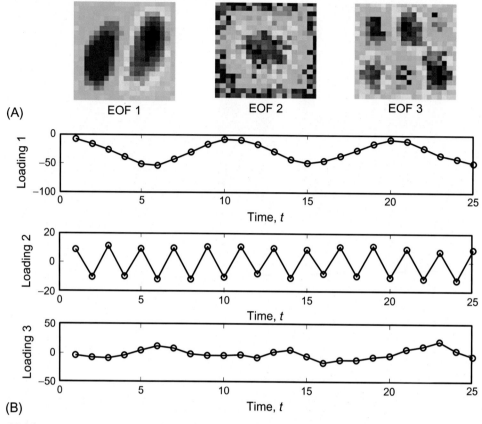

FIG. 10.15 (A) First three empirical orthogonal functions (EOFs) of the image sequence shown in Fig. 10.12. (B) Corresponding loadings as a function of time, t. *MatLab* script gda10_07.

(B) Compute and plot the variance $\sigma_{fA'}^2 + \sigma_{fB'}^2$ as a function of angle θ. (C) Note the angle of the minimum variance and verify that the angle is the one predicted by the varimax formula. (D) Verify that the factors corresponding to the angle are as stated in Eq. (10.12).

10.3 Suppose that a data set represents a function of three spatial dimensions; that is, with samples $S(t, x, y, z)$ on an evenly spaced three-dimensional grid. How can these samples be unfolded into a matrix, **S**?

References

Kaiser, H.F., 1958. The varimax criterion for analytic rotation in factor analysis. Psychometrika 23, 187–200.
Menke, W., Menke, J., 2011. Environmental Data Analysis with MatLab. Academic Press, Elsevier Inc, Oxford. 263 pp.

Continuous Inverse Theory and Tomography

11.1 THE BACKUS-GILBERT INVERSE PROBLEM

While continuous inverse problems are not the main subject of this book, we will cover them briefly to illustrate their relationship to discrete problems. Discrete and continuous inverse problems differ in their assumptions about the model parameters. Whereas the model parameters are treated as a finite-length vector in discrete inverse theory, they are treated as a

© 2018 Elsevier Inc. All rights reserved.

continuous function in continuous inverse theory. The standard form of the continuous inverse problem is

$$d_i = \int_a^b G_i(z)m(z)dz \tag{11.1}$$

when the model function $m(z)$ varies only with one parameter, such as depth z. When the model function depends on several—say L—variables, then Eq. (11.1) must be generalized to

$$d_i = \int_V G_i(\mathbf{x})m(\mathbf{x})d^L x \tag{11.2}$$

where $d^L x$ is the volume element in the space of $\mathbf{x}$.

The "solution" of a discrete problem can be viewed as either an estimate of the model parameter vector $\mathbf{m}^{\text{est}}$ or a series of weighted averages of the model parameters, $\mathbf{m}^{\text{avg}} = \mathbf{Rm}^{\text{true}}$, where $\mathbf{R}$ is the resolution matrix (see Section 4.3). If the discrete inverse problem is very underdetermined, then the interpretation of the solution in terms of weighted averages is most sensible, since a single model parameter is very poorly resolved. Continuous inverse problems can be viewed as the limit of discrete inverse problems as the number of model parameters becomes infinite, and they are inherently underdetermined. Attempts to estimate the model function $m(\mathbf{x})$ at a specific point $\mathbf{x} = \mathbf{x}'$ are futile. All determinations of the model function must be made in terms of local averages, which are simple generalizations of the discrete case, $m_i^{\text{avg}} = \sum_j G_{ij}^{-g}d_j = \sum_j R_{ij}m_j^{\text{true}}$ where $R_{ij} = \sum_k G_{ik}^{-g}G_{kj}$

$$m^{\text{avg}}(\mathbf{x}') = \sum_{i=1}^N G_i^{-g}(\mathbf{x}')d_i = \int R(\mathbf{x}', \mathbf{x})m^{\text{true}}(\mathbf{x})d^L x \tag{11.3}$$

where

$$R(\mathbf{x}', \mathbf{x}) = \sum_{i=1}^N G_i^{-g}(\mathbf{x}')G_i(\mathbf{x}) \tag{11.4}$$

Here, $G_i^{-g}(\mathbf{x}')$ is the continuous analogy to the generalized inverse G_{ij}^{-g} and the averaging function $R(\mathbf{x}', \mathbf{x})$ (often called the *resolving kernel*) is the analogy to the model resolution matrix R_{ij}. The average is localized near the target point $\mathbf{x}'$ if the resolving kernel is peaked near $\mathbf{x}'$. The solution of the continuous inverse problem involves constructing the most peaked resolving kernel possible with a given set of measurements, that is, with a given set of data kernels, $G_i(\mathbf{x})$. The spread of the resolution function is quantified by (compare with Eq. 4.23)

$$J(\mathbf{x}') = \int w(\mathbf{x}', \mathbf{x})R^2(\mathbf{x}', \mathbf{x})d^L x \tag{11.5}$$

Here, $w(\mathbf{x}', \mathbf{x})$ is a nonnegative function that is zero at the point $\mathbf{x}'$ and that grows monotonically away from that point. One commonly used choice is the quadratic function $w(\mathbf{x}', \mathbf{x}) = |\mathbf{x}' - \mathbf{x}|^2$. Other, more complicated functions can be meaningful if the elements of $\mathbf{x}$ have an interpretation other than spatial position. After inserting the definition of the resolving kernel (Eq. 11.3) into the definition of the spread (Eq. 11.5), we find

$$J(\mathbf{x}') = \int w(\mathbf{x}', \mathbf{x}) R(\mathbf{x}', \mathbf{x}) R(\mathbf{x}', \mathbf{x}) d^L x$$

$$= \int w(\mathbf{x}', \mathbf{x}) \sum_{i=1}^{N} G_i^{-g}(\mathbf{x}') G_i(\mathbf{x}) \sum_{j=1}^{N} G_j^{-g}(\mathbf{x}') G_j(\mathbf{x}) d^L x$$

$$= \sum_{i=1}^{N} \sum_{j=1}^{N} G_i^{-g}(\mathbf{x}') G_j^{-g}(\mathbf{x}') \int w(\mathbf{x}', \mathbf{x}) G_i(\mathbf{x}) G_j(\mathbf{x}) d^L x \qquad (11.6)$$

$$= \sum_{i=1}^{N} \sum_{j=1}^{N} G_i^{-g}(\mathbf{x}') G_j^{-g}(\mathbf{x}') [\mathbf{S}(\mathbf{x}')]_{ij}$$

where

$$[\mathbf{S}(\mathbf{x}')]_{ij} = \int w(\mathbf{x}', \mathbf{x}) G_i(\mathbf{x}) G_j(\mathbf{x}) d^L x \qquad (11.7)$$

Eq. (11.7) might be termed an *overlap integral,* since it is large only when the two data kernels overlap, that is, when they are simultaneously large in the same region of space. The continuous spread function has now been manipulated into a form completely analogous to the discrete spread function in Eq. (4.26). The generalized inverse that minimizes the spread of the resolution is the precise analogy of Eq. (4.34)

$$G_l^{-g}(\mathbf{x}') = \frac{\sum_{i=1}^{N} u_i [\mathbf{S}^{-1}(\mathbf{x}')]_{il}}{\sum_{i=1}^{N} \sum_{j=1}^{N} u_i [\mathbf{S}^{-1}(\mathbf{x}')]_{ij} u_j} \quad \text{where } u_i = \int G_i(\mathbf{x}) d^L x \qquad (11.8)$$

11.2 RESOLUTION AND VARIANCE TRADE-OFF

Since the data $\mathbf{d}$ are determined only up to some error quantified by the covariance matrix [cov $\mathbf{d}$], the localized average $m^{\mathrm{avg}}(\mathbf{x}')$ is determined up to some corresponding error

$$\mathrm{var}[m^{\mathrm{avg}}(\mathbf{x}')] = \sum_{i=1}^{N} \sum_{j=1}^{N} G_i^{-g}(\mathbf{x}') [\mathrm{cov}\,\mathbf{d}]_{ij} G_j^{-g}(\mathbf{x}') \qquad (11.9)$$

As in the discrete case, the generalized inverse that minimizes the spread of resolution may lead to a localized average with large error bounds. A slightly less localized average may be desirable because it may have much less error. This generalized inverse may be found by minimizing a weighted average of the spread of resolution and size of variance

$$\text{minimize } J'(\mathbf{x}') = \alpha \int w(\mathbf{x}', \mathbf{x}) R^2(\mathbf{x}', \mathbf{x}) d^L x + (1 - \alpha) \mathrm{var}[m^{\mathrm{avg}}(\mathbf{x}')] \qquad (11.10)$$

The parameter α, which varies between 0 and 1, quantifies the relative weight given to spread of resolution and size of variance. As in the discrete case (see Section 4.10), the

FIG. 11.1   Typical trade-off of resolution and variance for a linear continuous inverse problem. Note that the size (spread) function decreases monotonically with spread and that it is tangent to the two asymptotes at the endpoints $\alpha=1$ and $\alpha=0$.

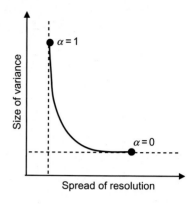

corresponding generalized inverse can be found by using Eq. (11.7), where all instances of $\mathbf{S}(\mathbf{x}')$ are replaced by

$$[\mathbf{S}'(\mathbf{x}')]_{ij} = \alpha \int w(\mathbf{x}',\mathbf{x})G_i(\mathbf{x})G_j(\mathbf{x})\mathrm{d}^L x + (1-\alpha)[\mathrm{cov}\,\mathbf{d}]_{ij} \qquad (11.11)$$

Backus and Gilbert (1968) prove a number of important properties of the trade-off curve (Fig. 11.1), which is a plot of size of variance against spread of resolution, including the fact that the variance decreases monotonically with spread.

11.3 APPROXIMATING CONTINUOUS INVERSE PROBLEMS AS DISCRETE PROBLEMS

A continuous inverse problem can be converted into a discrete one with the assumption that the model function can be represented by a finite number M of coefficients, that is

$$m(\mathbf{x}) \approx \sum_{j=1}^{N} m_j f_j(\mathbf{x}) \qquad (11.12)$$

The particular choice of the functions $f_j(\mathbf{x})$ implies particular prior information about the behavior of $m(\mathbf{x})$, so the solution one obtains is sensitive to the choice. One commonly used set of functions assumes that the model is constant within certain subregions V_j of the space of model parameters (e.g., voxels). In this case, $f_j(\mathbf{x})$ is unity inside V_j and zero outside it, and m_j is the value of the model in each voxel. In one dimension, where the x-axis is divided into intervals of width Δx

$$m(x) \approx \sum_{j=1}^{M} m_j f_j(x) \quad \text{with } f_j(x) = H(x-j\Delta x) - H(x-(j+1)\Delta x) \qquad (11.13)$$

Here $H(x-\xi)$ is the Heaviside step function, which is zero for $x<\xi$ and unity for $x>\xi$. Many other choices of $f_j(\mathbf{x})$ are encountered, based on polynomial approximations, splines, and truncated Fourier series representations of $m(\mathbf{x})$.

Inserting Eq. (11.12) into the standard form of the continuous problem (Eq. 11.2) leads to a discrete problem

$$d_i = \int G_i(\mathbf{x}) \sum_{j=1}^{M} m_j f_j(\mathbf{x}) \mathrm{d}^L x = \sum_{j=1}^{M} \left\{ \int G_i(\mathbf{x}) f_j(\mathbf{x}) \mathrm{d}^L x \right\} m_j \tag{11.14}$$

So the discrete form of the equation is $\mathbf{d} = \mathbf{Gm}$

$$d_i = \sum_{j=1}^{M} G_{ij} m_j \quad \text{with} \quad G_{ij} = \int G_i(\mathbf{x}) f_j(\mathbf{x}) \mathrm{d}^L x \tag{11.15}$$

In the special case of voxels of volume V_j centered on the points $\mathbf{x}^{(j)}$, Eq. (11.15) becomes

$$G_{ij} = \int_{V_j} G_i(\mathbf{x}) \mathrm{d}^L x \tag{11.16}$$

If the data kernel varies slowly with position so that it is approximately constant in the voxel, then we may approximate the integral as just its integrand evaluated at $\mathbf{x}^{(j)}$ times the volume V_j:

$$G_{ij} = \int_{V_j} G_i(\mathbf{x}) \mathrm{d}^L x \approx G_i\left(\mathbf{x}^{(j)}\right) V_j \tag{11.17}$$

In one dimension, this is equivalent to the Riemann summation approximation to the integral

$$d_i = \int G_i(x) m(x) \mathrm{d}x \approx \sum_{j=1}^{M} \{G_i(j\Delta x)\Delta x\} m_j = \sum_{j=1}^{M} G_{ij} m_j \tag{11.18}$$

Two different factors control the choice of the size of the voxels. The first is dictated by a prior assumption of the smoothness of the model. The second becomes important only when one uses Eq. (11.17) in preference to Eq. (11.16). Then the subregion must be small enough that the data kernels $G_i(\mathbf{x})$ are approximately constant in the voxel. This second requirement often forces the voxels to be much smaller than dictated by the first requirement, so Eq. (11.16) should be used whenever the data kernels can be integrated analytically. Eq. (11.17) fails completely whenever a data kernel has an integrable singularity within the subregion. This case commonly arises in problems involving the use of seismic rays to determine acoustic velocity structure.

11.4 TOMOGRAPHY AND CONTINUOUS INVERSE THEORY

The term "tomography" has come to be used in geophysics almost synonymously with the term "inverse theory." Tomography is derived from the Greek word *tomos*, that is, slice, and denotes forming an image of an object from measurements made from slices (or rays) through it. We consider tomography a subset of inverse theory, distinguished by a special form of the data kernel that involves measurements made along rays. The model function in tomography is a function of two or more variables and is related to the data by

$$d_i = \int_{C_i} m[x(s), y(s)] ds \tag{11.19}$$

Here, the model function is integrated along a curved ray C_i having arc length s. This integral is equivalent to the one in a standard continuous problem (Eq. (11.2)) when the data kernel is $G_i(x, y) = \delta\{x(s) - x_i[y(s)]\} ds/dy$, where $\delta(x)$ is the Dirac delta function:

$$d_i = \iint m(x, y) \delta\{x(s) - x_i[y(s)]\} \frac{ds}{dy} dx dy = \int_{C_i} m[x(s), y(s)] ds \tag{11.20}$$

Here x is supposed to vary with y along the curve C_i and y is supposed to vary with arc length s.

While the tomography problem is a special case of a continuous inverse problem, several factors limit the applicability of the formulas of the previous sections. First, the Dirac delta functions in the data kernel are not square integrable so that the overlap integrals S_{ij} (Eq. 11.7) have nonintegrable singularities at points where rays intersect. Further, in three-dimensional cases, the rays may not intersect at all so that all the S_{ij} may be identically zero. Neither of these problems is insurmountable, and they can be overcome by replacing the rays with tubes of finite cross-sectional width. (Rays are often an idealization of a finite-width process anyway, as in acoustic wave propagation, where they are an infinitesimal wavelength approximation.) Since this approximation is equivalent to some statement about the smoothness of the model function $m(x, y)$, it often suffices to discretize the continuous problem by dividing it into constant m subregions, where the subregions are large enough to guarantee a reasonable number containing more than one ray. The discrete inverse problem is then of the form $d_i = \sum_j G_{ij} m_j$, where the data kernel G_{ij} gives the arc length of the ith ray in the jth subregion. The concepts of resolution and variance, now interpreted in the discrete fashion of Chapter 4, are still applicable and of considerable importance.

11.5 TOMOGRAPHY AND THE RADON TRANSFORM

The simplest tomography problem involves straight-line rays and a two-dimensional model function $m(x, y)$ and is called Radon's problem. By historical convention, the straight-line rays C_i in Eq. (11.19) are parameterized by their perpendicular distance u from the origin and the angle θ (Fig. 11.2) that the perpendicular makes with the x-axis. Position (x, y) and ray coordinates (u, s), where s is arc length, are related by

$$\begin{bmatrix} x \\ y \end{bmatrix} = \begin{bmatrix} \cos\theta & -\sin\theta \\ \sin\theta & \cos\theta \end{bmatrix} \begin{bmatrix} u \\ s \end{bmatrix} \quad \text{and} \quad \begin{bmatrix} u \\ s \end{bmatrix} = \begin{bmatrix} \cos\theta & \sin\theta \\ -\sin\theta & \cos\theta \end{bmatrix} \begin{bmatrix} x \\ y \end{bmatrix} \tag{11.21}$$

The tomography problem is then

$$d(u, \theta) = \int_{-\infty}^{+\infty} m(x = u \cos\theta - s \sin\theta, y = u \sin\theta + s \cos\theta) ds \tag{11.22}$$

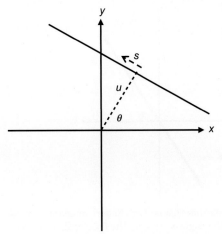

FIG. 11.2 The Radon transform is performed by integrating a function of (x,y) along straight lines *(bold)* parameterized by the arc length, s; perpendicular distance, u; and angle, θ.

In realistic experiments, $d(u, \theta)$ is sampled only at discrete points $d_i = d(u_i, \theta_i)$. Nevertheless, much insight can be gained into the behavior of Eq. (11.22) by regarding (u, θ) as continuous variables. Eq. (11.22) is then an integral transform that transforms variables (x,y) into two new variables (u,θ) and is called a *Radon* transform.

11.6 THE FOURIER SLICE THEOREM

The Radon transform is similar to another integral transform, the Fourier transform, which transforms spatial position x into spatial wave number k_x

$$\hat{f}(k_x) = \int_{-\infty}^{+\infty} f(x)\exp(ik_xx)dx \quad \text{and} \quad f(x) = \frac{1}{2\pi}\int_{-\infty}^{+\infty} \hat{f}(k_x)\exp(-ik_xx)dk_x \tag{11.23}$$

In fact, the two are quite closely related, as can be seen by Fourier transforming Eq. (11.22) with respect to $u \to k_u$:

$$\hat{d}(k_u, \theta) = \int_{-\infty}^{+\infty}\int_{-\infty}^{+\infty} m(u\cos\theta - s\sin\theta, u\sin\theta + s\cos\theta)ds\,\exp(ik_uu)du \tag{11.24}$$

We now transform the double integral from ds du to dx dy, using the fact that the Jacobian determinant $|\det[\partial(x, y)/\partial(u, s)]|$ is unity (see Eq. 11.21):

$$\hat{d}(k_u, \theta) = \int_{-\infty}^{+\infty}\int_{-\infty}^{+\infty} m(x, y)\exp(ik_ux\cos\theta + ik_uy\sin\theta)dxdy$$
$$= \hat{m}(k_x = k_u\cos\theta, k_y = k_u\sin\theta) \tag{11.25}$$

This result, called the *Fourier slice theorem*, provides a method of inverting the Radon transform. The Fourier-transformed quantity $\hat{d}(k_u, \theta)$ is simply the Fourier-transformed image $\hat{m}(k_x, k_y)$ evaluated along radial lines in the (k_x,k_y) plane (Fig. 11.3). If the Radon

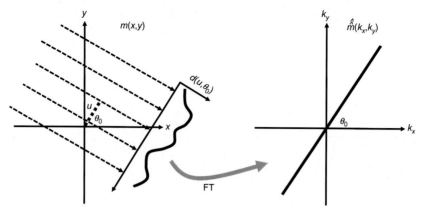

FIG. 11.3 (Left) The function $m(x,y)$ is integrated along a set of parallel lines *(dashed)* in a Radon transform to form the function $d(u,\theta_0)$. This function is called the *projection* of $m(x,y)$ at the angle θ_0. (Right) The Fourier slice theorem states that the Fourier transform (FT) of the projection is equal to the Fourier-transformed image evaluated along a line *(bold)* of angle θ_0 in the (k_x,k_y) plane.

transform is known for all values of (u,θ), then the Fourier-transformed image is known for all (k_x,k_y), and since the Fourier transform can be inverted uniquely, the image itself is known for all (x,y).

Since the Fourier transform and its inverse are unique, the Radon transform can be uniquely inverted if it is known for all possible (u,θ). Further, the Fourier slice theorem can be used to invert the Radon transform in practice by using discrete Fourier transforms in place of integral Fourier transforms. However, u must be sampled sufficiently evenly that the $u \rightarrow k_u$ transform can be performed and θ must be sampled sufficiently finely that $\hat{m}(k_x, k_y)$ can be sensibly interpolated onto a rectangular grid of (k_x,k_y) to allow the $k_x \rightarrow x$ and $k_y \rightarrow y$ transforms to be performed (see the discussion in *MatLab* script gda11_01). An example of the use of the Fourier slice theorem to invert tomography data is shown in Fig. 11.4.

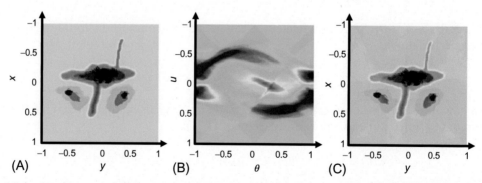

FIG. 11.4 (A) A test image $m^{\text{true}}(x,y)$ of 256×256 discrete values, or pixels. This synthetic image depicts a hypothetical magma chamber beneath a volcano. (B) The Radon transform $d(u,\theta)$ of the image in (A), also evaluated at 256×256 points. (C) The image $m^{\text{est}}(x,y)$ reconstructed from its Radon transform by direct application of the Fourier slice theorem. Small errors in the reconstruction arise from the interpolation of the Fourier-transformed image onto a rectangular grid. *MatLab* script gda11_02.

11.7 CORRESPONDENCE BETWEEN MATRICES AND LINEAR OPERATORS

The continuous function $m(x)$ is the continuous analog to the vector $\mathbf{m}$. What is the continuous analog to $\mathbf{Lm}$, where $\mathbf{L}$ is a matrix? Let us call it $\mathcal{L}m$, that is, $\mathcal{L}$ *operating* on the function $m(x)$. In the discrete case, $\mathbf{Lm}$ is another vector, so by analogy $\mathcal{L}m$ is another function. Just as $\mathbf{L}$ is linear in the sense that $\mathbf{L}(\mathbf{m}^A + \mathbf{m}^B) = \mathbf{Lm}^A + \mathbf{Lm}^B$, we would like to choose $\mathcal{L}$ so that it is linear in the sense that $\mathcal{L}(m^A + m^B) = \mathcal{L}m^A + \mathcal{L}m^B$. A hint to the identity of $\mathcal{L}$ can be drawn from the fact that a matrix can be used to approximate derivatives and integrals. Consider, for example,

$$\mathbf{L}^A = \frac{1}{\Delta x}\begin{bmatrix} -1 & 1 & 0 & \cdots & 0 & 0 \\ 0 & -1 & 1 & 0 & \cdots & 0 \\ & & & \ddots & & \\ 0 & 0 & 0 & \cdots & -1 & 1 \end{bmatrix} \quad \text{and} \quad \mathbf{L}^B = \Delta x \begin{bmatrix} 1 & 0 & 0 & \cdots & 0 & 0 \\ 1 & 1 & 0 & 0 & \cdots & 0 \\ & & & \ddots & & \\ 1 & 1 & 1 & \cdots & 1 & 1 \end{bmatrix} \tag{11.26}$$

Here, $\mathbf{L}^A\mathbf{m}$ is the finite difference approximation to the derivative dm/dx and $\mathbf{L}^B\mathbf{m}$ is the Riemann sum approximation to the indefinite integral

$$\int_0^x m(x')dx' \tag{11.27}$$

Thus, $\mathcal{L}$ can be any linear combination of integrals and derivatives (a *linear operator*, for short) (Lanczos, 1961). In the general multidimensional case, where $m(\mathbf{x})$ is a function of an N-dimensional position vector $\mathbf{x}$, $\mathcal{L}$ is built up of partial derivatives and volume integrals. However, for simplicity, we restrict ourselves to the one-dimensional case here.

Now let us consider whether we can define an inverse operator $\mathcal{L}^{-1}$ that is the continuous analog to the matrix inverse $\mathbf{L}^{-1}$. By analogy to $\mathbf{L}^{-1}\mathbf{Lm} = \mathbf{m}$, we want to choose $\mathcal{L}^{-1}$ so that $\mathcal{L}^{-1}\mathcal{L}m(x) = m(x)$. Actually, the derivative matrix $\mathbf{L}^A$ (Eq. 11.26) is problematical in this regard, for it is one row short of being square, and thus has no inverse. This corresponds to a function being determined only up to an integration constant when its derivative is known. This problem can be patched by adding to it a top row

$$\mathbf{L}^C = \frac{1}{\Delta x}\begin{bmatrix} 1 & 0 & 0 & 0 & & 0 & 0 \\ -1 & 1 & 0 & \cdots & & 0 & 0 \\ 0 & -1 & 1 & 0 & & \cdots & 0 \\ & & & \ddots & & & \\ 0 & 0 & 0 & \cdots & & -1 & 1 \end{bmatrix} \tag{11.28}$$

that fixes the value of m_1, that is, by specifying the integration constant. Thus, $\mathcal{L}$ is not just a linear operator, but rather a linear operator plus one or more associated *boundary conditions*. It is easy to verify that $\mathbf{L}^C\mathbf{L}^B = \mathbf{I}$. The analogous continuous result that the derivative is the inverse operator of the derivative

$$m(x) = \frac{d}{dx}\int_0^x m(x')dx' \tag{11.29}$$

is well known and is called the *fundamental theorem of calculus*. Note that a linear differential equation can be written $\mathcal{L}m(x)=f(x)$. Its solution in terms of the inverse operator is $m(x)=\mathcal{L}^{-1}f(x)$. But its solution can also be written as a Green function integral

$$m(x) = \int_{-\infty}^{+\infty} F(x,\xi)f(\xi)\mathrm{d}\xi = \mathcal{L}^{-1}f(x) \tag{11.30}$$

where $F(x,\xi)$ solves $\mathcal{L}F(x,\xi)=\delta(x-\xi)$. Hence, the inverse operator to a differential operator is the Green function integral.

Another important quantity ubiquitous in the formulas of inverse theory is the dot product between two vectors; written here generically as $s=\mathbf{a}^T\mathbf{b}=\sum_i a_i b_i$ where s is a scalar. The continuous analog is the integral

$$s = \int a(\mathbf{x})b(\mathbf{x})\mathrm{d}^N x = (a,b) \tag{11.31}$$

where s is a scalar and $\mathrm{d}^N x$ is the volume element, and the integration is over the whole space of $\mathbf{x}$. This integral is called the *inner product* of the functions $a(x)$ and $b(x)$ and is abbreviated $s=(a,b)$. As before, we restrict the discussion to the one-dimensional case:

$$s = \int_{-\infty}^{+\infty} a(x)b(x)\mathrm{d}x = (a,b) \tag{11.32}$$

Many of the dot products that we encountered earlier in this book contained matrices, for example, $[\mathbf{Aa}]^T\mathbf{b}$ where $\mathbf{A}$ is a matrix. The continuous analog is an inner product containing a linear operator, that is, $(\mathcal{L}a,b)$.

An extremely important property of the dot product $[\mathbf{Aa}]^T\mathbf{b}$ is that it can also be written as $\mathbf{a}^T[\mathbf{Bb}]$ with $\mathbf{B}=\mathbf{A}^T$. We can propose the analogous relationship for linear operators:

$$\left(\mathcal{L}^A a, b\right) = \left(a, \mathcal{L}^B b\right) \tag{11.33}$$

The question then is what is the relationship between the two linear operators $\mathcal{L}^A$ and $\mathcal{L}^B$, or put another way, what is the continuous analog of the transpose of matrix? As before, we will start off by merely giving the answer a name, the *adjoint*, and a symbol, $\mathcal{L}^\dagger$

$$\left(\mathcal{L}a, b\right) = \left(a, \mathcal{L}^\dagger b\right) \tag{11.34}$$

Several approaches are available for determining the $\mathcal{L}^\dagger$ corresponding to a particular $\mathcal{L}$. The most straightforward is to start with the definition of the inner product. For instance, if $\mathcal{L}=c(x)$, where $c(x)$ is an ordinary function, then $\mathcal{L}^\dagger=c(x)$, too, since

$$\int_{-\infty}^{+\infty} (ca)b\,\mathrm{d}x = \int_{-\infty}^{+\infty} a(cb)\,\mathrm{d}x \tag{11.35}$$

When $\mathcal{L}=\mathrm{d}/\mathrm{d}x$, we can use integration by parts to find $\mathcal{L}^\dagger$:

$$\int_{-\infty}^{+\infty} \frac{\mathrm{d}a}{\mathrm{d}x}b\,\mathrm{d}x = ab\Big|_{-\infty}^{+\infty} - \int_{-\infty}^{+\infty} a\frac{\mathrm{d}b}{\mathrm{d}x}\,\mathrm{d}x \tag{11.36}$$

So $\mathcal{L}^\dagger = -d/dx$, as long as the functions approach zero as x approaches $\pm\infty$. This same procedure, applied twice, can be used to show that $\mathcal{L} = d^2/dx^2$ is *self-adjoint*, that is, it is its own adjoint. As another example, we derive the adjoint of the indefinite integral $\int_{-\infty}^x d\xi$ by first writing it as

$$\int_{-\infty}^x a(\xi)d\xi = \int_{-\infty}^{+\infty} H(x-\xi)a(\xi)d\xi \tag{11.37}$$

where $H(x-\xi)$ is the Heaviside step function, which is unity if $x > \xi$ and zero if $x < \xi$. Then

$$\begin{aligned}
(\mathcal{L}a, b) &= \int_{-\infty}^{+\infty}\left\{\int_{-\infty}^x a(\xi)d\xi\right\}b(x)dx \\
&= \int_{-\infty}^{+\infty}\left\{\int_{-\infty}^{+\infty} H(x-\xi)a(\xi)d\xi\right\}b(x)dx \\
&= \int_{-\infty}^{+\infty} a(\xi)\left\{\int_{-\infty}^{+\infty} H(x-\xi)b(x)dx\right\}d\xi \\
&= \int_{-\infty}^{+\infty} a(\xi)\left\{\int_{\xi}^{+\infty} b(x)dx\right\}d\xi = (a, \mathcal{L}^\dagger b)
\end{aligned} \tag{11.38}$$

Hence, the adjoint of $\int_{-\infty}^x d\xi$ is $\int_x^{+\infty} d\xi$.

Another technique for computing an adjoint is to approximate the operator $\mathcal{L}$ as a matrix, transpose the matrix, and then to "read" the adjoint operator back by examining the matrix. In the cases of the integral and first derivative, the transposes are

$$\mathbf{L}^{AT} = \Delta x \begin{bmatrix} 1 & 1 & 1 & \cdots & 1 & 1 \\ 0 & 1 & 1 & 1 & \cdots & 1 \\ & & \ddots & & & \\ 0 & 0 & 0 & \cdots & 0 & 1 \end{bmatrix} \text{ and}$$

$$\mathbf{L}^{CT} = \frac{1}{\Delta x}\begin{bmatrix} 1 & -1 & 0 & \cdots & 0 & 0 \\ 0 & 1 & -1 & 0 & \cdots & 0 \\ & & \ddots & & & \\ 0 & 0 & 0 & \cdots & 1 & -1 \\ 0 & 0 & 0 & 0 & 0 & 1 \end{bmatrix} \tag{11.39}$$

$\mathbf{L}^{AT}$ represents the integral from progressively larger values of x to infinity and is equivalent to $\int_x^{+\infty} d\xi$. All but the last row of the $\mathbf{L}^{CT}$ represents $-d/dx$, and the last row is the boundary condition (which has moved from the top of $\mathbf{L}^C$ to the bottom of $\mathbf{L}^{CT}$). Hence, these results agree with those previously derived.

Adjoint operators have many of the properties of matrix transposes, including

$$\begin{aligned}
(\mathcal{L}^\dagger)^\dagger &= \mathcal{L} \text{ and } (\mathcal{L}^{-1})^\dagger = (\mathcal{L}^\dagger)^{-1} \\
(\mathcal{L}^A + \mathcal{L}^B)^\dagger &= (\mathcal{L}^B)^\dagger + (\mathcal{L}^A)^\dagger \text{ and } (\mathcal{L}^A\mathcal{L}^B)^\dagger = (\mathcal{L}^B)^\dagger(\mathcal{L}^A)^\dagger
\end{aligned} \tag{11.40}$$

11.8 THE FRÉCHET DERIVATIVE

Previously, we wrote the relationship between the model $m(x)$ and the data d_i as

$$d_i = \int G_i(x)m(x)\mathrm{d}x = (G_i, m) \tag{11.41}$$

But now, we redefine it in terms of perturbations around some reference model $m^{(0)}(x)$. We define $m(x) = m^{(0)}(x) + \delta m(x)$ where $m^{(0)}(x)$ is a reference function and $\delta m(x)$ is a perturbation. If we write $d_i = d_i^{(0)} + \delta d_i$ where $d_i^{(0)} = (G_i, m^{(0)})$ is the data predicted by the reference model, then

$$\delta d_i = \int G_i(x)\delta m(x)\mathrm{d}x = (G_i, \delta m) \tag{11.42}$$

This equation says that a perturbation $\delta m(x)$ in the model causes a perturbation δd_i in the data. This formulation is especially useful in linearized problems, since then the data kernel can be approximate, that is, giving results valid only when $\delta m(x)$ is small.

Eq. (11.42) is reminiscent of the standard derivative formula that we have used in linearized discrete problems:

$$\Delta d_i = \sum_{i=1}^{M} G_{ij}^{(0)} \Delta m_j \quad \text{with } G_{ij}^{(0)} = \frac{\partial d_i}{\partial m_j}\bigg|_{\mathbf{m}^{(0)}} \tag{11.43}$$

The only difference is that $\Delta \mathbf{m}$ is a vector, whereas $\delta m(x)$ is a continuous function. Thus, we can understand $G_i(x)$ in Eq. (11.42) an analog to a derivative. We might use the notation

$$G_i(x) = \frac{\delta d_i}{\delta m}\bigg|_{\mathbf{m}^{(0)}} \tag{11.44}$$

in which case it is called the *Fréchet derivative* of the datum d_i with respect to the model $m(x)$.

11.9 THE FRÉCHET DERIVATIVE OF ERROR

Fréchet derivatives of quantities other than the data are possible. One that is of particular usefulness is the Fréchet derivative of the error E with respect to the model, where the data $d(x)$ is taken to be a continuous variable.

$$E = (d^{\text{obs}} - d, d^{\text{obs}} - d) \quad \text{and} \quad \delta E = E - E^{(0)} = \left(\frac{\delta E}{\delta m}\bigg|_{\mathbf{m}^{(0)}}, \delta m\right) \tag{11.45}$$

Here the error is the continuous analog to the L_2 norm discrete error $E = (\mathbf{d}^{\text{obs}} - \mathbf{d})^{\mathrm{T}} (\mathbf{d}^{\text{obs}} - \mathbf{d})$. Suppose now that the data are related to the model via a linear operator, $d = \mathcal{L}m$. We compute the perturbation δE due to the perturbation δm as

$$\delta E = E - E^{(0)} = \left(d^{\mathrm{obs}} - d, d^{\mathrm{obs}} - d\right) - \left(d^{\mathrm{obs}} - d^{(0)}, d^{\mathrm{obs}} - d^{(0)}\right)$$

$$= -2\left(d, d^{\mathrm{obs}}\right) + (d, d) + 2\left(d^{(0)}, d^{\mathrm{obs}}\right) - \left(d^{(0)}, d^{(0)}\right)$$

$$= -2\left(d^{\mathrm{obs}} - d^{(0)}, d - d^{(0)}\right) + \left(d - d^{(0)}, d - d^{(0)}\right) \qquad (11.46)$$

$$= -2\left(d^{\mathrm{obs}} - d^{(0)}, \delta d\right) + (\delta d, \delta d) \approx -2\left(d^{\mathrm{obs}} - d^{(0)}, \delta d\right)$$

$$= -2\left(d^{\mathrm{obs}} - d^{(0)}, \mathcal{L}\delta m\right)$$

Note that we have ignored a term involving the second-order quantity $(\delta d)^2$. Using the adjoint operator $\mathcal{L}^\dagger$, we find

$$\delta E = \left(-2\mathcal{L}^\dagger\left(d^{\mathrm{obs}} - d^{(0)}\right), \delta m\right) \qquad (11.47)$$

which implies that the Fréchet derivative of the error E is

$$\left.\frac{\delta E}{\delta m}\right|_{\mathbf{m}^{(0)}} = -2\mathcal{L}^\dagger\left(d^{\mathrm{obs}} - d^{(0)}\right) \qquad (11.48)$$

This result can be useful when solving the inverse problem using a gradient method (see Section 9.8). As an example, consider the problem $d = \mathcal{L}m$ where

$$\mathcal{L}m(x) = a\frac{\mathrm{d}}{\mathrm{d}x}m(x) + b\int_{-\infty}^{x} m(x')\mathrm{d}x' \qquad (11.49)$$

where a and b are constants. Using the adjoint relationships derived in Section 11.7, we find

$$\mathcal{L}^\dagger d(x) = -a\frac{\mathrm{d}}{\mathrm{d}x}d(x) + b\int_{x}^{\infty} d(x')\mathrm{d}x' \qquad (11.50)$$

and hence the Fréchet derivative of the error E is

$$\left.\frac{\delta E}{\delta m}\right|_{\mathbf{m}^{(0)}} = 2a\frac{\mathrm{d}}{\mathrm{d}x}\left[d^{\mathrm{obs}}(x) - d^{(0)}(x)\right] - 2b\int_{x}^{\infty}\left[d^{\mathrm{obs}}(x') - d^{(0)}(x')\right]\mathrm{d}x' \qquad (11.51)$$

In order to use this result in a numerical scheme, one must discretize it; say by using voxels of width Δx and amplitude $\mathbf{m}$ for the model and $\mathbf{d}$ for the data, both of length $M = N$. Eq. (11.51) then yields the gradient $\partial E/m_i$, which can be used to minimize E via the gradient method (Fig. 11.5).

11.10 BACKPROJECTION

The formula for the Fréchet derivative of the error E (Eq. 11.48) is the continuous analog of the discrete gradient $\nabla E = -2\mathbf{G}^{\mathrm{T}}(\mathbf{d}^{\mathrm{obs}} - \mathbf{d}^{\mathrm{pre}}) = -2\mathbf{G}^{\mathrm{T}}(\mathbf{d}^{\mathrm{obs}} - \mathbf{G}\mathbf{m})$. In Section 3.4, the discrete gradient was set to zero, leading to the least-squares equation for the model parameters, $\mathbf{G}^{\mathrm{T}}\mathbf{G}\mathbf{m} = \mathbf{G}^{\mathrm{T}}\mathbf{d}^{\mathrm{obs}}$. A similar result is achieved in the continuous case:

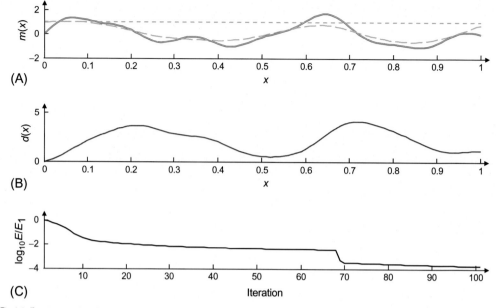

FIG. 11.5 Example of the solution of a continuous inverse problem using a gradient method to minimize the error E, where an adjoint method is used to compute ∇E. (A) A test function, $m^{\text{true}}(x)$ (*green*), trial function (*dotted green*) reconstructed function after 40 iterations (*dashed green*) and final reconstructed function after 15,890 iterations (*green*). (B) The data, $d(t)$, satisfies $d(t) = \mathcal{L}m(t)$, where $\mathcal{L}$ is the linear operator discussed in the text. (C) Error E is a function of iteration number, for the first 100 iterations. *MatLab* script gda11_03.

$$\left.\frac{\delta E}{\delta m}\right|_{\mathbf{m}^{(0)}} = 0 = -2\mathcal{L}^{\dagger}\left(d^{\text{obs}} - d\right) = -2\mathcal{L}^{\dagger}\left(d^{\text{obs}} - \mathcal{L}m\right) \quad \text{or}$$

$$\mathcal{L}^{\dagger}\mathcal{L}m = \mathcal{L}^{\dagger}d^{\text{obs}} \tag{11.52}$$

Now suppose that $\mathcal{J}$ represents the identity operator, that is, satisfying $m = \mathcal{J}m$ (in one dimension, this operator is $m(x) = \int_{-\infty}^{\infty} \delta(x - x')m(x')dx'$). Then we can write

$$\left(\mathcal{L}^{\dagger}\mathcal{L} + \mathcal{J} - \mathcal{J}\right)m = \mathcal{L}^{\dagger}d^{\text{obs}} \quad \text{or}$$

$$m = \mathcal{J}m = \mathcal{L}^{\dagger}d^{\text{obs}} - \left(\mathcal{L}^{\dagger}\mathcal{L} - \mathcal{J}\right)m \tag{11.53}$$

In the special case that $\mathcal{L}^{\dagger}\mathcal{L} = \mathcal{J}$ (i.e., $\mathcal{L}^{\dagger} = \mathcal{L}^{-1}$), the solution is very simple, $m = \mathcal{J}m = \mathcal{L}^{\dagger}d^{\text{obs}}$. However, even in other cases the equation is still useful, since it can be viewed as a recursion relating an old estimate of the model parameters $m^{(i)}$ to a new one, $m^{(i+1)}$

$$m^{(i+1)} = \mathcal{L}^{\dagger}d^{\text{obs}} - \left(\mathcal{L}^{\dagger}\mathcal{L} - \mathcal{J}\right)m^{(i)} \tag{11.54}$$

If the recursion is started with $m^{(0)} = 0$, the new estimate is

$$m^{(1)} = \mathcal{L}^{\dagger}d^{\text{obs}} \tag{11.55}$$

As an example, suppose that $\mathcal{L}$ is the indefinite integral, so that

$$d^{\mathrm{obs}}(x) = \mathcal{L}m(x) = \int_{-\infty}^{x} m(x')\mathrm{d}x' \qquad (11.56a)$$

and

$$m^{(1)}(x) = \mathcal{L}^{\dagger}d^{\mathrm{obs}}(x) = \int_{x}^{\infty} d^{\mathrm{obs}}(x')\mathrm{d}x' \qquad (11.56b)$$

Eq. (11.56b) may seem crazy, since an indefinite integral is inverted by taking a derivative, not another integral. Yet this result, while approximate, is nevertheless quite good in some cases, at least up to an overall multiplicative factor (Fig. 11.6).

Eq. (11.56a) might be considered an ultrasimplified one-dimensional tomography problem, relating acoustic slowness $m(x)$ to traveltime $d^{\mathrm{obs}}(x)$. Note that the ray associated with traveltime $d^{\mathrm{obs}}(x)$ starts at $-\infty$ and ends at x; that is, the traveltime at a point x depends only upon the slowness to the left of x. The formula for $m^{(1)}$ has a simple interpretation: the slowness at x is estimated by summing up all the traveltimes for rays ending at points $x' > x$; that is, summing up traveltimes only for those rays that sample the slowness at x. This process, which carries over to multidimensional tomography, is called *backprojection*. The inclusion, in a model parameter's own estimate, of just those data that are affected by it, is intuitively appealing.

Note that the accuracy of the backprojection depends upon the degree to which $\mathcal{L}^{\dagger}\mathcal{L} - \mathcal{J} = 0$ in Eq. (11.54). Some insight into this issue can be gained by examining the singular value decomposition of a discrete data kernel $\mathbf{G} = \mathbf{U}_p\mathbf{\Lambda}_p\mathbf{V}_p^{\mathrm{T}}$, which has transpose $\mathbf{G}^{\mathrm{T}} = \mathbf{V}_p\mathbf{\Lambda}_p\mathbf{U}_p^{\mathrm{T}}$ and generalized inverse $\mathbf{G}^{-g} = \mathbf{V}_p\mathbf{\Lambda}_p^{-1}\mathbf{U}_p^{\mathrm{T}}$. Clearly, the accuracy of the approximation

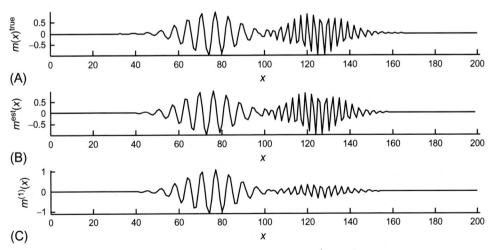

FIG. 11.6   (A) True one-dimensional model $m^{\mathrm{true}}(x)$. The data satisfy $d^{\mathrm{obs}} = \mathcal{L}m^{\mathrm{true}}$, where $\mathcal{L}$ is the indefinite integral. (B) Estimated model, using $m^{\mathrm{est}} = \mathcal{L}^{-1}d^{\mathrm{obs}}$ where $\mathcal{L}^{-1}$ is the first derivative. Note that $m^{\mathrm{est}} = m^{\mathrm{true}}$. (C) Backprojected model $m^{(1)} = \mathcal{L}^{\dagger}d^{\mathrm{obs}}$, where $\mathcal{L}^{\dagger}$ is the adjoint of $\mathcal{L}$. Note that, up to an overall multiplicative factor, $m^{(1)} \approx m^{\mathrm{true}}$. *MatLab* script gda11_04.

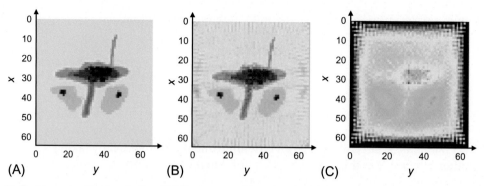

FIG. 11.7 Example of backprojection. (A) The true two-dimensional model, for which traveltimes associated with a dense and well-distributed set of rays are measured. (B) The estimated model, using damped least squares. (C) The estimated model, using backprojection. *MatLab* script gda11_05.

$\mathbf{G}^{-g} = \mathbf{G}^{T}$ will depend upon the degree to which $\mathbf{\Lambda}_p = \mathbf{\Lambda}_p^{-1}$, that is, to the degree to which the singular values have unit amplitude. The correspondence might be improved by scaling the rows of $\mathbf{G}$ and corresponding elements of $\mathbf{d}^{obs}$ by carefully chosen constants c_i: $G_{ij} \rightarrow c_i G_{ij}$ and $d_i \rightarrow c_i d_i$ so that the singular values are of order unity. This analysis suggests a similar scaling of the continuous problem, $\mathcal{L} \rightarrow c(x)\mathcal{L}$ and $d^{obs}(x) \rightarrow c(x)d^{obs}(x)$, where $c(x)$ is some function. In tomography, a commonly used scaling is $c(x) = 1/L(x)$, where $L(x)$ is the length of ray x. The transformed data d^{obs}/L represents the average slowness along the ray. The backprojection process sums up the average slowness of all rays that interact with the model parameter at point x. This is a bit counterintuitive; one might expect that averaging the averages, as contrasted to summing the averages, would be more appropriate. Remarkably, the use of the summation introduces only long-wavelength errors into the image. An example of two-dimensional back projection is shown in Fig. 11.7.

11.11 FRÉCHET DERIVATIVES INVOLVING A DIFFERENTIAL EQUATION

Eq. (11.42) links model parameters directly to the data. Some inverse problems are better analyzed when the link is indirect, through a field, $u(x)$. The model parameters are linked to the field, and the field to the data. Consider, for example, a problem in ocean circulation, where the model parameters $m(x)$ represent the force of the wind, the field $u(x)$ represents the velocity of the ocean water, and the data d_i represent the volume of water transported from one ocean to another. Wind forcing is linked to water velocity through a fluid-mechanical differential equation, and water transport is related to water velocity through an integral of the velocity across the ocean-ocean boundary. Such a relationship has the form

$$\mathcal{L}u(x) = m(x) \quad \text{and} \quad \mathcal{L}\delta u(x) = \delta m(x) \tag{11.57}$$

$$d_i = (h_i(x), u(x)) \quad \text{and} \quad \delta d_i = (h_i(x), \delta u(x)) \tag{11.58}$$

Here, Eq. (11.57) is a differential equation with known boundary conditions and Eq. (11.58) is an inner product involving a known function $h_i(t)$. Symbolically, we can write the solution of the differential equation as

$$u(x) = \int F(x, \xi)m(\xi)d\xi = \mathcal{L}^{-1}m(x)$$

$$\delta u(x) = \int F(x, \xi)\delta m(\xi)d\xi = \mathcal{L}^{-1}\delta m(x)$$

(11.59)

where $F(x,\xi)$ is the Green function. Note that $\mathcal{L}^{-1}$ is a linear integral operator, whereas $\mathcal{L}$ is a linear differential operator. We now combine Eqs. (11.57), (11.58):

$$\delta d_i = (h_i, \delta u) = \left(h_i, \mathcal{L}^{-1}\delta m\right) = \left(\left(\mathcal{L}^{-1}\right)^{\dagger}h_i, \delta m\right) = \left(\left(\mathcal{L}^{\dagger}\right)^{-1}h_i, \delta m\right)$$

(11.60)

Comparing to Eq. (11.42), we find

$$G_i(x) = \left(\mathcal{L}^{\dagger}\right)^{-1}h_i(x) \quad \text{or} \quad \mathcal{L}^{\dagger}G_i(x) = h_i(x)$$

(11.61)

Thus, the scalar field satisfies the equation $\mathcal{L}u(x) = m(x)$ and the data kernel satisfies the adjunct equation $\mathcal{L}^{\dagger}G_i(x) = h_i(x)$. In most applications, the scalar field $u(t)$ must be computed by numerical solution of its differential equation. Thus, the computational machinery is typically already in place to solve the adjoint differential equation and thus to construct the data kernel.

An important special case is when the data is the field $u(x)$ itself, that is, $d_i = u(x_i)$. This choice implies that the weighting function in Eq. (11.58) is a Dirac delta function, $h_i(x) = \delta(x - x_i)$ and that the data kernel is the Green function, say $Q(x,x_i)$, to the adjoint equation. Then

$$d_i = (Q(x, x_i), m(x)) \quad \text{with} \quad \mathcal{L}^{\dagger}Q(x, x_i) = \delta(x - x_i)$$

(11.62)

As an example, we consider an object that is being heated by a flame. We will use time t instead of position x in this example; the object is presumed to be of a spatially uniform temperature $u(t)$ that varies with time, t. Suppose that temperature obeys the simple Newtonian heat flow equation

$$\mathcal{L}u(t) = \left\{\frac{d}{dt} + c\right\}u(t) = m(t)$$

(11.63)

where the function $m(t)$ represents the heating and c is a thermal constant. This equation implies that in the absence of heating, the temperature of the object decays away toward zero at a rate determined by c. We assume that the heating is restricted to some finite time interval, so that temperature satisfies the boundary condition $u(t \to -\infty) = 0$.

The solution to the heat flow equation can be constructed from its Green function $F(t,\tau)$, which represents the temperature at time t due to an impulse of heat at time τ. It solves the equation

$$\left\{\frac{d}{dt} + c\right\}F(t, \tau) = \delta(t - \tau)$$

(11.64)

The solution to this equation can be shown to be

$$F(t, \tau) = H(t - \tau)\exp\{-c(t - \tau)\}$$

(11.65)

Here the function $H(t-\tau)$ is the Heaviside step function, which is zero when $t < \tau$ and unity when $t > \tau$. This result can be verified by first noting that it satisfies the boundary condition and then by checking, through direct differentiation, that it solves the differential equation. The temperature of the object is zero before the heat pulse is applied, jumps to a maximum at the moment of application, and then exponentially decays back to zero at a rate determined by the constant, c (Fig. 11.8A). A hypothetical heating function $m(x)$ and resulting temperature $u(t)$ are shown in Fig. 11.8B and C, respectively.

We now need to couple the temperature function $u(t)$ to observations, in order to build a complete forward problem. Suppose that a chemical reaction occurs within the object, at a rate that is proportional to temperature. We observe the amount of chemical product $d_i = P(t_i)$, which is given by the integral

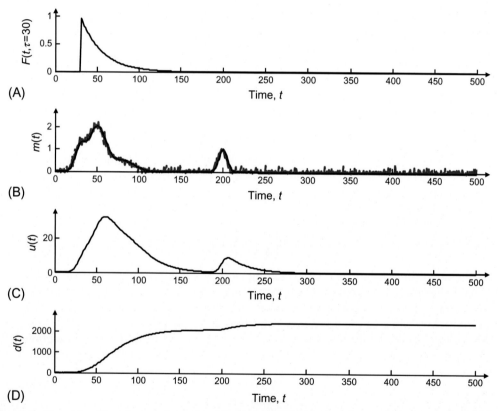

FIG. 11.8 Example of the solution of a continuous inverse problem involving a differential equation. (A) Green function $F(t,\tau)$ for $\tau=30$. (B) True *(black)* and estimated *(red)* heat production function $m(t)$. (C) Temperature $u(t)$, which solves $\mathcal{L}u=m$. (D) Observed data $d(t)$, which is proportional to the integral of $u(t)$. *MatLab* script gda11_06.

$$d_i = P(t_i) = b \int_0^{t_i} u(t)dt = b \int_0^{\infty} H(t_i - t)u(t)dt = (bH(t_i - t), u(t)) \tag{11.66}$$

where b is the proportionality factor. Thus, $h_i(t) = bH(t_i - t)$, where H is the Heaviside step function. The adjoint equation is

$$\mathcal{L}^{\dagger} G_i(t) = \left\{ -\frac{d}{dt} + c \right\} G_i(t) = h_i(t) \tag{11.67}$$

since, as noted in Section 11.7, the adjoint of d/dt is $-d/dt$ and the adjunct of a constant c is itself. The Green function of the adjoint equation can be shown to be

$$Q(t, \tau) = H(\tau - t)\exp\{ +c(t - \tau) \} \tag{11.68}$$

Note that $Q(t, \tau)$ is just a *time-reversed* version of $F(t, \tau)$, a relationship that arises because the two differential equations differ only by the sign of t. The data kernel $G_i(t)$ is given by the Green function integral

$$
\begin{aligned}
G_i(t) &= \int_0^{\infty} Q(t, \tau)h_i(\tau)d\tau \\
&= \int_0^{\infty} H(\tau - t)\exp\{ c(t - \tau) \}bH(t_i - \tau)d\tau \\
&= b \int_0^{\infty} H(\tau - t)H(t_i - \tau)\exp\{ c(t - \tau) \}d\tau \\
&= b \int_t^{t_i} \exp\{ -c(\tau - t) \}d\tau
\end{aligned}
\tag{11.69}
$$

The integral is nonzero only in the case where $t_i > t$:

$$G_i(t) = \begin{cases} 0 & t_i \le t \\ -\dfrac{b}{c}[\exp\{ -c(t_i - t) \} - 1] & t_i > t \end{cases} \tag{11.70}$$

The data kernel $G_i(t)$ (Fig. 11.9) quantifies the effect of heat applied at time t on an observation made at time t_i. It being zero for times $t > t_i$ is a manifestation of causality; heat applied in the future of a given observation cannot affect it.

The problem that we have been discussing is completely linear; neither the differential equation relating $m(t)$ to $u(t)$ nor the equation relating $u(t)$ to d_i contains any approximations.

FIG. 11.9 Data kernel $G_i(t)$ for the continuous inverse problem involving a differential equation. See text for further discussion. *MatLab* script gda11_06.

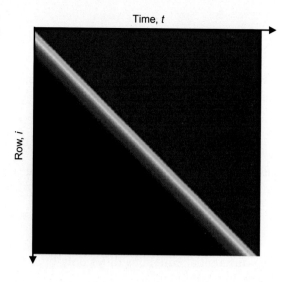

The data kernel is therefore accurate for perturbations of any size and hence holds for $m(t)$ and d_i, as well as for the perturbations $\delta m(t)$ and δd_i:

$$d_i = (G_i(t), m(t)) \tag{11.71}$$

We solve the inverse problem—determining the heating through observations of the chemical product—by first discretizing all quantities with a time increment Δt

$$m(t) \rightarrow m_i = m(i\Delta t) \quad \text{and} \quad G_i(t) \rightarrow G_{ij} = \Delta t G_i(j\Delta t) \tag{11.72}$$

so that

$$d_i = (G_i(t), m(t)) \rightarrow \mathbf{d} = \mathbf{Gm} \tag{11.73}$$

Note that a factor of Δt has been included in the definition of $\mathbf{G}$ to account for the one that arises when the inner product is approximated by its Riemann sum. We assume that data is measured at all times, and then solve Eq. (11.73) by damped least squares (Fig. 11.8B). Some noise amplification occurs, owing to the very smooth data kernel. Rapid fluctuations in heating $m(t)$ have very little effect on the temperature $u(t)$, owing to the diffusive nature of heat, and thus are poorly constrained by the data. Adding smoothness constraints would lead to a better solution.

11.12 DERIVATIVE WITH RESPECT TO A PARAMETER IN A DIFFERENTIAL EQUATION

The previous section was concerned with the effect on the data of a perturbation in *forcing* (meaning the function on the right hand side of a differential equation). This section considers the effect on the data of a perturbation of a *parameter* in the differential operator $\mathcal{L}$.

As an example, suppose that the constant c in Eq. (11.63) is replaced with a function $c_0 + mc_1(t)$, where c_0 is a known constant, $c_1(t)$ is a known function, and m is an unknown parameter:

$$\left\{\frac{d}{dt} + [c_0 + mc_1(t)]\right\}u(t) = f(t) \tag{11.74}$$

The forcing $f(t)$ is now considered known. The partial derivative $\partial u/\partial m$ quantifies how a small change Δm in the model parameter m away from a reference value m_0 leads to a small perturbation δu of the solution away from its corresponding value u^0:

$$\delta u = \frac{\partial u}{\partial m}\bigg|_{m_0} \Delta m \tag{11.75}$$

We derive an expression for this partial derivative, starting from the general case of the differential equation:

$$\mathcal{L}(m)u(t) = f(t) \tag{11.76}$$

First, the field is represented as $u = u^0 + \delta u$, where the reference field u^0 solves the differential equation for $m = m_0$. Second, the operator $\mathcal{L}(m)$ is expanded in a Taylor series around $m = m_0$ and terms higher than first-order are discarded:

$$\mathcal{L}(m) \approx \mathcal{L}(m_0) + \frac{\partial\mathcal{L}}{\partial m}\bigg|_{m_0} \Delta m \tag{11.77}$$

Third, these two approximations are inserted into Eq. (11.76) and only first-order terms are retained:

$$f \approx \left(\mathcal{L}(m_0) + \frac{\partial\mathcal{L}}{\partial m}\bigg|_{m_0} \Delta m\right)(u^0 + \delta u) \approx \mathcal{L}(m_0)u^0 + \frac{\partial\mathcal{L}}{\partial m}\bigg|_{m_0} u^0 \Delta m + \mathcal{L}(m_0)\delta u \tag{11.78}$$

Fourth, the reference equation $\mathcal{L}(m_0)u^0 = f$ is subtracted out and the remaining terms are rearranged:

$$\mathcal{L}(m_0)\delta u = -\frac{\partial\mathcal{L}}{\partial m}\bigg|_{m_0} u^0 \Delta m \tag{11.79a}$$

and

$$\delta u = \left(-\mathcal{L}^{-1}(m_0)\frac{\partial\mathcal{L}}{\partial m}\bigg|_{m_0} u^0\right)\Delta m \tag{11.79b}$$

Eq. (11.79a) is known as the *Born approximation*. It indicates that the field perturbation δu solves a differential equation with a *source term* that depends on the reference field u^0 and the perturbation δm of the model parameter. The reference field u^0 *interacts* with Δm to produce a *scattered* field δu. Comparison of Eqs. (11.75), (11.79b) indicates that the quantity in parentheses is the desired partial derivative:

$$\frac{\partial u}{\partial m}\bigg|_{m_0} = -\mathcal{L}^{-1}(m_0)\frac{\partial\mathcal{L}}{\partial m}\bigg|_{m_0} u^0 \tag{11.80}$$

While the derivative $\partial\mathcal{L}/\partial m$ of an operator $\mathcal{L}$ might at first seem like a mysterious quantity, it is calculated simply by applying the $\partial/\partial m$ to functions of m and treating all other components of the operator as constants. For example, the operator in Eq. (11.74) has derivative:

$$\frac{\partial\mathcal{L}}{\partial m}\bigg|_{m_0} = \frac{\partial}{\partial m}\left\{\frac{d}{dt} + c_0 + mc_1(t)\right\}\bigg|_{m_0} = [0+0+c_1(t)]|_{m_0} = c_1(t) \tag{11.81}$$

We now return to the case examined in the previous section where the data d_i are related to the field $u_i(t)$ through an inner product with a known functions $h_i(t)$; that is $d_i = (h_i, u_i)$. Following Eq. (11.58), we have:

$$\delta d_i = (h_i, \delta u_i) = -\left(h_i, \mathcal{L}^{-1}(m_0)\frac{\partial\mathcal{L}}{\partial m}\bigg|_{m_0} u^0\right)\Delta m \tag{11.82}$$

Eq. (11.82) is of the form $\delta d_i = G_i \Delta m$, where G_i is the data kernel. We now use adjoint methods to manipulate the inner product:

$$G_i(m_0) = \frac{\partial d_i}{\partial m}\bigg|_{m_0} = -\left(h_i, \mathcal{L}^{-1}(m_0)\frac{\partial\mathcal{L}}{\partial m}\bigg|_{m_0} u^0\right) = -\left(\mathcal{L}^{-1\dagger}(m_0)h_i, \frac{\partial\mathcal{L}}{\partial m}\bigg|_{m_0} u^0\right) = (\lambda_i, \xi)$$

with $\lambda_i(t) \equiv \mathcal{L}^{-1\dagger}(m_0)h_i(t)$ or $\mathcal{L}^{\dagger}(m_0)\lambda_i(t) = h_i(t)$ $\qquad(11.83)$

and $\xi(t) \equiv -\dfrac{\partial\mathcal{L}}{\partial m}\bigg|_{m_0} u^0(t)$

Here we have introduced an adjoint field $\lambda_i(t)$ that solves the adjoint differential equation with a source term $h_i(t)$. The data kernel is calculated in the following four steps:

(1) Solve the unperturbed equation $\mathcal{L}(m_0)u^0(t) = f(t)$ for the reference field $u^0(t)$;
(2) Form the quantity $\xi(t) = -\partial\mathcal{L}/\partial m|_{m_0}u^0(t)$;
(3) Solve the adjoint equation $\mathcal{L}^{\dagger}(m_0)\lambda_i(t) = h_i(t)$ for the adjoint field $\lambda_i(t)$; and
(4) Take the inner product of $\lambda_i(t)$ and $\xi(t)$, which yields the data kernel $G_i(m_0)$

Most of the work is in solving the two differential equations. The other two steps require minimal effort.

As an example, consider the temperature equation (Eq. 11.74) when the forcing $f(t) = \delta(t)$ is a spike of heat at time zero and the perturbation in the material parameter $c_1(t) = \delta(t - t_0)$ is a spike at time t_0. We calculate the data kernel $G_i(m_0)$ when the data are integrals of the field (as in Eq. 11.66), so that $h_i = bH(t_i - t)$, and for the choice $m_0 = 0$. The reference field is given by Eq. (11.65), with $\tau = 0$ and with c replaced by c_0:

$$u^0(t) = H(t)\exp(-c_0 t) \tag{11.84}$$

where $H(t)$ is the Heaviside step function. The function $\xi(t)$ is then:

$$\xi(t) = -H(t)\exp(-c_0 t)\delta(t - t_0) \tag{11.85}$$

The Green function $Q(t - \tau)$ of the adjoint equation is given by Eq. (11.68), with c replaced by c_0. The adjoint field is given by the integral:

$$\lambda_i(t) = \int_{-\infty}^{+\infty} Q(t-\tau)h_i(\tau)\,\mathrm{d}\tau$$

$$= \int_{-\infty}^{+\infty} H(\tau-t)\exp\{c_0(t-\tau)\}bH(t_i-\tau)\,\mathrm{d}\tau$$

$$= b\,\exp(c_0 t)\int_t^{t_i}\exp\{-c_0\tau\}\,\mathrm{d}\tau \tag{11.86}$$

$$= -\frac{b}{c_0}H(t_i-t)\exp(c_0 t)[\exp(-c_0 t_i) - \exp\{-c_0 t\}]$$

$$= \frac{b}{c_0}H(t_i-t)[1 - \exp\{-c_0(t_i-t)\}]$$

As in the previous section, the adjoint field is "backward in time." The data kernel is then:

$$G_i(m=0) = (\lambda_i,\,\xi) = \int_{-\infty}^{+\infty}\lambda_i(t)\xi(t)\,\mathrm{d}t$$

$$= -\frac{b}{c_0}\int_{-\infty}^{+\infty}H(t_i-t)[1-\exp\{-c_0(t_i-t)\}]H(t)\exp(-c_0 t)\delta(t-t_0)\,\mathrm{d}t \tag{11.87}$$

$$= -\frac{b}{c_0}H(t_i-t_0)H(t_0)[1-\exp\{-c_0(t_i-t_0)\}]\exp(-c_0 t_0)$$

The data kernel's dependence on $H(t_0)$ implies it is zero when the spike in the material parameter occurs before the source time (at time zero). Its dependence on $H(t_i-t_0)$ implies that is zero when the spike in the material parameter occurs after the upper limit t_i of the data integral. This data kernel is illustrated in Fig. 11.10.

While the derivation in Eq. (11.86) was for an operator that depended on a single model parameter, it can be adapted trivially to the general case of an operator $\mathcal{L}(\mathbf{m})$ that depends on a vector $\mathbf{m}$ of model parameters. One merely considers each model parameter m_j separately, so that Eq. (11.83) becomes:

$$G_{ij}(\mathbf{m}_0) = \frac{\partial d_i}{\partial m_j}\bigg|_{\mathbf{m}_0} = \left(\lambda_i,\,\xi_j\right)$$

$$\text{with } \mathcal{L}(\mathbf{m}_0)u^0(t) = f(t) \text{ and } \mathcal{L}^\dagger(\mathbf{m}_0)\lambda_i(t) = h_i(t) \text{ and } \xi_j(t) = -\frac{\partial\mathcal{L}}{\partial m_j}\bigg|_{\mathbf{m}_0}u^0(t) \tag{11.88}$$

Here $\mathbf{m}_0$ is the reference value of the model parameter vector. This data kernel can be used in a linearized least squares inversion to estimate the value of $\mathbf{m}$ that best matches the predicted data d_i to observations. However, it must be recalculated during each iteration of the algorithm, since $\mathbf{m}_0$ is updated before every iteration.

Also of interest is the derivative $\partial E/\partial m|_{\mathbf{m}_0}$ of the total error $E=(e,e)$, where $e(t)=u^{obs}(t)-u(t)$ is the mismatch between the observed and predicted field. It is derived following steps similar to those used for the data kernel:

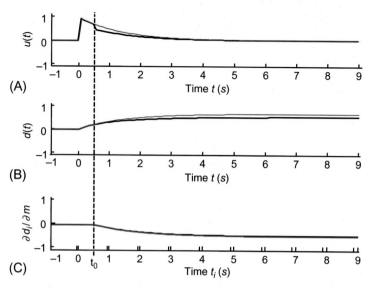

FIG. 11.10   Data kernel for a perturbation in the material parameter in the heat flow problem. (A) Reference field $u_0(t)$ *(red curve)* and perturbed field $u(t) = u_0(t) + \delta u(t)$ *(red curve)* for a spike in the material parameter $c_1(t) = m\delta(t - t_0)$ at time t_0 *(dashed line)*. (B) Corresponding unperturbed *(black)* and perturbed data *(red)*. (C) Data kernel calculated by finite differences *(black curve)* and adjoint methods *(red curve)* match exactly. *MatLab* script gda11_07.

$$\left.\frac{\partial E}{\partial m}\right|_{m_0} = 2\left(e, \left.\frac{de}{dm}\right|_{m_0}\right) = -2\left(e, \left.\frac{du}{dm}\right|_{m_0}\right)$$

$$= 2\left(e, \mathcal{L}^{-1}(m_0)\left.\frac{\partial \mathcal{L}}{\partial m}\right|_{m_0} u^0\right) = \left(\mathcal{L}^{-1\dagger}(m_0)e, 2\left.\frac{\partial \mathcal{L}}{\partial m}\right|_{m_0} u^0\right) = (\lambda, \xi) \tag{11.89}$$

with $\lambda(t) \equiv \mathcal{L}^{-1\dagger}(m_0)e(t)$ or $\mathcal{L}^{\dagger}(m_0)\lambda(t) = e(t)$

and $\xi(t) \equiv 2\left.\frac{\partial \mathcal{L}}{\partial m}\right|_{m_0} u^0(t)$

As with the data kernel, the error derivative can be adapted trivially to the general case of a vector $\mathbf{m}$ of model parameters:

$$\left.\frac{\partial E}{\partial m_j}\right|_{\mathbf{m}_0} = (\lambda, \xi_j)$$

with $\mathcal{L}(\mathbf{m}_0)u^0(t) = f(t)$ and $\mathcal{L}^{\dagger}(\mathbf{m}_0)\lambda(t) = e(t)$ and $\xi_j(t) = 2\left.\frac{\partial \mathcal{L}}{\partial m_j}\right|_{\mathbf{m}_0} u^0(t)$ $\tag{11.90}$

Calculation of the error derivative requires a four-step process similar to the one used for the data kernel. The error derivative can be used in a gradient descent algorithm to estimate the value of $\mathbf{m}$ that best matches the predicted field $u(t)$ to observations. However, it must be recalculated during each iteration of the algorithm, since $\mathbf{m}_0$ is updated before every iteration.

Adjoint techniques have had extensive application in seismology (Dahlen et al., 2000; Tromp et al., 2005) and in atmospheric and ocean science, where they are associated with the term *data assimilation* (Hall and Cacuci, 1983; Moore et al., 2004).

11.13 PROBLEMS

11.1 This inverse problem is due to Robert Parker. Consider a radially stratified sphere of radius $R=1$ with unknown density $\rho(r)$. Its mass $M=4\pi \int \rho(r)r^2dr$ and its moment of inertia $I = \left(\frac{8\pi}{3}\right) \int \rho(r)r^4dr$ are measured. What is the best averaging kernel that can be designed that is centered about $\frac{1}{2}R$? Since all the integrals involve only rational functions and all the matrices are 2×2, the problem can be solved analytically. Do as much of the problem as you can, analytically; do the parts that you cannot, numerically.

11.2 Suppose that a function $m(x)$ is discretized in each of two ways: (A) by dividing it into M rectangles, each of height m_i and width Δx; and (B) by representing it as a sum of M overlapping Gaussians, spaced at regular intervals Δx, and with coefficients

$$m(x) = \sum_{i=1}^{M} m'_i \frac{\Delta x}{(2\pi)\sigma} \exp\left\{-\frac{(x-i\Delta x)^2}{2\sigma^2}\right\}$$

Here, σ^2 is a prescribed variance. Discuss the advantages and disadvantages of each representation. Include the respective effect of damping in the two cases.

11.3 How sensitive are the results of tomography to noise in the data (u, θ)? Run experiments with the Fourier slice script. Try two different types of noise: (A) uncorrelated random noise drawn from a Gaussian distribution and (B) a few large outliers. Comment upon the results.

11.4 Modify the example in Section 11.9 to use a truncated Fourier sine series representation of the model:

$$m^{(0)}(x) = \sum_{n=1}^{K} m_n^{(0)} \sin\left(\frac{n\pi x}{L}\right) \quad \text{and} \quad \delta m(x) = \sum_{n=1}^{K} \delta m_n^{(0)} \sin\left(\frac{n\pi x}{L}\right)$$

where m_n are scalar coefficients, $K=15$ and $0<x<L$. You will need to use the chain rule

$$\left.\frac{\delta E}{\delta m_n}\right|_{\mathbf{m}^{(0)}} = \left(\left.\frac{\delta E}{\delta m}\right|_{\mathbf{m}^{(0)}}, \left.\frac{\delta m}{\delta m_n}\right|_{\mathbf{m}^{(0)}}\right)$$

11.5 Suppose that the Green function $F(t,\tau)$ of a particular differential equation $\mathcal{L}^F F(t,\tau) = \delta(t-\tau)$ depends only upon time differences; that is, $F(t,\tau) = F(t-\tau)$. Denote the Green function integral as the linear operator $(\mathcal{L}^F)^{-1}$. (A) Use the formula for the inner product to derive $(\mathcal{L}^F)^{-1\dagger}$. Call the function within this operator F'. (B) Suppose the differential equation $\mathcal{L}^{Q\dagger}Q(t,\tau)=\delta(t-\tau)$ has Green function $Q(t,\tau)$. Denote the Green function integral as the linear operator $(\mathcal{L}^{Q\dagger})^{-1}$. By comparing $(\mathcal{L}^F)^{-1\dagger}$ and $(\mathcal{L}^{Q\dagger})^{-1}$, establish the relationship between F' and Q. Explain your results in words.

11.6 The file `prob1106_data.txt`, which was generated by `gda11_08.m`, contains three columns, corresponding to $\mathbf{t}$, $\delta\mathbf{m}^{true}$, and $\mathbf{d}^{true}$ (all of length N) in the temperature problem discussed in Section 11.12. The parameters associated with this problem include a source time of zero, $c_0=0.7$, $b=0.4$, $\mathbf{m}_0=0$, and:

$$c_1(t) = \sum_{i=1}^{M} \delta m_i^{true}\delta(t-t_i)$$

where $M=N$. Construct a set of synthetic data $\mathbf{d}^{obs}$ by adding random noise to $\mathbf{d}^{true}$ and then use the data kernel in Eq. (11.87) to invert for $\delta\mathbf{m}^{est}$. Compare your result with $\delta\mathbf{m}^{true}$. You will need to add prior information of smallness or smoothness to the inversion, in order to prevent amplification of high-frequency noise.

References

Backus, G.E., Gilbert, J.F., 1968. The resolving power of gross earth data. Geophys. J. Roy. Astron. Soc. 16, 169–205.

Dahlen, F.A., Hung, S.-H., Nolet, G., 2000. Fréchet kernels for finite frequency travel times—I. Theory. Geophys. J. Int. 141, 157–174.

Hall, M.C.G., Cacuci, D.G., 1983. Physical interpretation of the adjoint functions for sensitivity analysis of atmospheric models. J. Atmos. Sci. 40, 2537–2546.

Lanczos, C., 1961. Linear Differential Operators. Van Nostrand-Reinhold, Princeton, NJ.

Moore, A.M., Arango, H.G., Di Lorenzo, E., Cornuelle, B.D., Miller, A.J., Neilson, D.J., 2004. A comprehensive ocean prediction and analysis system based on the tangent linear and adjoint of a regional ocean model. Ocean Model. 7, 227–258.

Tromp, J., Tape, C., Liu, Q., 2005. Seismic tomography, adjoint methods, time reversal and banana-doughnut kernels. Geophys. J. Int. 160, 195–216.

12

Sample Inverse Problems

12.1 AN IMAGE ENHANCEMENT PROBLEM

Suppose that a camera moves slightly during the exposure of an image, so that the picture is blurred. Also suppose that the amount and direction of motion are known. Can the image be "unblurred?"

The optical sensor in the camera consists of rows and columns of light-sensitive elements that measure the total amount of light received during the exposure. For simplicity, we shall consider that the camera moves parallel to a row of pixels. The data d_i are a set of numbers that represent the amount of light recorded at each of N pixels. Because the scene's brightness varies continuously, this is properly a problem in continuous inverse theory. We shall discretize it, however, by assuming that the scene can be adequately approximated by a row of small square elements, each with a constant brightness. These elements form M model parameters m_i. Since the camera's motion is known, it is possible to calculate each scene

Geophysical Data Analysis
https://doi.org/10.1016/B978-0-12-813555-6.00012-5

© 2018 Elsevier Inc. All rights reserved.

element's relative contribution to the light recorded at a given camera pixel. For instance, if the camera moves through three scene elements during the exposure, then each camera element records the average of three neighboring scene brightness

$$d_i = \frac{1}{3}[m_{i-1} + m_i + m_{i+1}] \tag{12.1}$$

Note that this is a linear equation and can be written in the form $\mathbf{Gm} = \mathbf{d}$, where

$$\mathbf{G} = \frac{1}{3}\begin{bmatrix} 1 & 1 & 1 & 0 & 0 & \cdots & 0 \\ 0 & 1 & 1 & 1 & 0 & \cdots & 0 \\ & & & \ddots & & & \\ 0 & \cdots & 0 & 0 & 1 & 1 & 1 \end{bmatrix} \tag{12.2}$$

In general, there will be several more model parameters than data, so the problem will be underdetermined. In this case $M = N + 2$, so there will be at least two null vectors. In fact, the problem is purely underdetermined, and there are only two null vectors, which can be identified by inspection as

$$\begin{aligned} \mathbf{m}^{(1)\text{null}} &= [1 \quad 0 \quad -1 \quad 1 \quad 0 \quad -1 \quad \cdots \quad 1 \quad 0 \quad -1]^{\mathrm{T}} \\ \mathbf{m}^{(2)\text{null}} &= [0 \quad 1 \quad -1 \quad 0 \quad 1 \quad -1 \quad \cdots \quad 0 \quad 1 \quad -1]^{\mathrm{T}} \end{aligned} \tag{12.3}$$

These null vectors have rapidly fluctuating elements, which indicates that at best only the longer wavelength features can be recovered. To find a solution to the inverse problem, we must add prior information. We shall use a simplicity constraint and find the scene of shortest length. If the data are normalized so that zero represents gray (with white negative and black positive), then this procedure in effect finds the scene of least contrast that fits the data. We therefore estimate the solution with the minimum length generalized inverse as

$$\mathbf{m}^{\text{est}} = \mathbf{G}^{\mathrm{T}}[\mathbf{GG}^{\mathrm{T}}]^{-1}\mathbf{d}^{\text{obs}} \tag{12.4}$$

As an example, we shall solve the problem for the data kernel given earlier, for a blur width of 100 and with an image $M = 2000$ pixels wide. We first select a true scene (Fig. 12.1A) and blur it by multiplication with the data kernel to produce synthetic data (Fig. 12.1B). Note that the blurred image is much smoother than the true scene. We now try to invert back for the true scene by premultiplying by the generalized inverse. The result (Fig. 12.1C) has not only correctly captured some of the sharp features in the original scene but also contains a short wavelength oscillation not present in the true scene. This error results from creating a reconstructed scene containing an incorrect combination of null vectors.

In *MatLab*, the inverse problem for each row of the image is solved separately, with Eq. (12.5) evaluated in a loop over rows:

```
epsilon=1.0e-6; % damping, just in case GGT is singular
GGT= G*G'+epsilon*speye(J,J);
for i= [1:I]dobsrow= dobs(i,:)';
mestrow= G'*(GGT\dobsrow);
mest(i,:)=mestrow';
end
```

(*MatLab* script gda12_01)

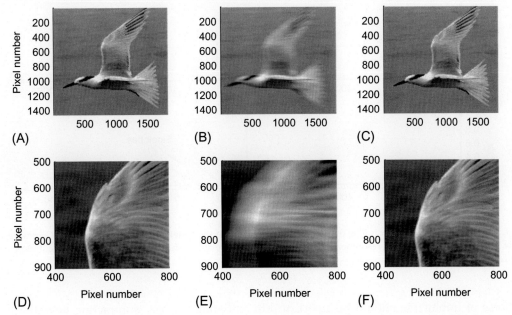

FIG. 12.1 Example of removing blur from image. (A) True image. (B) Image blurred with 100 pixel-wide boxcar filter. (C) Estimated image, unblurred using the minimum length generalized inverse. (D–F) Enlargement of portion of images (A–C). *MatLab* script gda12_01.

The matrix **G** is defined as sparse, and the matrix product [**GG**^{T}] is calculated by simple matrix multiplication.

We note that the matrix **G** is simple enough for the matrix product [**GG**^{T}] to be computed analytically

$$\mathbf{GG}^T = \frac{1}{9}\begin{bmatrix} 3 & 2 & 1 & 0 & 0 & 0 & \cdots & 0 \\ 2 & 3 & 2 & 1 & 0 & 0 & \cdots & 0 \\ 1 & 2 & 3 & 2 & 1 & 0 & \cdots & 0 \\ \vdots & \vdots & \vdots & \vdots & \vdots & \vdots & \ddots & \vdots \\ 0 & 0 & 0 & 0 & 0 & 1 & 2 & 3 \end{bmatrix} \qquad (12.5)$$

In this problem, which deals with moderate-sized matrices, the effort saved in using the analytic version over a numerical computation is negligible. In larger inverse problems, however, considerable saving can result from a careful analytical treatment of the structures of the various matrices.

Each row of the generalized inverse states how a particular model parameter is constructed from the data. We might expect that this blurred image problem would have a localized generalized inverse, meaning that each model parameter would be constructed mainly from a few neighboring data. By examining **G**$^{-g}$, we see that this is not the case (Fig. 12.2A). We also note that the process of blurring is an integrating process; it sums neighboring data. The

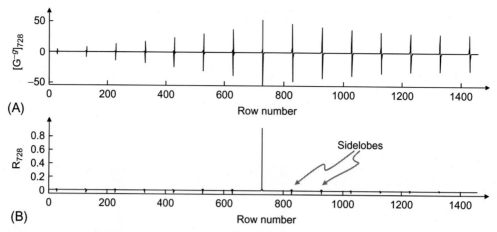

FIG. 12.2   (A) Central row (row 728) of the generalized inverse of the image deblurring problem. (B) Central row (row 728) of the corresponding resolution matrix. The sidelobes are about 6% of the amplitude of the central peak. *MatLab* script gda12_01.

process of unblurring should be a sort of differentiating process, subtracting neighboring data. The generalized inverse is, in fact, just that.

The resolution matrix for this problem is seen to be quite "spiky" (Fig. 12.2B); the diagonal elements are several orders of magnitude larger than the off-diagonal elements. On the other hand, the off-diagonal elements are all of uniform size, indicating that if the estimated model parameters are interpreted as localized averages, they are in fact not completely localized. It is interesting to note that the Backus-Gilbert inverse (not shown), which generally gives very localized resolution kernels, returns only the blurred image itself. The solution to this problem contains an unavoidable trade-off between the width of resolution and the presence of sidelobes.

12.2 DIGITAL FILTER DESIGN

Suppose that two signals $d(t)$ and $g(t)$ are known to be related by *convolution* with a filter $m(t)$:

$$d(t) = m(t) * g(t) = \int g(t - \tau) m(\tau) \, d\tau \tag{12.6}$$

where τ is a dummy integration variable. Can $m(t)$ be found if $g(t)$ and $d(t)$ are known?

Since the signals and filter are continuous functions, this is a problem in continuous inverse theory. We shall analyze it, however, by approximating the functions as *time series*. Each function will be represented by its value at a set of points spaced equally in time (with interval Δt). We shall assume that the signals are transient (have a definite beginning and end) so that $d(t)$ and $g(t)$ can be represented by time series of length N. Typically, the advantage of relating two signals by a filter is realized only when the filter length M is shorter than either signal, so $M < N$ is presumed. The convolution integral can then be approximated by the sum

$$d_i = \Delta t \sum_{j=1}^{M} g_{i-j+1} m_j \tag{12.7}$$

where $g_i = 0$, if $i < 1$ or $i > N$. This equation is linear in the unknown filter coefficients and can be written in the form $\mathbf{Gm} = \mathbf{d}$, where

$$\mathbf{G} = \Delta t \begin{bmatrix} g_1 & 0 & 0 & \cdots & 0 \\ g_2 & g_1 & 0 & \cdots & 0 \\ \vdots & \vdots & \vdots & \ddots & \vdots \\ g_N & g_{N-1} & g_{N-2} & \cdots & g_{N-M+1} \end{bmatrix} \tag{12.8}$$

The time series d_i is identified with the data and the filter m_i with the model parameters. The equation is therefore an overdetermined linear system for $M < N$ filter coefficients. For this problem to be consistent with the tenets of probability theory, however, $\mathbf{g}$ must be known exactly, while $\mathbf{d}$ must contain uncorrelated Gaussian noise of uniform variance.

Many approaches are available for solving this inverse problem. The simplest is to use the least squares equation $[\mathbf{G}^T\mathbf{G}]\mathbf{m}^{\text{est}} = \mathbf{G}^T\mathbf{d}$. This formulation is especially attractive because the matrices $[\mathbf{G}^T\mathbf{G}]$ and $\mathbf{G}^T\mathbf{d}$ can be computed analytically as

$$\mathbf{G}^T\mathbf{G} = (\Delta t)^2 \begin{bmatrix} \sum_{i=1}^{N} g_i^2 & \sum_{i=2}^{N} g_i g_{i-1} & \cdots \\ \sum_{i=2}^{N} g_i g_{i-1} & \sum_{i=1}^{N-1} g_i^2 & \cdots \\ \vdots & \vdots & \ddots \end{bmatrix} \quad \text{and} \quad \mathbf{G}^T\mathbf{d} = \Delta t \begin{bmatrix} \sum_{i=1}^{N} d_i g_i \\ \sum_{i=2}^{N} d_i g_{i-1} \\ \vdots \end{bmatrix} \tag{12.9}$$

Furthermore, the summations along each diagonal of $\mathbf{G}^T\mathbf{G}$ usually can be approximated adequately as all having the same limits, so that the resulting matrix is *Toeplitz* (meaning that it has constant diagonals). Then $\mathbf{G}^T\mathbf{G}$ matrix contains the *autocorrelation* of $\mathbf{g}$ (denoted $\mathbf{g} \star \mathbf{g}$) and $\mathbf{G}^T\mathbf{d}$ the *cross-correlation* of $\mathbf{g}$ with $\mathbf{d}$ (denoted $\mathbf{g} \star \mathbf{d}$)

$$[\mathbf{G}^T\mathbf{G}]_{ij} = (\Delta t)^2 [\mathbf{g} \star \mathbf{g}]_{|i-j|+1} \quad \text{and} \quad [\mathbf{G}^T\mathbf{d}]_i = \Delta t [\mathbf{g} \star \mathbf{d}]_i,$$
$$\text{where } [\mathbf{a} \star \mathbf{b}]_i = \sum_{j=1}^{N} a_j b_{i+j-1} \tag{12.10}$$

However, one often finds that the least squares solution is very rough, because the high frequency components of $m(t)$ are poorly constrained (i.e., the problem is really mixed-determined).

An alternative approach is to incorporate prior information of smoothness using weighted damped least squares $\mathbf{F}^T\mathbf{Fm} = \mathbf{F}^T\mathbf{d}$, where the matrix $\mathbf{F}$ contains both $\mathbf{G}$ and a smoothness matrix $\mathbf{H}$, scaled by a damping parameter ε^2 (see Eq. 3.51). If the biconjugate gradient method is used to solve this equation (see Section 3.9.3), then the only quantity that need be computed is the product $\mathbf{F}^T(\mathbf{Fv}) = \mathbf{G}^T(\mathbf{Gv}) + \varepsilon^2\mathbf{H}^T(\mathbf{Hv})$, where $\mathbf{v}$ is an arbitrary vector. The $\mathbf{G}^T(\mathbf{Gv})$ product can be computed extremely efficiently starting with $\mathbf{g}$ and using *MatLab* `xcorr()` cross-correlation function (see the accompanying `filterfun()` function for details).

As an example of filter construction, we shall consider a time series $g(t)$, which represents a recording of the sound emitted by a seismic exploration airgun (Fig. 12.3A). Signals of this sort are used to detect layering at depth in the earth through echo sounding. Ideally, a very spiky sound from the airgun is best because it allows echoes from layers at depth to be most easily detected. Engineering constraints, however, limit the airgun signal to a series of pulses. We shall attempt, therefore, to find a filter that, when applied to the airgun pulse, produces a signal spike or delta function $\mathbf{d} = [0, 0, 0, \ldots, 0, 1, 0, \ldots, 0]^T$ centered on the largest pulse in the original signal. This filter can then be applied to the recorded echo soundings to remove the reverberation of the airgun and reveal the layering of the earth. The least squares filter $m(t)$ (computed for this example using weighted damped least squares with both smoothness and length constraints and solved with the biconjugate gradient method) is shown in Fig. 12.3B and the resulting signal $d^{\text{pre}}(t) = m(t) * g(t)$ in Fig. 12.3C. Note that, although the reverberations are reduced in amplitude, they are by no means completely removed.

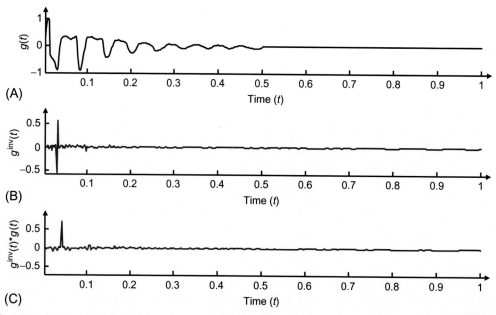

FIG. 12.3 (A) An airgun signal $g(t)$, after Smith (1975). Ideally, the inverse filter $g^{\text{inv}}(t)$ when convolved with $g(t)$ should produce the spike $\delta(t - t_0)$, centered at time t_0. (B) Estimate of the inverse filter $g^{\text{inv}}(t)$ for $t_0 = 0.04$, computed via generalized least squares with prior information on solution size and smoothness. (C) The convolution of $g(t)$ with the estimated $g^{\text{inv}}(t)$. While not a perfect spike, the result is significantly spikier than the airgun signal, $g(t)$. *MatLab* script gda12_02.

12.3 ADJUSTMENT OF CROSSOVER ERRORS

Consider a set of radar altimetry data from a remote-sensing satellite. These data consist of measurements of the distance from the satellite to the surface of the earth directly below the satellite. If the altitude of the satellite with respect to the earth's center were known, then these data could be used to measure the elevation of the surface of the earth. Unfortunately, while the height of the satellite during each orbit is approximately constant, its exact value is unknown. Since the orbits crisscross the earth, one can try to solve for the satellite height in each orbit by minimizing the overall crossover error (Kaula, 1966).

Suppose that there are M orbits and that the unknown altitude of the satellite during the ith orbit is m_i. We shall divide these orbits into two groups (Fig. 12.4), the ascending orbits (when the satellite is traveling north) and the descending orbits (when the satellite is traveling south). The ascending and descending orbits intersect at N points. At one such point ascending orbit number A_i intersects with descending orbit D_i (where the numbering refers to the ordering in $\mathbf{m}$). At this point, the two orbits have measured a satellite-to-earth distance of, say, S_{A_i} and S_{D_i}, respectively. The elevation of the ground is $m_{A_i} - S_{A_i}$ according to the data collected on the ascending orbit and $m_{D_i} - S_{D_i}$, according to the data from the descending orbit. The crossover error at the ith intersection is $e_i = (m_{A_i} - S_{A_i}) - (m_{D_i} - S_{D_i}) = (m_{A_i} - m_{D_i}) - (S_{A_i} - S_{D_i})$. The assertion that the crossover error should be zero leads to a linear equation of the form $\mathbf{Gm} = \mathbf{d}$, where

$$G_{ij} = \delta_{jA_i} - \delta_{jD_i} \quad \text{and} \quad d_i = S_{A_i} - S_{D_i} \tag{12.11}$$

Each row of the data kernel contains one 1, one -1, and $M - 2$ zeros. Note that the matrix $\mathbf{G}$ is extremely sparse; we would be well advised to declare it as such in a *MatLab* script.

Initially, we might assume that we can use simple least squares to solve this problem. We note, however, that the solution is always to a degree underdetermined. Any constant can be added to all the ms without changing the crossover error since the error depends only on the

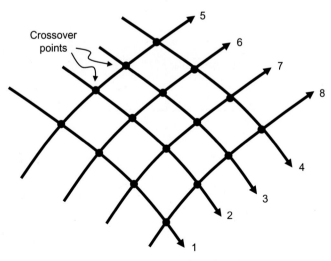

FIG. 12.4 Descending tracks 1–4 intersect ascending tracks 5–8 at 16 points. The height of the satellite along each track is determined by minimizing the crossover error at the intersections.

Crossover points

difference between the elevations of the satellite during the different orbits. This problem is therefore mixed-determined. We should therefore impose the prior constraint that $\sum_i m_i = 0$. While this constraint is not physically realistic (implying as it does that the satellite has on average zero altitude), it serves to remove the underdeterminacy. Any desired constant can subsequently be added to the solution.

This constraint can be approximately implemented with damped least squares

$$\mathbf{m}^{est} = \left[\mathbf{G}^T\mathbf{G} + \varepsilon^2 \mathbf{I}\right]^{-1} \mathbf{G}^T\mathbf{d} \tag{12.12}$$

As long as damping parameter ε^2 is chosen carefully, the solution will very closely approximate the exact one. An example is shown in Fig. 12.5.

As with the previous cases that we have studied, the matrices $[\mathbf{G}^T\mathbf{G}]$ and $[\mathbf{G}^T\mathbf{d}]$ can be computed analytically; furthermore, there is good reason for doing so. A realistic problem may

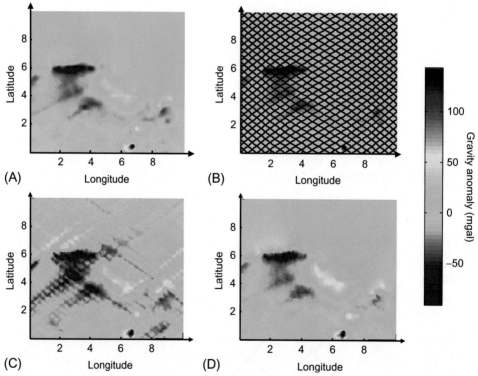

FIG. 12.5 Example of crossover error adjustment of satellite gravity data. (A) True gravity anomaly data for the equatorial Atlantic Ocean. It reflects variations in the depth of the seafloor and density variations within the oceanic crust. (B) Hypothetical satellite tracks, along which the gravity is measured. The measurements along each track have a constant offset reflecting errors in the assumed altitude of the satellite. (C) Reconstructed gravity anomaly without crossover correction. Artifacts parallel to the tracks are clearly visible. (D) Reconstructed gravity anomaly with crossover correction. The artifacts are eliminated. Data courtesy of Bill Haxby, Lamont-Doherty Earth Observatory. *MatLab* script gda12_03.

have thousands of orbits that intersect at millions of points. The data kernel will therefore be very large, with dimensions on the order of $1{,}000{,}000 \times 1000$. We find

$$[\mathbf{G}^\mathrm{T}\mathbf{G}]_{rs} = \sum_{i=1}^{N} G_{ir}G_{is} = \sum_{i=1}^{N} (\delta_{rA_i} - \delta_{rD_i})(\delta_{sA_i} - \delta_{sD_i})$$

$$= \sum_{i=1}^{N} (\delta_{rA_i}\delta_{sA_i} - \delta_{rA_i}\delta_{sD_i} - \delta_{rD_i}\delta_{sA_i} + \delta_{rD_i}\delta_{sD_i})$$

(12.13)

The diagonal elements of $[\mathbf{G}^\mathrm{T}\mathbf{G}]$ are

$$[\mathbf{G}^\mathrm{T}\mathbf{G}]_{rr} = \sum_{i=1}^{N} (\delta_{rA_i}\delta_{rA_i} - 2\delta_{rA_i}\delta_{rD_i} + \delta_{rD_i}\delta_{rD_i})$$

(12.14)

The first term contributes to the sum whenever the ascending orbit is r, and the third term contributes whenever the descending orbit is r. The second term is zero since an orbit never intersects itself. The rth element of the diagonal is the number of times the rth orbit is intersected by other orbits.

Only the two middle terms of the sum in the expression for $[\mathbf{G}^\mathrm{T}\mathbf{G}]_{rs}$ contribute to the off-diagonal elements. The second term contributes whenever $A_i = r$ and $D_i = s$, and the third when $A_i = s$ and $D_i = r$. The (r, s) off-diagonal element is the number of times the rth and sth orbits intersect, multiplied by -1.

The other matrix product is

$$[\mathbf{G}^\mathrm{T}\mathbf{d}]_r = \sum_{i=1}^{N} G_{ir}d_i = \sum_{i=1}^{N} (\delta_{rA_i} - \delta_{rD_i})d_i$$

(12.15)

We note that the delta functions can never both equal 1 since an orbit can never intersect itself. Therefore $[\mathbf{G}^\mathrm{T}\mathbf{d}]_r$ is the sum of all the ds that have ascending orbit number $A_i = r$ minus the sum of all the ds that have descending orbit number $D_i = r$.

We can then compute the matrix products. We first prepare a table that gives the ascending orbit number A_i, descending orbit number D_i, and elevation difference d_i for each of the N orbital intersections. We then start with $[\mathbf{G}^\mathrm{T}\mathbf{G}]$ and $[\mathbf{G}^\mathrm{T}\mathbf{d}]$ initialized to zero and, for each ith row of the table, execute the following steps:

1. Add 1 to the $r = A_i$, $s = A_i$ element of $[\mathbf{G}^\mathrm{T}\mathbf{G}]_{rs}$.
2. Add 1 to the $r = D_i$, $s = D_i$ element of $[\mathbf{G}^\mathrm{T}\mathbf{G}]_{rs}$.
3. Subtract 1 from the $r = A_i$, $s = D_i$ element of $[\mathbf{G}^\mathrm{T}\mathbf{G}]_{rs}$.
4. Subtract 1 from the $r = D_i$, $s = A_i$ element of $[\mathbf{G}^\mathrm{T}\mathbf{G}]_{rs}$.
5. Add d_i to the $r = A_i$ element of $[\mathbf{G}^\mathrm{T}\mathbf{d}]_r$.
6. Subtract d_i from the $r = D_i$ element of $[\mathbf{G}^\mathrm{T}\mathbf{d}]_r$.

The final form for $\mathbf{G}^\mathrm{T}\mathbf{G}$ is relatively simple. If two orbits intersect at most once, then it will contain only zeros and ones on its off-diagonal elements. As an alternative to damped least squares, the $\sum_i m_i = 0$ constraint can be implemented exactly using the Lagrange multiplier method (see Section 3.10).

$$\left[\begin{matrix} [\mathbf{G}^\mathrm{T}\mathbf{G}] & \mathbf{1} \\ \mathbf{1}^\mathrm{T} & 0 \end{matrix}\right] = \left[\begin{matrix} \mathbf{m} \\ \lambda \end{matrix}\right] = \left[\begin{matrix} [\mathbf{G}^\mathrm{T}\mathbf{d}] \\ 0 \end{matrix}\right] \tag{12.16}$$

Here, $\mathbf{1}$ is a length M column vector of ones, λ is a Lagrange multiplier, and the matrix on the left-hand side of the equation is $M+1 \times M+1$.

12.4 AN ACOUSTIC TOMOGRAPHY PROBLEM

An acoustic tomography problem was discussed previously in Section 1.1.3. In that simple case, travel times d_i of sound rays are measured through the rows and columns of a square 4×4 grid of bricks, each having height and width h and acoustic slowness m_i. The data kernel G_{ij} represents the length of ray i in brick j. Since the sound ray paths are either exactly horizontal or exactly vertical, each of the lengths is equal to h and the data kernel is

$$\left[\begin{matrix} d_1 \\ d_2 \\ \vdots \\ d_8 \end{matrix}\right] = h \left[\begin{matrix} 1 & 1 & 1 & 1 & 0 & 0 & 0 & 0 & 0 & 0 & 0 & 0 & 0 & 0 & 0 & 0 \\ 0 & 0 & 0 & 0 & 1 & 1 & 1 & 1 & 0 & 0 & 0 & 0 & 0 & 0 & 0 & 0 \\ \vdots & \vdots & \vdots & \vdots & \vdots & \vdots & \vdots & \vdots & \vdots & \vdots & \vdots & \vdots & \vdots & \vdots & \vdots & \vdots \\ 0 & 0 & 1 & 0 & 0 & 0 & 1 & 0 & 0 & 0 & 1 & 0 & 0 & 0 & 1 \end{matrix}\right] \left[\begin{matrix} m_1 \\ m_2 \\ \vdots \\ d_{16} \end{matrix}\right] \tag{12.17}$$

The matrix $\mathbf{G}$ is sparse, since the typical ray crosses just a small fraction of the total number of bricks. In the general case, where the image is divided up into rectangular pixels and where the ray paths are slanted and/or curved, the data kernel G_{ij} still represents the length of ray i in pixel j, but now that length is no longer constant. These variable lengths may be difficult to calculate analytically; instead, one must resort to numerical approximations.

One commonly used technique begins by initializing the data kernel to zero and then stepping along each ray i in arc length increments of Δs, chosen to be much smaller than the size of the pixels. The pixel index j of the center of each increment is determined, and the whole increment is added to the corresponding element of the data kernel; that is, $G_{ij} \rightarrow G_{ij} + \Delta s$.

We examine a test case where a square object is divided into a 256×256 grid of pixels (Fig. 12.6A), with the model parameter m_i representing the acoustic slowness within the pixels. A set of evenly distributed source and receiver points are placed on the four edges of the square and connected with straight-line rays (Fig. 12.6A). The data kernel is constructed using the approximate ray-stepping method described earlier and a data set of synthetic travel times is constructed via $\mathbf{d} = \mathbf{Gm}^{\mathrm{true}}$. Having an effective way of plotting travel time data is important, for instance, in detecting outliers. We parameterize each ray by its perpendicular distance r from the center of the object and by its angle θ from the horizontal and form an (r, θ) image of the data (Fig. 12.6C) for this purpose. The estimated image (Fig. 12.6D), computed using damped least squares solved with the biconjugate gradient method, recovers most of the long-wavelength features of the image, but contains faint streaks along ray paths. These streaks can be eliminated by increasing the number of rays (not shown).

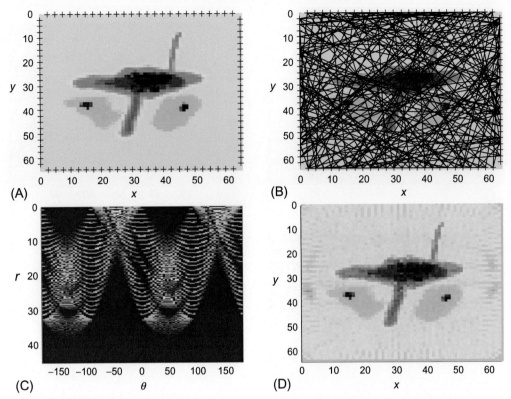

FIG. 12.6   Acoustic tomography problem. (A) True image. (B) Ray paths (only a few percent of rays are shown, else the image would be black). (C) Travel time data, organized by the distance r and angle θ of the ray from the midpoint of the image. (D) Reconstructed image. See text for further discussion. *MatLab* script gda12_04.

The success of tomography is critically dependent upon the ray coverage, which must not only be spatially dense but must—at every point—cover a 90 degree suite of angles (from horizontal to vertical). Tomographic reconstructions based on poorer ray coverage (Fig. 12.7) generally have poor resolution.

12.5 ONE-DIMENSIONAL TEMPERATURE DISTRIBUTION

As a hot slab within a uniform whole space cools, heat flows from the slab to the surrounding material, slowly warming it. The width of the warm zone slowly grows with time and the boundary between the slab and the whole space, initially distinct, slowly fades away. If several hot slabs are present, the temperatures from each blend together as time increases, making them hard to distinguish. The question that we pose is how well the initial pattern of temperatures can be reconstructed, given the temperature profile measured at some later time.

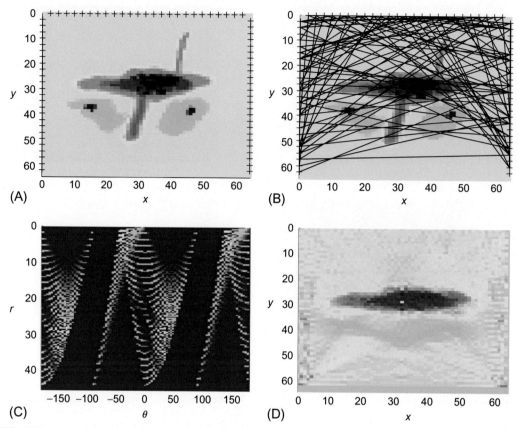

FIG. 12.7 Acoustic tomography problem with deficient distribution of rays. (A) True image. (B) Ray paths (only a few percent of rays are shown, else the image would be black). (C) Travel time data, organized by the distance r and angle θ of the ray from the midpoint of the image. (D) Reconstructed image. See text for further discussion. *MatLab* script gda12_05.

For a single slab, the temperature $T(x, t)$ at position x and time t can be shown to be (Menke and Abbott, 1990, their Section 6.3.3)

$$T(x, t) = \frac{1}{2}T_0\left\{ \text{erf}\left[\frac{x - (\xi - {}^1/_2 h)}{\sqrt{t}}\right] - \text{erf}\left[\frac{x - (\xi + {}^1/_2 h)}{\sqrt{t}}\right]\right\} = T_0 g(x, t, \xi) \qquad (12.18)$$

Here x is the distance measured perpendicular to the face of the slab, h is the thickness of the slab, and ξ is its position (Fig. 12.8A). The slab has initial temperature T_0 while the whole space is initially at zero temperature. The thermal diffusivity of both materials is taken, for simplicity, to be unity. The special function erf() is called the *error function*; *MatLab*'s implementation of it has the same name.

If several slabs of different temperature m_i are placed face-to-face within the whole space, the temperature at position x_i and time t is

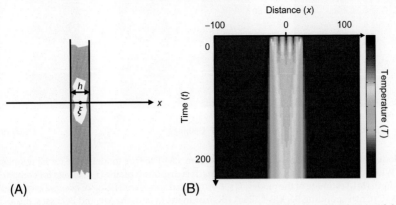

FIG. 12.8 (A) Single hot slab of thickness, h, located at position, $x=\xi$. (B) Temporal evolution of the temperature $T(x, t)$ of 100 adjacent slabs. The initial temperature distribution of the slabs, $T(x, t=0)$, is taken to be the model parameter vector, **m**. It is nonzero only for slabs near $|x| \leq 20$. The temperature, $T(x, t=0)$, at subsequent times, t, can be computed from the initial temperature distribution, since the data kernel can be calculated from the physics of heat transport. Note that the band of hot temperatures widens with increasing time, and that fine scale temperature fluctuations are preferentially attenuated. *MatLab* script gda12_06.

$$T(x_i, t) = \sum_{j=1}^{M} g\left(x_i, t, \xi_j\right) m_j \qquad (12.19)$$

The inverse problem that we consider is how well the temperature distribution m_i can be reconstructed by making measurements of the temperature at all positions, but at a fixed time, t. We might anticipate that the reconstruction will be well resolved at short times, since heat will not have much time to flow and the initial pattern of temperature will still be partly preserved. On the other hand, it will be poorly resolved at long times, since the initial pattern of temperature will have faded away.

Our test scenario has $M=100$ slabs located in the distance interval $|x| \leq 20$, which have a sinusoidal temperature pattern with five oscillations, overall (Fig. 12.8B). The details of this pattern fade with time, so that after about $t \approx 20$ only a broad warm zone is present. The width of this warm zone slowly grows with time, roughly doubling by $t \approx 250$.

We reconstruct the initial temperature using $N=100$ observations of temperature $T(x_i, t)$ from equally spaced between $-100 < x < 100$ all made at a fixed time t. Two different inversion methods are used, minimum length and Backus-Gilbert (Fig. 12.9). As hypothesized, their quality falls off rapidly with observation time t, with little detail being present after $t \approx 50$. The minimum length inversion does better at recovering the sinusoidal pattern, but it also contains artifacts, especially at long time, where two instead of five oscillations appear to be present. In contrast, the Backus-Gilbert inversion provides a poorer reconstruction, but one that is artifact free. These differences can be understood by examining the model resolution matrices of the two methods (Fig. 12.10). The minimum length inversion has the narrower resolution, but also the stronger sidelobes.

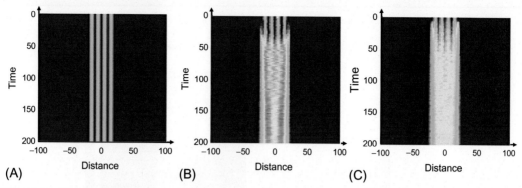

FIG. 12.9 Model for the temperature distribution problem. (A) The true model is the initial temperature distribution, $T(x, t=0)$. While this function is not a function of time, it is displayed on the (x, t) image for comparison purposes. (B) Minimum length (ML) estimate of the model, $T(x, t=0)$, for a data set consisting of observations at all distances at a single time $t > 0$. (C) Corresponding Backus-Gilbert (BG) estimate. In both the ML and BG cases, the ability of the data to resolve fine details declines with time, with the BG case declining fastest. *MatLab* script gda12_06.

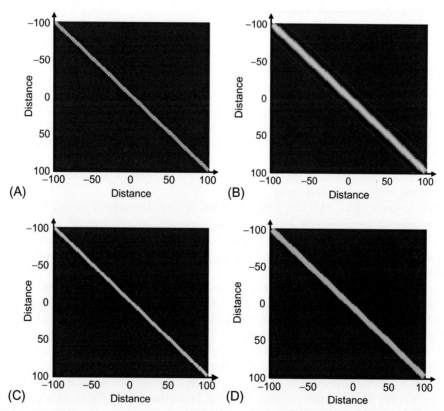

FIG. 12.10 Model resolution matrices for the temperature distribution problem. (A) Minimum length (ML) solution for data at time $t=10$. (B) ML solution for data at time $t=40$. (C) Backus-Gilbert (BG) solution for data at time $t=10$. (D) BG solution for data at time $t=40$. Note that the BG resolution matrix has smaller sidelobes than the corresponding ML case. *MatLab* script gda12_06.

12.6 L_1, L_2, AND L_∞ FITTING OF A STRAIGHT LINE

The L_1, L_2, and L_∞ problem is to fit the straight line $d_i = m_1 + m_2 z_i$ to a set of (z, d) pairs by minimizing the prediction error under a variety of norms. This is a linear problem with an $N \times 2$ data kernel

$$\mathbf{G} = \begin{bmatrix} 1 & z_1 \\ 1 & z_2 \\ \vdots & \vdots \\ 1 & z_N \end{bmatrix} \tag{12.20}$$

The L_2 norm is the simplest to implement. It implies that the error follows a Gaussian probability density function with $[\text{cov } \mathbf{d}] = \sigma_d^2 \mathbf{I}$. The simple least squares solution $\mathbf{m}^{\text{est}} = [\mathbf{G}^T \mathbf{G}]^{-1} \mathbf{G}^T \mathbf{d}$ is adequate since the problem typically is very overdetermined. Since the L_2 problem has been discussed, we shall not treat it in detail here. However, it is interesting to compute the data resolution matrix $\mathbf{N} = \mathbf{G} \mathbf{G}^{-g}$

$$\mathbf{N} = \begin{bmatrix} 1 & z_1 \\ 1 & z_2 \\ \vdots & \vdots \\ 1 & z_N \end{bmatrix} \frac{1}{N \sum z_i^2 - \left(\sum z_i \right)^2} \begin{bmatrix} \sum_{k=1}^{N} z_k^2 & -\sum_{k=1}^{N} z_k \\ -\sum_{k=1}^{N} z_k & N \end{bmatrix} \begin{bmatrix} 1 & 1 & 1 & \cdots & 1 \\ z_1 & z_2 & z_3 & \cdots & z_N \end{bmatrix} \tag{12.21}$$

$$N_{ij} = \frac{\sum z_k^2 - (z_i + z_j) \sum z_k + z_i z_j N}{N \sum z_i^2 - \left(\sum z_i \right)^2} = A_i + B_i z_j$$

Each row of the resolution matrix N_{ij} is a linear function of z_j, so that the elements with the largest absolute value are at an edge of the matrix and not along its main diagonal. The resolution is not at all localized; instead, the points with most extreme z_i control the fit of the straight line.

The L_1 and L_∞ estimates can be determined by using the transformation to a linear programming problem described in Chapter 8. Although more efficient algorithms exist, we shall set up the problems so that they can be solved with a standard linear programming algorithm. This algorithm determines a vector $\mathbf{y}$ that minimizes $\mathbf{c}^T \mathbf{y}$ subject to $\mathbf{A} \mathbf{y} = \mathbf{b}$ and $\mathbf{y} \geq 0$. The first step is to define two new variables $\mathbf{m}'$ and $\mathbf{m}''$ such that $\mathbf{m} = \mathbf{m}' - \mathbf{m}''$. This definition relaxes the positivity constraints on the model parameters. For the L_1 problem, we define three additional vectors $\boldsymbol{\alpha}$, $\mathbf{x}$, and $\mathbf{x}'$ and then arrange them in the form of a linear programming problem in $2M + 3N$ variables and $2N$ constraints as

$$\mathbf{y}^T = \left[[m'_1, \dots m'_M], [m''_1, \dots, m''_M], [\alpha_1, \dots, \alpha_N], [x_1, \dots, x_N], [x'_1, \dots, x'_N] \right]$$

$$\mathbf{c}^T = \left[[0, \dots, 0], [0, \dots, 0], [1, \dots, 1], [0, \dots, 0], [0, \dots, 0] \right]$$

$$\mathbf{A} = \begin{bmatrix} \mathbf{G}_{N \times M} & -\mathbf{G}_{N \times M} & -\mathbf{I}_{N \times N} & \mathbf{I}_{N \times N} & \mathbf{O}_{N \times N} \\ \mathbf{G}_{N \times M} & -\mathbf{G}_{N \times M} & \mathbf{I}_{N \times N} & \mathbf{O}_{N \times N} & -\mathbf{I}_{N \times N} \end{bmatrix} \tag{12.22}$$

$$\mathbf{b}^T = \left[[d_1, \dots, d_N][d_1, \dots, d_N] \right]$$

FIG. 12.11 Fitting a straight line to data $d(z)$. The true model *(black line)* is the straight line, $d^{true}(z) = 1 + 3z$. The $N = 10$ observations d_i^{obs} are the true data perturbed with exponentially distributed noise with variance $\sigma_d^2 = (0.4)^2$. Three different fits have been computed by minimizing the L_1, L_2, and L_∞ norms of the error. The corresponding predicted data are shown in *green*, *blue*, and *red*, respectively. *MatLab* script gda12_07.

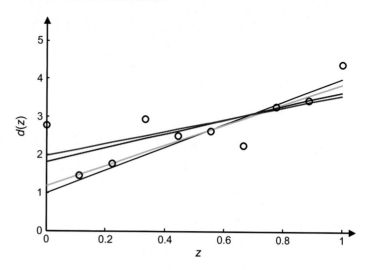

The L_∞ problem is transformed into a linear programming problem with additional variables **x** and **x'** and a scalar parameter α. The transformed problem in $2M + 2N + 1$ unknowns and $2N$ constraints is

$$\mathbf{y}^T = \left[\left[m'_1, \ldots m'_M \right], \left[m''_1, \ldots, m''_M \right], [\alpha], [x_1, \ldots, x_N], [x'_1, \ldots, x'_N] \right]$$

$$\mathbf{c}^T = \left[[0, \ldots, 0], [0, \ldots, 0], [1], [0, \ldots, 0], [0, \ldots, 0] \right]$$

$$\mathbf{A} = \begin{bmatrix} \mathbf{G}_{N \times M} & -\mathbf{G}_{N \times M} & -1 & \mathbf{I}_{N \times N} & \mathbf{O}_{N \times N} \\ \mathbf{G}_{N \times M} & -\mathbf{G}_{N \times M} & 1 & \mathbf{O}_{N \times N} & -\mathbf{I}_{N \times N} \end{bmatrix} \tag{12.23}$$

$$\mathbf{b}^T = \left[[d_1, \ldots, d_N][d_1, \ldots, d_N] \right]$$

We illustrate the results of using these three different norms to data with exponentially distributed error (Fig. 12.11). Note that the L_1 line is by far the best; it is designed to give little weight to outliers, which are common in data with exponentially distributed noise. Both the L_1 and L_∞ fits may be nonunique. Most versions of the linear programming algorithm will find only one solution, so the process of identifying the complete range of minimum solutions may be difficult.

12.7 FINDING THE MEAN OF A SET OF UNIT VECTORS

Suppose that a set of measurements of direction (defined by unit vectors in a three-dimensional Cartesian space) are thought to scatter randomly about a mean direction (Fig. 12.12). How can the mean vector be determined?

This problem is similar to that of determining the mean of a group of scalar quantities (Sections 5.1 and 8.2) and is solved by direct application of the principle of maximum likelihood. In the scalar mean problems, we assume that the data possess a Gaussian or exponential probability density function and then apply the principle of maximum

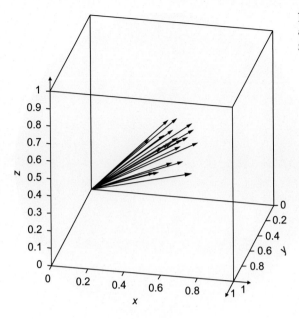

FIG. 12.12   Several unit vectors *(black)* scattering about a central direction *(red)*. *MatLab* script gda12_08.

likelihood to estimate a single model parameter, the mean. Neither of these probability density functions is applicable to directional data because they are defined on the wrong interval ($[-\infty, +\infty]$, instead of $[0, \pi]$). A better choice is the Fisher probability density function (Fisher, 1953). Its vectors are clumped near the mean direction with no preferred azimuthal direction. It proposes that the probability of finding a vector in an increment of solid angle $d\Omega = \sin(\theta)d\theta d\varphi$, located at an angle with inclination θ and azimuth φ from the mean direction (Fig. 12.13), is

$$p(\theta, \phi) = \frac{\kappa}{4\pi\sinh(\kappa)} \exp\left[\kappa \, \cos(\theta)\right] \tag{12.24}$$

This distribution is peaked near the mean direction $\theta = 0$, and its width depends on the value of the *precision parameter* κ (Fig. 12.14). The reciprocal of this parameter serves a role similar to the variance in the Gaussian distribution. When $\kappa = 0$ the distribution is completely white or random on the sphere, but when $\kappa \gg 1$ it becomes very peaked near the mean direction.

If the data are specified by N Cartesian unit vectors (x_i, y_i, z_i) and the mean by its unit vector (m_1, m_2, m_3), then the cosine of the inclination for any data is just the dot product $\cos(\theta_i) = x_i m_1 + y_i m_2 + z_i m_3$. The joint probability density function for the data $p(\theta, \varphi)$ is then the product of N distributions of the form $p(\theta_i, \varphi_i) \sin(\theta_i)$. The $\sin(\theta_i)$ term must be included since we are using Cartesian coordinates, as contrasted to polar coordinates, to represent directions. The joint distribution is then

$$p(\theta, \phi) = \left[\frac{\kappa}{4\pi\sinh(\kappa)}\right]^N \exp\left[\kappa \sum_{i=1}^{N} \cos(\theta_i)\right] \prod_{i=1}^{N} \sin(\theta_i) \tag{12.25}$$

FIG. 12.13 Unit sphere, showing coordinate system used in Fisher distribution. *MatLab* script gda12_09.

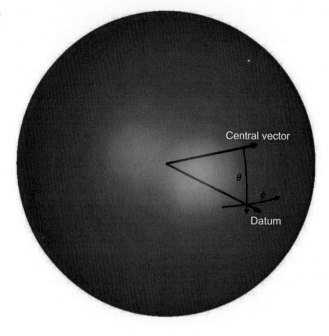

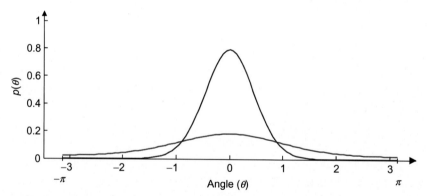

FIG. 12.14 Fisher distribution, $p(\theta, \varphi)$, for a large precision parameter ($\kappa=5$, *blue curve*) and a small precision parameter ($\kappa=1$, *red curve*). *MatLab* script gda12_10.

where $\cos(\theta_i) = \mathbf{x}^{\mathrm{T}}\mathbf{m} = [x_i m_1 + y_i m_2 + z_i m_3]$. The likelihood function is

$$L = \log(p) = N\log(\kappa) - N\log(4\pi) - N\log[\sinh(\kappa)]$$
$$+ \kappa \sum_{i=1}^{N} [x_i m_1 + y_i m_2 + z_i m_3] + \sum_{i=1}^{N} \log[\sin(\theta_i)] \tag{12.26}$$

The best estimate of model parameters occurs when the likelihood is maximized. However, since the solution is assumed to be a unit vector, L must be maximized under the constraint that $\mathbf{m}$ represents a unit vector; that is, $\sum_i m_i^2 = 1$. We first simplify this maximization by making an approximation. As long as κ is reasonably large (say, $\kappa > 5$), any variations in the magnitude of the joint probability that are caused by varying the m_i will come mainly from the exponential, since it varies much faster than the sine. We therefore ignore the last term in the likelihood function when computing the derivatives $\partial L/\partial m_q$. The Lagrange multiplier equations for the problem are then approximately

$$\kappa \sum_i x_i - 2\lambda m_1 = 0$$

$$\kappa \sum_i y_i - 2\lambda m_2 = 0$$

$$\kappa \sum_i z_i - 2\lambda m_3 = 0 \tag{12.27}$$

$$\frac{N}{\kappa} - N \frac{\cosh(\kappa)}{\sinh(\kappa)} + \sum_{i=1}^{N} [x_i m_1 + y_i m_2 + z_i m_3] = 0$$

where λ is the Lagrange multiplier. The first three equations can be solved simultaneously along with the constraint equation for the model parameters as

$$[m_1, m_2, m_3]^{\mathrm{T}} = \frac{\left[\sum_i x_i, \sum_i y_i, \sum_i z_i \right]^{\mathrm{T}}}{\left\{ \left(\sum_i x_i \right)^2 + \left(\sum_i y_i \right)^2 + \left(\sum_i z_i \right)^2 \right\}^{1/2}} \tag{12.28}$$

Note that the mean vector is the normalized vector sum of the individual observed unit vectors. The fourth equation is an implicit transcendental equation for κ. Since we have assumed that $\kappa > 5$, we can use the approximation $\cosh(\kappa)/\sinh(\kappa) \approx 1$, and the fourth equation yields

$$\kappa \approx \frac{N}{N - \sum_i \cos(\theta_i)} \tag{12.29}$$

An example is shown in Fig. 12.15.

12.8 GAUSSIAN AND LORENTZIAN CURVE FITTING

Many types of spectral data consist of several overlapping peaks, each of which has either Gaussian or *Lorentzian* shape. (Both the Gaussian and the Lorentzian have a single maximum, but the Lorentzian is much longer tailed.) The problem is to determine the location, area, and width of each peak through least squares curve fitting.

Suppose that the data consist of $N(z, d)$ pairs, where the auxiliary variable z represents spectral frequency. Each of, say, q peaks is parameterized by its center frequency f_i, area

FIG. 12.15 Lower hemisphere stereonet showing P-axes of deep (300–600 km) earthquakes in the Kurile-Kamchatka subduction zone. The axes mostly dip to the west, parallel to the direction of subduction, indicating that the subducting slab is in down-dip compression. *(Black circles)* Axes of individual earthquakes. *(Red circle)* Central axis, computed by maximum-likelihood technique. *(Blue dots)* Scatter of central axis determined by bootstrapping. Data courtesy of the Global CMT Project. *MatLab* script gda12_11.

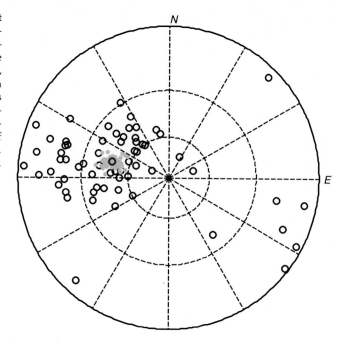

A_i, and width c_i. There are then $M = 3q$ model parameters $\mathbf{m} = [A_1, f_1, c_1, \ldots, A_q, f_q, c_q]^{\mathrm{T}}$. The model is nonlinear and of the form $\mathbf{d} = \mathbf{g}(\mathbf{m})$

$$\text{Gaussian}: \quad d_i = \sum_{j=1}^{q} \frac{A_j}{(2\pi)^{1/2} c_j} \exp\left[-\frac{(z_i - f_j)^2}{2c_j^2}\right]$$

$$\text{Lorentzian}: \quad d_i = \sum_{j=1}^{q} \frac{A_j c_j^2}{(z_i - f_j)^2 + c_j^2}$$

(12.30)

Note that in the case of the Gaussian, the quantity c_i^2 has the interpretation of variance, but in the case of the Lorentzian (which has infinite variance), it does not. If the data have Gaussian error, it is appropriate to use Newton's method to solve this problem. Furthermore, the problem will typically be overdetermined, at least if $N > M$ and if the peaks do not overlap completely. The equation is linearized around an initial guess using Taylor's theorem, as in Section 9.3. This linearization involves computing a matrix $\mathbf{G}$ of derivatives with rows

$$[\partial g_i/\partial A_1 \;\; \partial g_i/\partial f_1 \;\; \partial g_i/\partial c_1 \;\; \ldots \;\; \partial g_i/\partial A_p \;\; \partial g_i/\partial f_p \;\; \partial g_i/\partial c_p] \qquad (12.31)$$

In this problem the Gaussian and Lorentzian models are simple enough for the derivatives to be computed analytically as

Gaussian:

$$\partial g_i/\partial A_j = \left[1/(2\pi)^{1/2}c_j\right]\exp\left[-(z_i-f_j)^2/2c_j^2\right]$$

$$\partial g_i/\partial f_j = \left[\frac{A_j}{(2\pi)^{1/2}c_j}\right]\left[(z_i-f_j)/c_j^2\right]\exp\left[-(z_i-f_j)^2/2c_j^2\right]$$ (12.32)

$$\partial g_i/\partial c_j = \left[\frac{A_j}{(2\pi)^{1/2}c_j^2}\right]\left[\left((z_i-f_j)^2/c_j^2\right)-1\right]\exp\left[-(z_i-f_j)^2/2c_j^2\right]$$

Lorentzian:

$$\partial g_i/\partial A_j = c_j^2\bigg/\left[(z_i-f_j)^2+c_j^2\right]$$

$$\partial g_i/\partial f_j = 2A_jc_j^2(z_i-f_j)\bigg/\left[(z_i-f_j)^2+c_j^2\right]^2$$ (12.33)

$$\partial g_i/\partial c_j = 2A_jc_j\bigg/\left[(z_i-f_j)^2+c_j^2\right] - 2A_jc_j^3\bigg/\left[(z_i-f_j)^2+c_j^2\right]^2$$

These derivatives are evaluated at the trial solution $\mathbf{m}^{(p)}$ with the starting value (when $p=1$) chosen from visual inspection of the data. Improved estimates of the model parameters are found using the recursions $\mathbf{G}\Delta\mathbf{m}=\mathbf{d}-\mathbf{g}(\mathbf{m}^{(p)})$ and $\mathbf{m}^{(p+1)}=\mathbf{m}^{(p)}+\Delta\mathbf{m}$, where the matrix equation is solved with simple least squares. The iterations are terminated when the correction factor $\Delta\mathbf{m}$ becomes negligibly small (for instance, when the absolute value of each component becomes less than some given tolerance). An example is shown in Fig. 12.16.

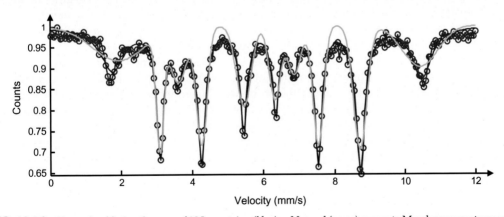

FIG. 12.16 Example of fitting the sum of 10 Lorentzian *(blue)* or Normal *(green)* curves to Mossbauer spectroscopic data *(red)*. The Lorentzian curves are better able to fit the shape of the curve, with the ratio of estimated variances being about 4. An *F*-test indicates that a null hypothesis that any difference between the two fits can be ascribed to random variation can be rejected to better than 99.99%. Data courtesy of NASA and the University of Mainz. *MatLab* script gda12_12.

Occasionally prior information requires that the separation between two peaks be equal to a known value $f_i - f_j = \langle s_{ij} \rangle$. This constraint can be implemented with the linearized version of weighted damped least squares (Eq. 9.27), with a very certain constraint $\mathbf{Hm} = \mathbf{h}$

$$
\begin{bmatrix} \mathbf{G}^{(p)} \\ \varepsilon \mathbf{H} \end{bmatrix} \mathbf{m}^{(p+1)} = \begin{bmatrix} \left\{ \mathbf{d} - \mathbf{g}(\mathbf{m}^{(p)}) + \mathbf{G}^{(p)}\mathbf{m}^{(p)} \right\} \\ \varepsilon \mathbf{h} \end{bmatrix} \tag{12.34}
$$

$$
\mathbf{H} = [\ldots\ 0\ 1\ 0\ \ldots\ 0,\ -1\ 0\ \ldots], \quad \text{and} \quad \mathbf{h} = [\langle s_{ij} \rangle]
$$

Here, ε is set to a very large number.

One of the drawbacks to these iterative methods is that if the initial guess is too far off, the solution may oscillate wildly or diverge from one iteration to the next. There are several possible remedies for this difficulty. One is to force the perturbation $\Delta \mathbf{m}$ to be less than a given length. This result can be achieved by examining the perturbation on each iteration and, if it is longer than some empirically derived limit, decreasing its length (but not changing its direction). This procedure will prevent the method from wildly "overshooting" the true minimum but will also slow the convergence. Another possibility is to constrain some of the model parameters to equal prior values for the first few iterations, thus allowing only some of the model parameters to vary. The constraints are relaxed after convergence, and the iteration is continued until the unconstrained solution converges.

12.9 EARTHQUAKE LOCATION

When a fault ruptures within the earth, seismic compressional P and shear S waves are emitted. These waves propagate through the earth and are recorded by instruments on the earth's surface. The earthquake location problem is to determine the *hypocenter* (location, $\mathbf{x}^{(0)} = [x_0, y_0, z_0]^T$) and *origin time* (time of occurrence, t_0) of the rupture on the basis of measurements of the arrival time of the P and S waves at the instruments.

The data are the arrival times t_i^P of P waves and the arrival time t_i^S of S waves, at a set of $i = 1, \ldots, N$ *receivers* (seismometers) located at $\mathbf{x}^{(i)} = [x_i, y_i, z_i]^T$. The model is the arrival time equals the origin time plus the travel time of the wave from the source (geologic fault) to the receiver. In a constant velocity medium, the waves travel along straight-line *rays* connecting source and receiver, so the travel time is the length of the line divided by the P or S wave velocity. In heterogeneous media, the rays are curved (Fig. 12.17) and the ray paths and travel times are much more difficult to calculate. We shall not discuss the process of *ray tracing* here,

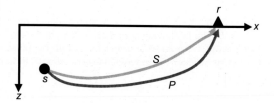

FIG. 12.17 Compressional P and shear S waves travel along rays from earthquake source s *(circle)* to receiver r *(triangle)*.

but merely assume that the travel times t_i^P and t_i^S can be calculated for arbitrary source and receiver locations. Then

$$t_i^P = T_i^P\left(\mathbf{x}^{(0)}, \mathbf{x}^{(i)}\right) + t_0 \quad \text{and} \quad t_i^S = T_i^S\left(\mathbf{x}^{(0)}, \mathbf{x}^{(i)}\right) + t_0 \tag{12.35}$$

These equations are nonlinear and of the form $\mathbf{d} = \mathbf{g}(\mathbf{m})$. If many observations are made so that the equations are overdetermined, an iterative least squares approach may be tried. This method requires that the derivatives $\nabla T = [\partial T_i/\partial x_0, \; \partial T_i/\partial y_0, \partial T_i/\partial y_0]^T$ be computed for various locations of the source. Unfortunately, there is no simple, differentiable analytic formula for travel time. One possible solution is to calculate this derivative numerically using the finite difference formula. For the P wave, we find

$$\frac{\partial T_i^P}{\partial x_0} \approx \frac{T_i^P(x_0 + \Delta x, y_0, z_0, x_i, y_i, z_i) - T_i^P(x_0, y_0, z_0, x_i, y_i, z_i)}{\Delta x}$$

$$\frac{\partial T_i^P}{\partial y_0} \approx \frac{T_i^P(x_0, y_0 + \Delta y, z_0, x_i, y_i, z_i) - T_i^P(x_0, y_0, z_0, x_i, y_i, z_i)}{\Delta y} \tag{12.36}$$

$$\frac{\partial T_i^P}{\partial z_0} \approx \frac{T_i^P(x_0, y, z_0 + \Delta z, x_i, y_i, z_i) - T_i^P(x_0, y_0, z_0, x_i, y_i, z_i)}{\Delta z}$$

and similarly for the S wave. Here, Δx, Δy, and Δz are small distance increments. These equations represent moving the location of the earthquake a small distance Δx along the directions of the coordinate axes and then computing the change in travel time. This approach has two disadvantages. First, if Δx is made very small so that the finite difference approximates a derivative very closely, the terms in the numerator become nearly equal and computer round-off error can become very significant. Second, this method requires that the travel time be computed for three additional earthquake locations and, therefore, is $4\times$ as expensive as computing travel time alone.

In some inverse problems, there is no alternative but to use finite element derivatives. Fortunately, it is possible in this problem to deduce the gradient of travel time by examining the geometry of a ray as it leaves the source (Fig. 12.18). If the earthquake is moved a small distance s parallel to the ray in the direction of the receiver, then the travel time is simply decreased by an amount s/v_0, where $v_0 = v(\mathbf{x}^{(0)})$ is the P or S wave velocity at the source. If it is moved a small distance perpendicular to the ray, then the change in travel time is negligible since the new ray path will have nearly the same length as the old. The gradient is therefore $\nabla T = -\mathbf{s}/v_0$, where $\mathbf{s}$ is a unit vector tangent to the ray at the source that points toward the

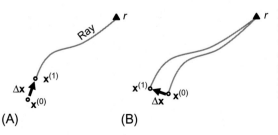

(A) (B)

FIG. 12.18 (A) Moving the source a distance Δx from $\mathbf{x}^{(0)}$ to $\mathbf{x}^{(1)}$ parallel to the ray path leads to a large change in travel time. (B) Moving the source perpendicular to the ray path leads to no (first order) change in travel time. The partial derivative of travel time with respect to distance can therefore be determined with minimal extra effort.

receiver. This insight is due to Geiger (1912) and linearized earthquake location is often referred to as *Geiger's method*. Since we must calculate the ray path to find the travel time, no extra computational effort is required to find the gradient. If $\mathbf{s}^{(i)P}$ is the direction of the P wave ray to the ith receiver (of a total of N receivers), and similarly $\mathbf{s}^{(i)S}$ is for the S wave, the linearized problem is

$$
\begin{bmatrix}
-s_1^{(1)P}/v_0^P & -s_2^{(1)P}/v_0^P & -s_3^{(1)P}/v_0^P & 1 \\
\cdots & \cdots & \cdots & \cdots \\
-s_1^{(N)P}/v_0^P & -s_2^{(N)P}/v_0^P & -s_3^{(N)P}/v_0^P & 1 \\
-s_1^{(1)S}/v_0^S & -s_2^{(1)S}/v_0^S & -s_3^{(1)S}/v_0^S & 1 \\
\cdots & \cdots & \cdots & \cdots \\
-s_1^{(N)S}/v_0^S & -s_2^{(N)S}/v_0^S & -s_3^{(N)S}/v_0^S & 1
\end{bmatrix}
\begin{bmatrix}
\Delta x_0 \\
\Delta y_0 \\
\Delta z_0 \\
t_0
\end{bmatrix}
=
\begin{bmatrix}
t_1^P - T_1^P \\
\cdots \\
t_N^P - T_N^P \\
t_1^S - T_1^S \\
\cdots \\
t_N^S - T_N^S
\end{bmatrix}
\tag{12.37}
$$

This equation is then solved iteratively using Newton's method. A sample problem is shown in Fig. 12.19.

There are some instances in which the matrix can become underdetermined and the least squares method will fail. This possibility is especially likely if the problem contains only P wave arrival times. The matrix equation consists of only the top half of Eq. (12.37). If all the rays leave the source in such a manner that one or more components of their unit vectors are all equal, then the corresponding column of the matrix will be proportional to the vector $[1, 1, 1, \ldots , 1]^T$ and will therefore be linearly dependent on the fourth column of the matrix. The earthquake location is then nonunique and can be traded off with origin time (Fig. 12.20). This problem occurs when the earthquake is far from all of the stations. The addition of S wave arrival times resolves the underdeterminacy (the columns are then proportional to $[1, 1, 1, \ldots , 1, v_S/v_P, \ldots]^T$ and are not linearly dependent on the fourth column). It is therefore

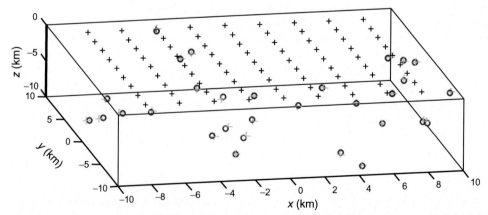

FIG. 12.19 Earthquake location example. Arrival times of P and S waves from earthquakes (*red circles*) are recorded on an array of 81 stations (*black crosses*). The observed arrival times include random noise with variance $\sigma_d^2 = (0.1)^2 s$. The estimated locations (*green crosses*) are computed using Geiger's method. *MatLab* script gda12_13.

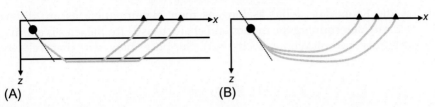

FIG. 12.20 (A) In layered media, rays follow "refracted" paths, such that the rays to all receivers *(triangles)* leave the source *(circle)* at the same angle. The position and travel time of the source on the dashed line therefore trade-off. (B) This phenomenon also occurs in nonlayered media if the source is far from the receivers.

wise to use singular-value decomposition and the natural inverse to solve earthquake location problems, permitting easy identification of underdetermined cases.

Another problem can occur if all the stations are on the horizontal plane. Then the solution is nonunique, with the solution $[x, y, z, t_0]^T$ and its mirror image $[x, y, -z, t_0]^T$ having exactly the same error. This problem can be avoided by including prior information that the earthquake occur beneath the ground and not in the air.

The earthquake location problem can also be extended to locate the earthquake and determine the velocity structure of the medium simultaneously. It then becomes similar to the tomography problem (Section 12.4) except that the orientations of the ray paths are unknown. To ensure that the medium is crossed by a sufficient number of rays to resolve the velocity structure, one must locate simultaneously a large set of earthquakes—on the order of 100 when the velocity structure is assumed to be one dimensional (Crosson, 1976) but thousands to millions when it is fully three dimensional.

12.10 VIBRATIONAL PROBLEMS

There are many inverse problems that involve determining the structure of an object from measurements of its *characteristic frequencies* of vibration (*eigenfrequencies*). For instance, the solar and terrestrial vibrational frequencies can be inverted for the density and elastic structure of those two bodies.

The forward problem of calculating the *modes* (spatial patterns) and eigenfrequencies of vibration of a body of a given structure is quite formidable. A commonly used approximate method is based on *perturbation theory*. The modes and eigenfrequencies are first computed for a simple *unperturbed* structure and then they are *perturbed* to match a more complicated structure. Consider, for instance, an acoustic problem in which the pressure modes $p_n(\mathbf{x})$ and corresponding eigenfrequencies ω_n satisfy the differential equation

$$-\omega_n^2 p_n(\mathbf{x}) = v^2(\mathbf{x})\nabla^2 p_n(\mathbf{x}) \tag{12.38}$$

with homogeneous boundary conditions. Many properties of the solutions to this equation can be deduced without actually solving it. For instance, two modes p_n and p_m can be shown to obey the orthogonality relationship

$$\int p_n(\mathbf{x})p_m(\mathbf{x})v^{-2}(\mathbf{x})\,\mathrm{d}^3x = \delta_{nm} \tag{12.39}$$

(where δ_{nm} is the Kronecker delta) as long as their eigenfrequencies are different; that is, $\omega_n \neq \omega_m$. Actually determining the modes and characteristic frequencies is a difficult problem for an arbitrary complicated velocity $v(\mathbf{x})$. Suppose, however, that velocity can be written as

$$v(\mathbf{x}) = v^{(0)}(\mathbf{x}) + \varepsilon v^{(1)}(\mathbf{x}) \tag{12.40}$$

where ε is a small number. Furthermore, suppose that the modes $p_n^{(0)}(\mathbf{x})$ and eigenfrequencies $\omega_n^{(0)}$ of the *unperturbed problem* (i.e., with $\varepsilon = 0$) are known. The idea of perturbation theory is to determine approximate versions of the *perturbed* modes $p_n(\mathbf{x})$ and eigenfrequencies ω_n that are valid for small ε.

The special case where all the eigenfrequencies are distinct (i.e., with numerically different values) is easiest (and the only one we solve here). We first assume that the eigenfrequencies and perturbed modes can be written as a series in powers in ε

$$\omega_n = \omega_n^{(0)} + \varepsilon \omega_n^{(1)} + \varepsilon^2 \omega_n^{(2)} + \cdots \quad \text{and}$$
$$p_n(\mathbf{x}) = p_n^{(0)}(\mathbf{x}) + \varepsilon p_n^{(1)}(\mathbf{x}) + \varepsilon^2 p_n^{(2)}(\mathbf{x}) + \cdots \tag{12.41}$$

Here, $\omega_n^{(i)}$ (for $i > 0$) are unknown constants and $p_n^{(i)}$ (for $i > 0$) are unknown functions. When ε is small, it suffices to solve merely for $\omega_n^{(1)}$ and $p_n^{(1)}$. We first represent $p_n^{(1)}$ as a sum of the other unperturbed modes

$$p_m^{(1)} = \sum_{\substack{m \\ \omega_m \neq \omega_n}}^{\infty} b_{nm} p_m^{(0)} \tag{12.42}$$

where b_{nm} are unknown coefficients. After substituting the Eqs. (12.41), (12.42) into Eq. (12.38), equating terms of equal power in ε, and applying the orthogonality relationship, the first-order quantities can be shown to be (see, e.g., Menke and Abbott, 1990, their Section 8.6.7)

$$\omega_n^{(1)} = \omega_n^{(0)} \int \left[p_n^{(0)}(\mathbf{x}) \right]^2 \left[v^{(0)}(\mathbf{x}) \right]^{-3} v^{(1)}(\mathbf{x}) \mathrm{d}^3 x$$
$$b_{nm} = \frac{2 \left(\omega_m^{(0)} \right)^2}{\left(\omega_m^{(0)} \right)^2 - \left(\omega_n^{(0)} \right)^2} \int p_n^{(0)}(\mathbf{x}) p_m^{(0)}(\mathbf{x}) \left[v^{(0)}(\mathbf{x}) \right]^{-3} v^{(1)}(\mathbf{x}) \mathrm{d}^3 x \tag{12.43}$$

The equation for $\omega_n^{(1)}$ is the relevant one for the discussion of the inverse problem. Notice that it is a linear function of the velocity perturbation $v^{(1)}(\mathbf{x})$,

$$\omega_n^{(1)} = \int G_n(\mathbf{x}) v^{(1)}(\mathbf{x}) \, \mathrm{d}^3 x \quad \text{with} \quad G_n(\mathbf{x}) = \omega_n^{(0)} \left[p_n^{(0)}(\mathbf{x}) \right]^2 \left[v^{(0)}(\mathbf{x}) \right]^{-3} \tag{12.44}$$

and so the problem of determining $v^{(1)}(\mathbf{x})$ from measurements of $\omega_n^{(1)}$ is a problem of linearized continuous inverse theory.

As an example, we will consider a one-dimensional organ pipe of length h, open at one end and closed at the other (corresponding to the boundary conditions $p|_{x=0} = \mathrm{d}p/\mathrm{d}x|_{x=h} = 0$).

We consider the unperturbed problem of a constant velocity structure, which has modes and eigenfrequencies given by

$$p_n^{(0)}(x) = \frac{2\left[v^{(0)}\right]^2}{h} \sin\left\{ \frac{\left(n - {}^1/_2\right)\pi}{h} x \right\} \quad \text{and}$$

$$\omega_n^{(0)} = \frac{\pi\left(n - {}^1/_2\right)v^{(0)}}{h} \quad \text{with } n = 1,2,3,\ldots$$

(12.45)

Note that the eigenfrequencies are evenly spaced in frequency (Fig. 12.21A). We now imagine that we measure the first N eigenfrequencies ω_n of an organ pipe whose velocity structure $v(x)$ is unknown (Fig. 12.21A). The data are the deviations of these frequencies from those of the unperturbed problem, $d_n = \omega_n^{(1)} = \omega_n - \omega_n^{(0)}$ The model parameters are the continuous function $m(x) = v^{(1)}(x) = v(x) - v^{(0)}$, which we will discretize into a vector $\mathbf{m}$ by sampling it at M evenly spaced values, with spacing Δx. The data kernel in the discrete equation $\mathbf{d} = \mathbf{Gm}$ is the discrete version of Eq. (12.44) (Fig. 12.21C)

$$G_{nm} = G_n(x_m)\Delta x = \omega_n^{(0)}\left[p_n^{(0)}(x_m)\right]^2\left[v^{(0)}\right]^{-3}\Delta x$$

(12.46)

The equation can be solved using damped least squares. Superficially, the estimated solution (red curve in Fig. 12.21B) looks reasonably good, except that it seems to have small steps

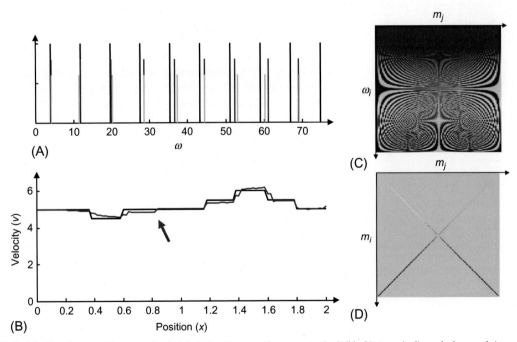

(B)

FIG. 12.21 Organ pipe example. (A) Ladder diagram for unperturbed *(black)*, true *(red)*, and observed *(green)* eigenfrequencies of the organ pipe. (B) True *(black)* and estimated *(red)* velocity structure. Arrow points to an artifact in the estimated solution. (C) Data kernel **G**. (D) Model resolution matrix **R**. *MatLab* script gda12_14.

(arrow in Fig. 12.21B) in locations where the true solution is smooth. However, a close inspection of the model resolution matrix (Fig. 12.21D), which is X-shaped, indicates a serious problem: two widely separated model parameters are trading off. The steps in the estimated solution are artifacts, steps from elsewhere in the model that are mapped into the wrong position. This is an inherent nonuniqueness of this and many other eigenfrequency-type problems. In seismology, it is addressed by combining eigenfrequency data with other data types (e.g., travel time measurements along rays) that do not suffer from this nonuniqueness.

12.11 PROBLEMS

12.1 Modify the *MatLab* script for the airgun problem to compute a shorter filter $m(t)$. How short can it be and still do a reasonable job at spiking the airgun pulse?

12.2 Modify the *MatLab* script for the tomography problem so that it retains receivers on the top and bottom of the model but omits them on the sides. Interpret the results.

12.3 Modify the *MatLab* script for the L_1 straight-line problem to allow for data of different accuracy. Test it against synthetic data, where the first $N/2$ data have variance $(\sigma^L)^2$ and the last $N/2$ have variance $(\sigma^R)^2$, both for the case where $\sigma^L = 10 \, \sigma^R$ and $\sigma^R = 10 \, \sigma^L$.

12.4 Modify the *MatLab* script for the earthquake location problem to include prior information that all the earthquakes occur near depth z_A. Adjust the true locations of the earthquakes to reflect this information. [Hint: You will need to use Eq. (9.27) to implement the prior information. It can be solved using `bicg()` with the weighted least squares function `weightedleastsquaresfcn()`.]

12.5 Modify the *MatLab* script for the earthquake location problem to use *differential P wave travel time data*, but no absolute travel time data. Each of these new data is the time difference between the arrival times of two earthquakes observed at a common station. Compare the locations to the results of the absolute-arrival time script, in the case where the variance of the differential data is four orders of magnitude smaller than the variance of the absolute data. [Hint: You may have to use a better starting guess than was used in the absolute-arrival time script.]

References

Crosson, R.S., 1976. Crustal structure modeling of earthquake data: 1. Simultaneous least squares estimation of hypocenter and velocity parameters. J. Geophys. Res. 81, 3036–3046.

Fisher, R.A., 1953. Dispersion on a sphere. Philos. Trans. R. Soc. Lond. A 217, 295–305.

Geiger, L., 1912. Probability method for the determination of earthquake epicenters from the arrival time only (translated from Geiger's 1910 German article). Bull. St. Louis Univ. 8 (1), 56–71.

Kaula, W.M., 1966. Theory of Satellite Geodesy. Ginn (Blaisdel), Boston, MA.

Menke, W., Abbott, D., 1990. Geophysical Theory. Columbia University Press, New York. 457 pp.

Smith, S.G., 1975. Measurement of airgun waveforms. Geophys. J. R. Astron. Soc. 42, 273–280.

13

Applications of Inverse Theory to Solid Earth Geophysics

13.1 EARTHQUAKE LOCATION AND DETERMINATION OF THE VELOCITY STRUCTURE OF THE EARTH FROM TRAVEL TIME DATA

The problem of determining the *hypocentral* parameters of an earthquake (i.e., its location and its origin time) and the problem of determining the velocity structure of the earth from travel time data are very closely coupled in geophysics, because earthquakes, which occur naturally at initially unknown locations, are a primary source of structural information.

Geophysical Data Analysis
https://doi.org/10.1016/B978-0-12-813555-6.00013-7

© 2018 Elsevier Inc. All rights reserved.

When the wavelength of the seismic waves is smaller than the scale of the velocity heterogeneities, the propagation of seismic waves can be described with *ray theory*, an approximation in which energy is assumed to propagate from source to observer along a curved path called a *ray*. In this approximation, the wave field is completely determined by the pattern of rays and the travel time along them.

In areas where the velocity structure is already known, earthquakes can be located using Geiger's (1912) method (see Section 12.9). The travel time $T(\mathbf{x})$ of an earthquake with location $\mathbf{x}$ and its ray tangent $\mathbf{t}(\mathbf{x})$ (pointing from earthquake to receiver) is calculated using ray theory. The arrival time is $t = \tau + T(\mathbf{x})$, where τ is the origin time of the earthquake. The perturbation in arrival time due to a change in the earthquake location from $\mathbf{x}_0$ to $\mathbf{x}_0 + \delta\mathbf{x}$ and a change in the origin time from τ_0 to $\tau_0 + \delta\tau$ is

$$\delta t = t - [\tau_0 + T(\mathbf{x}_0)] \approx \delta\tau - s(\mathbf{x}_0)\mathbf{t}(\mathbf{x}_0) \cdot \delta\mathbf{x} \tag{13.1}$$

where s is the slowness (reciprocal velocity). Note that this linearized equation has four model parameters, τ, and the three components of $\mathbf{x}$.

The forward problem, that of finding the ray path connecting earthquake and station and its travel time, is computationally intensive. Two alternative strategies have been put forward, one based on first finding the ray path using a shooting or bending strategy and then calculating the travel time by an integral of slowness along the ray (Cerveny, 2001), or conversely, by first finding the travel time by solving its partial differential equation (the *Eikonal* equation) and then calculating the ray path by taking the gradient of the travel time (Vidale, 1990).

The velocity structure can be represented with varying degrees of complexity. The simplest is a vertically stratified structure consisting of a stack of homogeneous layers or, alternatively, a continuously varying function of depth. Analytic formula for the travel time and its derivative with source parameters can be derived in some cases (Aki and Richards, 2002, their Section 9.3). The most complicated cases are fully three-dimensional velocity models represented with voxels or splines. Compromises between these extremes are also popular and include two-dimensional models (in cases where the structure is assumed to be uniform along the strike of a linear tectonic feature) and regionalized models (which assign a different vertically stratified structure to each tectonically distinct region but average them when computing travel times of rays that cross from one region to another).

In the simplest formulation, each earthquake is located separately. The data kernel of the linearized problem is an $N \times 4$ matrix, where N is the number of travel time observations, and is typically solved by singular-value decomposition. A very commonly used computer program that uses a layered velocity model is HYPOINVERSE (Klein, 1985). However, locations can often be improved by simultaneously solving for at least some aspect of the velocity model, which requires the earthquakes to be located simultaneously. The simplest modification is to solve for a *station correction* for each station; that is, a time increment that can be added to predicted travel times to account for near-station velocity heterogeneity not captured by the velocity model. More complicated formulations solve for a vertically stratified structure (Crosson, 1976) and for fully three-dimensional models (Menke, 2005; Zelt, 1998; Zelt and Barton, 1998). In the three-dimensional case, the inverse problem is properly one of tomographic inversion, as the perturbation in arrival time δt includes a line integral

$$\delta t \approx \delta \tau - s(\mathbf{x}_0)\mathbf{t}(\mathbf{x}_0) \cdot \delta \mathbf{x} + \int_{\text{unperturbed ray}} \delta s \, d\boldsymbol{\ell} \tag{13.2}$$

Here, δs is the perturbation in slowness with respect to the reference model and $d\boldsymbol{\ell}$ is the arc-length along the unperturbed ray. This problem is $N \times (4K+L)$, where N is the number of arrival time observations, K is the number of earthquakes, and L is the number of unknown parameters in the velocity model. One of the limitations of earthquake data is that earthquakes tend to be spatially clustered. Thus, the ray paths often poorly sample the model volume, and prior information, usually in the form of smoothness information, is required to achieve useful solutions. Even so, earthquake location (and especially earthquake depth) will often trade off strongly with velocity structure.

A very useful algorithm due to Pavlis and Booker (1980) simplifies the inversion of the very large matrices that result from simultaneously inverting for the source parameter of many thousands of earthquakes and an even larger number of velocity model parameters. Suppose that the linearized problem has been converted into a standard discrete linear problem $\mathbf{d} = \mathbf{Gm}$, where the model parameters can be arranged into two groups $\mathbf{m} = [\mathbf{m}_1, \mathbf{m}_2]^T$, where $\mathbf{m}_1$ is a vector of the earthquake source parameters and $\mathbf{m}_2$ is a vector of the velocity parameters. Then the inverse problem can be written

$$\mathbf{d} = \mathbf{Gm} = [\mathbf{G}_1 \ \mathbf{G}_2]\begin{bmatrix} \mathbf{m}_1 \\ \mathbf{m}_2 \end{bmatrix} = \mathbf{G}_1\mathbf{m}_1 + \mathbf{G}_2\mathbf{m}_2 \tag{13.3}$$

Now suppose that $\mathbf{G}_1$ has singular-value decomposition $\mathbf{G}_1 = \mathbf{U}_{1p}\Lambda_{1p}\mathbf{V}_{1p}^T$, with p nonzero singular values, so that $\mathbf{U}_1 = [\mathbf{U}_{1p}, \mathbf{U}_{10}]$ where $\mathbf{U}_{10}^T \mathbf{U}_{1p} = 0$. Premultiplying the inverse problem (Eq. 13.3) by $\mathbf{U}_{10}^T$ yields

$$\mathbf{U}_{10}^T\mathbf{d} = \mathbf{U}_{10}^T\mathbf{G}_1\mathbf{m}_1 + \mathbf{U}_{10}^T\mathbf{G}_2\mathbf{m}_2 = \mathbf{U}_{10}^T\mathbf{G}_2\mathbf{m}_2 \quad \text{or} \quad \mathbf{d}' = \mathbf{G}'\mathbf{m}_2 \quad \text{with} \quad \mathbf{d}' = \mathbf{U}_{10}^T\mathbf{d} \quad \text{and} \quad \mathbf{G}' = \mathbf{U}_{10}^T\mathbf{G}_2 \tag{13.4}$$

Here, we have used the fact that $\mathbf{U}_{10}^T\mathbf{G}_1\mathbf{m}_1 = \mathbf{U}_{10}^T\mathbf{U}_{1p}\Lambda_{1p}\mathbf{V}_{1p}^T\mathbf{m}_1 = 0$. Premultiplying the inverse problem (Eq. 13.3) by $\mathbf{U}_{1p}^T$ yields

$$\mathbf{U}_{1p}^T\mathbf{d} = \mathbf{U}_{1p}^T\mathbf{G}_1\mathbf{m}_1 + \mathbf{U}_{1p}^T\mathbf{G}_2\mathbf{m}_2 = \Lambda_{1p}\mathbf{V}_{1p}^T\mathbf{m}_1 + \mathbf{U}_{1p}^T\mathbf{G}_2\mathbf{m}_2$$

$$\text{or} \quad \mathbf{d}'' = \mathbf{G}''\mathbf{m}_1 \quad \text{with} \quad \mathbf{d}'' = \mathbf{U}_{1p}^T\mathbf{d} - \mathbf{U}_{1p}^T\mathbf{G}_2\mathbf{m}_2 \quad \text{and} \quad \mathbf{G}'' = \Lambda_{1p}\mathbf{V}_{1p}^T \tag{13.5}$$

The inverse problem has been partitioned into two equations, an equation for $\mathbf{m}_2$ (the velocity parameters) that can be solved first, and an equation for $\mathbf{m}_1$ (the source parameters) that can be solved once $\mathbf{m}_2$ is known (see *MatLab* script gda13_01). When the data consist of only absolute travel times (as contrasted to differential travel times, described later), the source parameter data kernel $\mathbf{G}_1$ is block diagonal, since the source parameters of a given earthquake affect only the travel times associated with that earthquake. The problem of computing the singular-value decomposition of $\mathbf{G}_1$ can be broken down into the problem of computing the singular-value decomposition of the submatrices.

Signal processing techniques based on waveform cross-correlation are able to determine the difference $\Delta t^{A,B} = t^A - t^B$ in arrival times of two earthquakes (say, labeled A and B), observed at a common station, to several orders of magnitude better precision than the absolute arrival times t^A and t^B can be determined (e.g., Menke and Menke, 2015). Thus, earthquake locations can be vastly improved, either by supplementing absolute arrival time data t with differential arrival time data $\Delta t^{A,B}$ or by relying on differential data alone. The perturbation in differential arrival time for two earthquakes, A and B, is formed by differencing two versions of Eq. (13.1)

$$\delta \Delta t^{A,B} \approx \delta \tau^A - s\left(\mathbf{x}_0^A\right) \mathbf{t}\left(\mathbf{x}_0^A\right) \cdot \delta \mathbf{x}^A - \delta \tau^B + s\left(\mathbf{x}_0^B\right) \mathbf{t}\left(\mathbf{x}_0^B\right) \cdot \delta \mathbf{x}^B \tag{13.6}$$

Each equation involves eight unknowns. Early versions of this method focused on using differential data to improve the *relative* location between earthquakes (e.g., Spence and Alexander, 1968). The modern formulation, known as the *double-difference method*, acknowledges that the relative locations are better determined than the absolute locations but solves for both (e.g., Menke and Schaff, 2004; Slunga et al., 1995; Waldhauser and Ellsworth, 2000). Double-difference methods have also been extended to include the tomographic estimation of velocity structure (Zhang and Thurber, 2003).

At the other extreme is the case where the origin times and locations of the seismic sources are known and the travel time T of elastic waves through the earth is used to determine the slowness $s(\mathbf{x})$ in the earth

$$\delta T = \int_{\text{perturbed ray}} \delta s \; d\boldsymbol{\ell} \approx \int_{\text{unperturbed ray}} \delta s \; d\boldsymbol{\ell} \tag{13.7}$$

The approximation relies on *Fermat's Principle*, which states that perturbations in ray path cause only second-order changes in travel time. Tomographic inversion based on Eq. (13.7) is now a standard tool in seismology and has been used to image the earth at many scales, from kilometer-scale geologic structures to the whole earth. An alternative to model estimation is presented by Vasco (1986) who uses extremal inversion to determine quantities such as the maximum difference between the velocities in two different parts of the model, assuming that the velocity perturbation is everywhere within a given set of bounds (see Chapter 6).

Tomographic inversions are often performed separately from earthquake location but then are degraded by whatever error is present in the earthquake locations and origin times. In the case where the earthquakes are all very distant from the volume being imaged, the largest error arises from the uncertainty in origin time, since errors in location do not change the part of the ray path in the imaged volume by very much. Suppose that the slowness in the imaged volume is represented as $s(x,y,z) = s_1(z) + s_2(x,y,z)$, with (x,y) horizontal and z vertical position, and with the mean of $s_2(x,y,z)$ zero at every depth, so that it represents horizontal variations in velocity, only. Poor knowledge of origin times affects only the estimates of $s_1(z)$ and not $s_2(x,y,z)$ (Aki et al., 1976; Menke et al., 2006). Estimates of the horizontal variations in velocity contain important information about the geologic structures that are being imaged, such as their (x,y) location and dip direction. Consequently, the method, called *teleseismic tomography*, has been applied in many areas of the world.

13.2 MOMENT TENSORS OF EARTHQUAKES

Seismometers measure the motion or *displacement* of the ground, a three-component vector $\mathbf{u}(\mathbf{x},t)$ that is a function of position $\mathbf{x}$ and time t. At wavelengths longer than the spatial scale of a seismic source, such as a geologic fault, an earthquake can be approximated by a system of nine *force-couples* acting at a point $\mathbf{x}_0$, the *centroid* of the source. The amplitude of these force-couples is described by a 3×3 symmetric matrix $\mathbf{M}(t)$, called the *moment tensor*, which is a function of time, t (Aki and Richards, 2002, their Section 3.3). Ground displacement depends on the time derivative $d\mathbf{M}/dt$, called *the moment-rate tensor*. Its six independent elements constitute six continuous model parameters $m_i(t)$. The ground displacement at $(\mathbf{x},t)$ due to a set of force-couples at $\mathbf{x}_0$ is linearly proportional to the elements of the moment-rate tensor via

$$u_i^{\text{pre}}(\mathbf{x}, t) = \sum_{j=1}^{6} \int G_{ij}(\mathbf{x}, \mathbf{x}_0, t - t_0) m_j(t_0) dt_0 \tag{13.8}$$

Here, $G_{ij}(\mathbf{x}, \mathbf{x}_0, t - t_0)$ are data kernels that describe the ith component displacement at $(\mathbf{x},t)$ due to the jth force couple at $(\mathbf{x}_0, t_0)$. The data kernels depend upon earth structure and, though challenging to calculate, are usually assumed to be known. If the location $\mathbf{x}_0$ of the source is also known, then the problem of estimating the moment-rate tensor from observations of ground displacement is completely linear (Dziewonski and Woodhouse, 1983; Dziewonski et al., 1981). The time variability of the moment-rate tensor is usually parameterized, with a triangular function of fixed width (or several overlapping such functions) being popular, so that the total number of unknowns is $M = 6L$, where L is the number of triangles. The error $e_i^{(j)}(t_k)$ associated with component i, observation location j, and observation time k is then

$$e_i^{(j)}(t_k) = u_i^{\text{obs}}\left(\mathbf{x}^{(j)}, t_k\right) - u_i^{\text{pre}}\left(\mathbf{x}^{(j)}, t_k\right) \tag{13.9}$$

When the total error E is defined as the usual L_2 prediction error

$$E = \sum_i \sum_j \sum_k \left[e_i^{(j)}(t_k)\right]^2 \tag{13.10}$$

the problem can be solved with simple least squares. Most natural sources, such as earthquakes, are thought to occur without causing volume changes, so some authors add the linear constraint $\dot{M}_{11} + \dot{M}_{22} + \dot{M}_{33} = 0$ (at all times).

Sometimes, the location $\mathbf{x}_0 = [x_0, y_0, z_0]^{\text{T}}$ of the source (or just its depth z_0) is also assumed to be unknown. The inverse problem is then nonlinear. This problem can be solved using a grid search over source location, that is, by solving the linear inverse problem for a grid of fixed source locations and then choosing the one with smallest total error E. The Fréchet derivatives of E with respect to hypocentral parameters are known; so alternately, Newton's method can be used. Adjoint methods are also applicable to this problem (Kim et al., 2011).

Earthquake locations are now routinely calculated by the U.S. Geological Survey (http://earthquake.usgs.gov/earthquakes/recenteqsww) and seismic moment tensors by the Global Centroid Moment Tensor (CMT) Project (http://www.globalcmt.org). They are standard data sets used in earthquake and tectonic research.

13.3 ADJOINT METHODS IN SEISMIC IMAGING

The fields most important to seismology depend upon both position $\mathbf{x}$ and time t. They include the displacement $\mathbf{u}(\mathbf{x},t)$, which quantifies the motion of particles in the Earth, the corresponding particle velocity $\dot{\mathbf{u}}(\mathbf{x}, t)$ and particle acceleration $\ddot{\mathbf{u}}(\mathbf{x}, t)$, and the body force $\mathbf{f}(\mathbf{x}, t)$, which describes seismic sources such as earthquakes.

The displacement field $\mathbf{u}(\mathbf{x}, t)$ satisfies a partial differential equation of the form:

$$\mathcal{L}(\mathbf{m})\,\mathbf{u} = \mathbf{f} \tag{13.11}$$

Here $\mathcal{L}(\mathbf{m})$ is a 3×3 second-order matrix partial differential operator that depends on material parameters, such as density ρ, Lamé parameter λ, and shear modulus μ, which we lump together into a model parameter vector $\mathbf{m}$. The seismic imaging problem is to estimate these model parameters using observations of the displacement $\mathbf{u}(\mathbf{x}, t)$, itself, or of parameters derived from it (such as finite-frequency travel time, discussed in Section 13.5). In some applications, the force $\mathbf{f}$ is known, but in others it is an additional model parameter that must be determined as the inverse problem is solved.

In general, $\mathcal{L}(\mathbf{m})$ can be very complicated. The simplest example is for an isotropic, homogeneous, elastic medium, where it has the form:

$$\mathcal{L} = \rho \frac{\partial^2}{\partial t^2} \mathbf{I} - \begin{bmatrix} (\lambda + 2\mu)\dfrac{\partial^2}{\partial x^2} + \mu \dfrac{\partial^2}{\partial y^2} + \mu \dfrac{\partial^2}{\partial z^2} & (\lambda + \mu)\dfrac{\partial^2}{\partial x \partial y} & (\lambda + \mu)\dfrac{\partial^2}{\partial x \partial z} \\[3ex] (\lambda + \mu)\dfrac{\partial^2}{\partial x \partial y} & \mu \dfrac{\partial^2}{\partial x^2} + (\lambda + 2\mu)\dfrac{\partial^2}{\partial y^2} + \mu \dfrac{\partial^2}{\partial z^2} & (\lambda + \mu)\dfrac{\partial^2}{\partial y \partial z} \\[3ex] (\lambda + \mu)\dfrac{\partial^2}{\partial x \partial z} & (\lambda + \mu)\dfrac{\partial^2}{\partial y \partial z} & \mu \dfrac{\partial^2}{\partial x^2} + \mu \dfrac{\partial^2}{\partial y^2} + (\lambda + 2\mu)\dfrac{\partial^2}{\partial z^2} \end{bmatrix} \tag{13.12}$$

Here $\mathbf{I}$ is the 3×3 identity matrix. In this case, the partial derivative of the operator with respect to density is also very simple. Suppose the density is $\rho(x) = \rho_0(x) + m\,\delta(\mathbf{x} - \mathbf{x}_H)$, which contains a point density heterogeneity of unknown amplitude m and known position $\mathbf{x}_H$. After inserting this formula into Eq. (13.12) and differentiating, we find:

$$\frac{\partial \mathcal{L}}{\partial m} = \delta(\mathbf{x} - \mathbf{x}_H)\frac{\partial^2}{\partial t^2}\mathbf{I} \tag{13.13}$$

Because seismic fields vary in both space and time, the inner product, previously defined as a spatial integral in Eq. (11.31), must be modified to include a time integral. The inner product between two vector, $\mathbf{a}(\mathbf{x}, t)$ and $\mathbf{b}(\mathbf{x}, t)$, is defined as:

$$\begin{aligned} (\mathbf{a}, \mathbf{b}) &= \iiint_V \int_{-\infty}^{+\infty} \mathbf{a}^{\mathrm{T}}(\mathbf{x}, t)\mathbf{b}(\mathbf{x}, t)\,dt\,d^3x \\ &= \iiint_V \int_{-\infty}^{+\infty} \sum_{i=1}^{3} a_i(\mathbf{x}, t)b_i(\mathbf{x}, t)\,dt\,d^3x \end{aligned} \tag{13.14}$$

Here $\mathbf{a}^{\mathrm{T}}\mathbf{b} = \sum_i a_i b_i$ is the dot product and V is the volume of the Earth.

Adjoints can be used to manipulate inner products, using the rule $(\mathbf{a}, \mathcal{L}\mathbf{b}) = (\mathcal{L}^\dagger \mathbf{a}, \mathbf{b})$. It can be shown that the adjoint of a matrix operator is the matrix transpose of the adjoint of its elements; that is

$$[\mathcal{L}^\dagger]_{ij} = [\mathcal{L}_{ji}]^\dagger \tag{13.15}$$

The wave equation operator in Eq. (13.12) is self-adjoint; that is, $\mathcal{L}^\dagger = \mathcal{L}$. This property is shared by many other second-order wave equations, including more complicated forms of the seismic wave equation.

A characteristic of seismic observations is that sampling in space is often irregular and sparse (because each observation point requires a separate geophone), whereas sampling in time is always regular and closely spaced (because seismic recording systems use very high-speed electronics). Consequently, seismic data are often treated as continuous functions of time made at discrete points in space. For example, $u_i^{(k)} = u_i(\mathbf{x}^{(k)}, t)$ is the ith component of displacement measured by the kth receiver. Here, $i=1,2,3$ and $k=1,\ldots,K$ when three-component observations are made at K receivers.

A datum, similar to the one defined in Eq. (11.58), can be derived from a single component i of displacement measured at a single receiver k, using the rule:

$$d_i^{(k)} = \int_{-\infty}^{+\infty} h(t) u_i\left(\mathbf{x}^{(k)}, t\right) dt \tag{13.16}$$

Here $h(t)$ is a known function. This equation is not in the form of an inner product, as defined in Eq. (13.14), because it omits integration over space. It can be manipulated into that form by inserting a Dirac impulse function and by defining a vector function $\mathbf{h}^{(i)}$ that is zero, except for its jth component, which is $h(t)$; that is, $[\mathbf{h}^{(i)}]_j = h(t)\delta_{ij}$. Then:

$$\left(\mathbf{h}^{(i)}\delta\left(\mathbf{x} - \mathbf{x}^{(k)}\right), \mathbf{u}(\mathbf{x}, t)\right) = \iiint_V \int_{-\infty}^{+\infty} \sum_{j=1}^{3} [h(t)\delta_{ij}] u_j\left(\mathbf{x}^{(k)}, t\right) \delta\left(\mathbf{x} - \mathbf{x}^{(k)}\right) dt\, d^3x$$

$$= \int h(t) u_i\left(\mathbf{x}^{(k)}, t\right) dt = d_i^{(k)} \tag{13.17}$$

The data kernel, evaluated for a reference model $\mathbf{m}_0$ can then be found using adjoint methods:

$$\frac{\partial}{\partial m} d_i^{(k)}\bigg|_{\mathbf{m}_0} = \left(\mathbf{h}^{(i)}\delta\left(\mathbf{x} - \mathbf{x}^{(k)}\right), \frac{\partial \mathbf{u}}{\partial m}\bigg|_{\mathbf{m}_0}\right)$$

$$= \left(\mathbf{h}^{(i)}\delta\left(\mathbf{x} - \mathbf{x}^{(k)}\right), -[\mathcal{L}(\mathbf{m}_0)]^{-1}\frac{\partial \mathcal{L}}{\partial m}\bigg|_{\mathbf{m}_0} \mathbf{u}_0(\mathbf{x}, t)\right) \tag{13.18}$$

$$= -\left(\mathcal{L}^{-1\dagger}\mathbf{h}^{(i)}\delta\left(\mathbf{x} - \mathbf{x}^{(k)}\right), \frac{\partial \mathcal{L}}{\partial m}\bigg|_{\mathbf{m}_0} \mathbf{u}_0(\mathbf{x}, t)\right)$$

Here, we have used the vector form of the Born approximation, which is analogous to Eq. (11.79) (and is derived in a similar manner). The expression can be simplified by defining an adjoint field λ:

$$\frac{\partial}{\partial m} d_i^{(k)}\bigg|_{\mathbf{m}_0} = (\lambda, \xi)$$

with

$$\mathcal{L}(\mathbf{m}_0)\,\mathbf{u}_0 = \mathbf{f} \text{ and } [\mathcal{L}(\mathbf{m}_0)]^\dagger \boldsymbol{\lambda}(\mathbf{x}, t) = \mathbf{h}^{(i)} \delta\left(\mathbf{x} - \mathbf{x}^{(k)}\right) \text{ and } \boldsymbol{\xi}(\mathbf{x}, t) = -\frac{\partial \mathcal{L}}{\partial m}\bigg|_{\mathbf{m}_0} \mathbf{u}_0(\mathbf{x}, t) \qquad (13.19)$$

The reference field $\mathbf{u}_0$ emanates from the source and propagates forward in time. The adjoint field emanates from the receiver and propagates backward in time. When the version of $\partial\mathcal{L}/\partial m$ corresponding to a point density heterogeneity (Eq. 13.13) is substituted into Eq. (13.19), the spatial part of the inner product can be performed trivially:

$$\frac{\partial}{\partial m} d_i^{(k)}\bigg|_{\mathbf{m}_0} = \left(\boldsymbol{\lambda}, -\frac{\partial \mathcal{L}}{\partial m}\bigg|_{\mathbf{m}_0} \mathbf{u}(\mathbf{x}, t)\right) = -(\boldsymbol{\lambda}(\mathbf{x}, t), \delta(\mathbf{x} - \mathbf{x}_H)\ddot{\mathbf{u}}_0(\mathbf{x}, t))$$

$$= -\int_{-\infty}^{+\infty} \boldsymbol{\lambda}^{\mathrm{T}}(\mathbf{x}_H, t)\ddot{\mathbf{u}}_0(\mathbf{x}_H, t)\,\mathrm{d}t \qquad (13.20)$$

The data kernel is constructed by taking the dot product of the adjoint field and the reference acceleration, both evaluated at the heterogeneity, and then time-integrating the result. This process is often referred to as *correlating* $\boldsymbol{\lambda}$ and $\ddot{\mathbf{u}}_0$ (since it corresponds to the zero-lag cross-correlation the two fields).

A similar process can be used to construct the derivative of the wave field error associated with the ith component of the kth receiver. This error is defined as:

$$E_i^{(k)} = \int \left(e_i^{(k)}\right)^2 \mathrm{d}t \text{ with } e_i^{(k)} = e_i\left(\mathbf{x}^{(k)}, t\right) = u_i^{\mathrm{obs}}\left(\mathbf{x}^{(k)}, t\right) - u_i^{(k)} \qquad (13.21)$$

It is equivalent to:

$$E_i^{(k)} = \iiint_V \int_{-\infty}^{+\infty} \sum_{j=1}^{3} \delta_{ij} \left(e_j^{(k)}\right)^2 \delta\left(\mathbf{x} - \mathbf{x}^{(k)}\right) \mathrm{d}t\,\mathrm{d}^3 x \qquad (13.22)$$

Differentiating with respect to the model yields:

$$\frac{\partial}{\partial m} E_i^{(k)}\bigg|_{\mathbf{m}_0} = -2 \iiint_V \int_{-\infty}^{+\infty} \sum_{j=1}^{3} \delta_{ij} e_j^{(k)} \frac{\partial u_j}{\partial m} \delta\left(\mathbf{x} - \mathbf{x}^{(k)}\right) \mathrm{d}t\,\mathrm{d}^3 x$$

$$= \left(\mathbf{r}^{(k)} \delta\left(\mathbf{x} - \mathbf{x}^{(k)}\right), -2\frac{\partial}{\partial m} \mathbf{u}^{(k)}\right) \qquad (13.23)$$

Here $[\mathbf{r}^{(k)}(t)]_i = e_j^{(k)} \delta_{ij}$ is a vector that is zero, except for its jth component, which is $e_j(t)$. After inserting the Born approximation, manipulating the inner product with adjoint methods, and defining an adjoint field $\boldsymbol{\lambda}$, we find:

$$\frac{\partial}{\partial m} E_i^{(k)}\bigg|_{\mathbf{m}_0} = (\boldsymbol{\lambda}, \boldsymbol{\xi})$$

with

$$\mathcal{L}(\mathbf{m}_0)\,\mathbf{u}_0 = \mathbf{f}(\mathbf{x}, t) \text{ and } [\mathcal{L}(\mathbf{m}_0)]^\dagger \boldsymbol{\lambda} = \mathbf{r}^{(k)}(t)\delta\left(\mathbf{x} - \mathbf{x}^{(k)}\right) \text{ and } \boldsymbol{\xi} = 2\frac{\partial \mathcal{L}}{\partial m}\bigg|_{\mathbf{m}_0} \mathbf{u}_0(\mathbf{x}, t) \qquad (13.24)$$

The total error and its derivative can then be constructed by summing over components and receivers:

$$E = \sum_{i=1}^{3} \sum_{k=1}^{K} E_i^{(k)} \quad \text{and} \quad \frac{\partial E}{\partial m}\bigg|_{\mathbf{m}_0} = \sum_{i=1}^{3} \sum_{k=1}^{K} \frac{\partial}{\partial m} E_i^{(k)}\bigg|_{\mathbf{m}_0} \tag{13.25}$$

Both the calculation of the data kernel (Eq. 13.20) and the error derivative (Eq. 13.25) require the numerical solution of differential equations—a very computationally intensive process. The reference field must be propagated from the source, forward in time, throughout the entire Earth model (or at least to a volume large enough to encompass all the heterogeneities). Similarly, Eq. (13.25) would seem to require that an adjoint field be propagated for all three components of every receiver, backward in time, throughout the entire source volume (or at least to a volume large enough to encompass all heterogeneities). Performed this way, $3K + 1$ differential equations must be solved, where K is the number of receivers. Fortunately, a very simple modification of Eqs. (13.24), (13.25) can reduce the number of required solutions to *two*—a very substantial savings! The idea is to create a *combined adjoint source* $\mathbf{s}^C$, which is the sum of the sources of the individual fields:

$$[\mathbf{s}^C]_i = \sum_{j=1}^{3} \sum_{k=1}^{K} r_j^{(k)}(t) \delta_{ij} \delta\left(\mathbf{x} - \mathbf{x}^{(k)}\right) \quad \text{or} \quad \mathbf{s}^C = \sum_{k=1}^{K} \mathbf{r}^{(k)}(t) \delta\left(\mathbf{x} - \mathbf{x}^{(k)}\right) \tag{13.26}$$

and to use it as the source of a *combined adjoint field* $\boldsymbol{\lambda}^C$. Thus, the number of required solutions is reduced from $3K$ to one! The combined adjoint source $\mathbf{s}^C$ is spatially distributed, but this complication presents no problem for most schemes for numerically solving a differential equation. The error derivative is then just the combined adjoint field $\boldsymbol{\lambda}^C$ correlated with the $\boldsymbol{\xi}$:

$$\frac{\partial E}{\partial m}\bigg|_{\mathbf{m}_0} = (\boldsymbol{\lambda}^C, \boldsymbol{\xi})$$

with

$$\mathcal{L}(\mathbf{m}_0) \mathbf{u}_0 = \mathbf{f}(\mathbf{x}, t) \quad \text{and} \quad \boldsymbol{\xi} = 2 \frac{\partial \mathcal{L}}{\partial m}\bigg|_{\mathbf{m}_0} \mathbf{u}_0(\mathbf{x}, t) \quad \text{and} \quad [\mathcal{L}(\mathbf{m}_0)]^\dagger \boldsymbol{\lambda}^C = \mathbf{s}^C \tag{13.27}$$

This manipulation relies on the linearity of the adjoint differential equation and the correlation process; the result of summing the outputs is the same a summing the inputs.

In addition to solving the two differential equations, M correlations must be performed, one for each heterogeneity. However, correlation requires negligible computational effort compared to the solution of a differential equation. Most seismic imaging problems have many, say N, sources. These processes must be repeated for each source, so overall, $2N$ solutions of differential equations and $N \times M$ correlations must be performed.

13.4 WAVEFIELD TOMOGRAPHY

In order to study the essential character of the error derivative, we step away from the complications of vector wave fields and consider the simple case where a "displacement" $u(\mathbf{x}, t)$ satisfies the scalar wave equation:

$$\mathcal{L}u \equiv \left(s^2 \frac{\partial^2}{\partial t^2} - \nabla^2 \right) u = f \tag{13.28}$$

The wave equation has one material parameter, the *slowness s*. We consider a point source at $\mathbf{x}_S$ with source time function $w(t)$, so that $f(\mathbf{x},t)=w(t)\delta(\mathbf{x}-\mathbf{x}_S)$. We consider the special case of a constant background slowness s_0 and a small perturbation δs, so that $s=s_0+\delta s$ and $s^2 \approx s_0^2 + 2s_0\delta s$. Furthermore, we assume a point heterogeneity of the form $2s_0\delta s = m\delta(\mathbf{x}-\mathbf{x}_H)$, with position $\mathbf{x}_H$ and unknown amplitude m. The reference solution, with $m_0=0$, can be shown to be:

$$u_0(\mathbf{x}, t) = \frac{w(t - T_{SX})}{4\pi R_{SX}} \text{ with } R_{SX} \equiv |\mathbf{x} - \mathbf{x}_S| \text{ and } T_{SX} \equiv s_0 R_{SX} \tag{13.29}$$

Here R_{SX} is the distance from the source to $\mathbf{x}$ and T_{SX} is the corresponding travel time. The solution is a spherical wave diverging from the source, with an amplitude proportional to $1/R_{SX}$ and with a pulse shape that is a delayed version of the source time function $w(t)$.

When the slowness has a point heterogeneity of the form $2s_0\delta s = m\delta(\mathbf{x}-\mathbf{x}_H)$, the derivative of the operator in Eq. (13.16) is:

$$\frac{\partial \mathcal{L}}{\partial m} = \delta(\mathbf{x} - \mathbf{x}_H) \frac{\partial^2}{\partial t^2} \tag{13.30}$$

Consequently,

$$\xi(\mathbf{x}, t) = 2 \frac{\partial \mathcal{L}}{\partial m} u_0(\mathbf{x}, t) = 2\delta(\mathbf{x} - \mathbf{x}_H)\ddot{u}_0(\mathbf{x}, t) = 2\delta(\mathbf{x} - \mathbf{x}_H)\frac{\ddot{w}(t - T_{SX})}{4\pi R_{SX}} \tag{13.31}$$

The operator in the adjoint equation is the same as in the wave equation, since the latter is self-adjoint. The adjoint field emanates from the receiver and has a time function given by the wave field error; that is, $f(\mathbf{x},t)=e_0(\mathbf{x},t)\delta(\mathbf{x}-\mathbf{x}_R)$. The solution to the adjoint equation is analogous to Eq. (13.29), except that it is backward in time:

$$\lambda(\mathbf{x}, t) = \frac{e_0(t + T_{RX})}{4\pi R_{RX}} \tag{13.32}$$

The error derivative for a single receiver R is then:

$$\frac{\partial E_R}{\partial m} = (\lambda(\mathbf{x}, t), \xi(\mathbf{x}, t)) = 2 \frac{1}{4\pi R_{SH}} \frac{1}{4\pi R_{RH}} \int_{-\infty}^{+\infty} e_0(\mathbf{x}_R, t + T_{RH})\ddot{w}(t - T_{SH})\,dt \tag{13.33}$$

The spatial pattern of the derivative is axially symmetric about a line drawn from source to receiver, since R_{SH}, R_{RH}, T_{RH}, and T_{SH} are depend only on the perpendicular distance of the heterogeneity from the S-R line. When $\ddot{w}(t)$ is oscillatory, the time integral is oscillatory and the derivative consists of a series of concentric ellipses of alternating sign, with foci at the source and receiver (Fig. 13.1). The ellipses represent surfaces of equal travel time from source to heterogeneity to receiver. The fields scattered field from heterogeneities on a given ellipse all arrive at the receiver at the same time and all contribute to changing the error at that time; they have no affect on the error at other times. The amplitude of the derivative varies across the surface of an ellipse, because it depends upon the product of the source-to-heterogeneity and heterogeneity-to-receiver distances, rather than their sum.

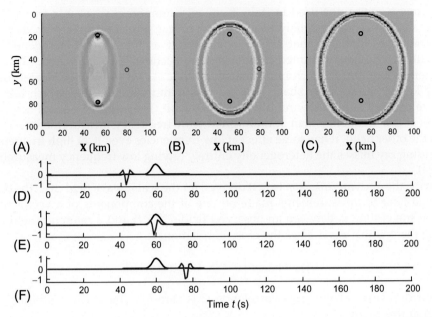

FIG. 13.1 Partial derivative $\partial E/\partial m$ of the wave field error $E(\mathbf{x}_R)$ for an acoustic wave propagation problem. The source $\mathbf{x}_S$ and receiver $\mathbf{x}_R$ *(back circles)* are separated by a distance of $R=60$ km. The perturbation in acoustic velocity is $\delta v = (m/2v_0)\,\delta(\mathbf{x}-\mathbf{x}_h)$, where $v_0 = 1$ km/s is a constant reference velocity and x_h is the heterogeneity's location. The wave field error $e(t)$ is a Gaussian pulse at time $R/v_0 + \tau$. (A–C) Partial derivative *(colors)* for a suite of $\mathbf{x}_h$'s covering the (x,y) plane, for $\tau=25,40$, and 57 s, respectively. Case (B) corresponds to the case when the reference field emanating from source and the adjoint field emanating from the receiver arrive at a particular heterogeneity $\mathbf{x}_H$ *(red circle)* at the same time, leading to a large value of the derivative. (D–F) Reference field $u^0(\mathbf{x}_H,t)$ *(black)* and adjoint field $\lambda(\mathbf{x}_H,t)$ *(red)* for $\tau=25,40$, and 57 s, respectively. *MatLab* script `gda13_02.m`.

13.5 FINITE-FREQUENCY TRAVEL TIME TOMOGRAPHY

Traditionally, seismic tomography has used travel times based on "picks" of the onset of motion of a seismic phase on a seismogram, either determined "by eye" or by an algorithm that detects a sudden increase in the amplitude of the ground motion. Such travel times are easy to measure on short-period seismograms but problematical at longer periods, owing to the emergent onset of the waveforms. A more suitable measurement technique for these data involves cross-correlating the observed seismic phase with a synthetic reference seismogram, because cross-correlation can accurately determine the small time difference, say τ, between two smooth pulses. However, the results of cross-correlation are dependent upon the frequency band of measurement; a phase that is observed to arrive earlier than the reference phase for one frequency band may well arrive later than it for another. Consequently, finite-frequency travel times must be interpreted in the context of the frequency band at which they are measured. Finite-frequency travel time tomography is based upon a derivative $\partial \tau/\partial m$ (where m is a model parameter) than incorporates the frequency-dependent behavior of cross-correlations.

At the long wavelengths used in finite-frequency tomography, seismic energy does not propagate along an infinitesimally thin ray connecting source and receiver, as it does in the high-frequency limits, so that $\partial\tau/\partial m$ is not simply an integral of heterogeneities along the ray path (as we have assumed previously, see Sections 1.3.3, 12.4, and 13.1). Instead, heterogeneities in entire Earth volume contribute to it. As we demonstrate later, the finite-frequency kernel has a diffuse shape that narrows with frequency, and which takes on the appearance of a ray only in the infinite frequency limit. The effect of a small heterogeneity on τ is less at low frequency than at high (giving low frequency measurements a disadvantage). On the other hand, it has some effect at low frequencies even if, at high frequencies, the corresponding ray misses the heterogeneity entirely (giving low frequency measurements an advantage).

For clarity, the derivative $\partial\tau/\partial m$ is first derived for the scalar wave field case; the result will be extended later to measurements made on one of the components of a vector field. The differential travel time τ_0 between an observed field $u^{obs}(\mathbf{x}, t)$ and a reference field $u_0(\mathbf{x}, t)$ is defined as the one that maximizes the cross-correlation:

$$C(\mathbf{x}, \tau; u_0) = \int u^{obs}(\mathbf{x}, t - \tau) u_0(\mathbf{x}, t) \, dt \tag{13.34}$$

The first derivative of the cross-correlation C is zero at τ_0, since by definition, C has a maximum at this point:

$$\left. \frac{dC(\mathbf{x}, \tau_0; u_0)}{d\tau} \right|_{\tau_0} = 0 \tag{13.35}$$

Now let the field be perturbed from u_0 to $u = u_0 + \delta u$. The cross-correlation experiences a corresponding perturbation:

$$C(\mathbf{x}, \tau; u) = C(\mathbf{x}, \tau; u_0) + \delta C(\mathbf{x}, \tau) \text{ with } \delta C(\mathbf{x}, \tau) = \int u^{obs}(\mathbf{x}, t - \tau) \delta u(\mathbf{x}, t) \, dt \tag{13.36}$$

The perturbed cross-correlation has a maximum at, say, $\tau = \tau_0 + \delta\tau$. Expanding $C(\mathbf{x}, \tau; u)$ in a Taylor series up to second order in small quantities yields:

$$C(\mathbf{x}, \tau; u) = C(\mathbf{x}, \tau_0; u_0) + 0 + \frac{1}{2} \left. \left(\frac{d^2}{d\tau^2} C(\mathbf{x}, \tau; u^0) \right) \right|_{\tau_0} (\delta\tau)^2 + \delta C(\mathbf{x}, \tau_0) + \left. \left(\frac{d}{dt} \delta C(\mathbf{x}, \tau) \right) \right|_{\tau_0} \delta\tau \tag{13.37}$$

As is shown in Eq. (13.35), the second term on the r.h.s. is zero. The maximum occurs where the derivative of Eq. (13.37) with respect to $\delta\tau$ is zero:

$$\frac{dC(\mathbf{x}, \tau)}{d\tau} = 0 \approx \left. \left(\frac{d^2}{d\tau^2} C(\mathbf{x}, \tau; u^0) \right) \right|_{\tau_0} \delta\tau + \left. \left(\frac{d}{d\tau} \delta C(\mathbf{x}, \tau) \right) \right|_{\tau_0} \tag{13.38}$$

Solving for $\delta\tau$ yields:

$$\delta\tau \approx -\frac{1}{D} \left. \left(\frac{d}{d\tau} \delta C(\mathbf{x}, \tau) \right) \right|_{\tau_0} \text{ with } D = \left. \left(\frac{d^2}{d\tau^2} C(\mathbf{x}, \tau) \right) \right|_{\tau_0} \tag{13.39}$$

This expression can be further simplified by differentiating the cross-correlations:

$$\frac{d\delta C}{d\tau}\bigg|_{\mathbf{x},\tau_0} = \frac{d}{d\tau}\int u^{obs}(\mathbf{x}, t-\tau)\,\delta u(\mathbf{x}, t)\,dt\bigg|_{\tau_0} = -\int \dot{u}^{obs}(\mathbf{x}, t-\tau_0)\delta u(\mathbf{x}, t)\,dt$$

$$D(\mathbf{x}, \tau_0) = \frac{d^2 C(\tau^0; u_B^0)}{d\tau^2}\bigg|_{\tau_0} = \frac{d^2}{d\tau^2}\int u^{obs}(\mathbf{x}, t-\tau)\,u_0(\mathbf{x}, t)dt\bigg|_{\tau_0} \tag{13.40}$$

$$= \int \ddot{u}^{obs}(\mathbf{x}, t-\tau_0)\,u_0(\mathbf{x}, t)\,dt$$

The scalar field $u(\mathbf{x},t)$ is now equated with the ith component of a vector field $\mathbf{u}(\mathbf{x},t)$. Eq. (13.39) can be manipulated into the form of an inner product:

$$\delta\tau^{(i)} = \left(\mathbf{h}^{(i)}\delta(\mathbf{x} - \mathbf{x}_R),\, \delta\mathbf{u}(\mathbf{x}, t)\right) \text{ with }$$

$$\left[\mathbf{h}^{(i)}\right]_j = \frac{-\dot{u}_j^{obs}\left(\mathbf{x}_R, t+\tau_0^{(j)}\right)}{D_j(\mathbf{x}_R, \tau_0)}\delta_{ij} \text{ and } D_j(\mathbf{x}, \tau_0) = \int \ddot{u}_j^{obs}\left(\mathbf{x}_R, t-\tau_0^{(j)}\right)u_j^{(0)}(\mathbf{x}_R, t)dt \tag{13.41}$$

Here $\mathbf{h}(t)$ is a vector with only one nonzero component (corresponding to the component of displacement for which travel time is being computed). The total travel time is $\tau^{(i)} = \tau_0^{(i)} + \delta\tau^{(i)}$ and its derivative with respect to the model is:

$$\frac{\partial}{\partial m}\tau^{(i)}\bigg|_{\mathbf{m}_0} = 0 + \frac{\partial}{\partial m}\delta\tau^{(i)}\bigg|_{\mathbf{m}_0} = \left(\mathbf{h}^{(i)}\delta(\mathbf{x} - \mathbf{x}_R),\, \frac{\partial}{\partial m}\mathbf{u}(\mathbf{x}, t)\bigg|_{\mathbf{m}_0}\right) \tag{13.42}$$

This equation, due to Marquering et al. (1999), is exactly the same form as the one for the data kernel in Eq. (13.17), so it can be calculated using the adjoint method of Eq. (13.18). Numerical tests (Fig. 13.2) show that the formula, though based on a first-order approximation, is nevertheless very accurate.

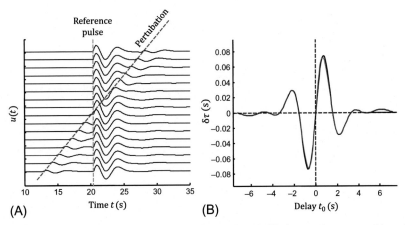

FIG. 13.2 Finite-frequency travel times. (A) A suite of signals $u(t)$ *(black curves)* are created by adding a small perturbation *(blue)* to a reference pulse u_0 *(red)*, where the signals differ only be the delay t_0 of the perturbation. (B) Cross-correlation *(black curve)* and the first-order approximation to it (Eq. (13.41), *red curve*) are used to estimate the travel time perturbation $\delta\tau$. Both yield similar results. Owing to wave interference, $\delta\tau(t_0)$ behaves like $\delta\tau \propto t_0$ only for very small delays. It is oscillatory at larger delays. *MatLab* script `gda13_03.m`.

13.6 BANANA-DOUGHNUT KERNELS

Once again, we step away from the complications of vector wave fields and use the scalar wave equation (Eq. 13.28) to study the behavior finite-frequency travel times. In keeping with the idea that the wave field is narrow-band, we assume a band-limited source time function $w(t)$:

$$w(t) = \mathrm{sinc}(\omega_1 t) - \mathrm{sinc}(\omega_2 t) \tag{13.43}$$

This function has a spectrum that is nonzero only angular frequencies between ω_1 and ω_2. The reference field $u_0(\mathbf{x}, t)$ and the related quantity $\xi(\mathbf{x}, t)$ are the same as given by Eqs. (13.29), (13.31), respectively. The adjoint equation and its solution are:

$$\mathcal{L}^\dagger \lambda = \frac{-\ddot{u}^{obs}(\mathbf{x}_R, t)}{D(\mathbf{x}_R)} \delta(\mathbf{x} - \mathbf{x}_R) \quad \text{and} \quad \lambda(\mathbf{x}, t) = \frac{-1}{D(\mathbf{x}_R)} \frac{\dot{u}^{obs}(\mathbf{x}_R, t + T_{RX})}{4\pi R_{RX}} \tag{13.44}$$

Assuming $\mathbf{m}_0 = 0$ for the reference value of the model parameters, the data kernel is:

$$\frac{\partial}{\partial m} \tau^{(i)} \Big|_0 = (\lambda, \xi) = \frac{-2}{D(\mathbf{x}_R)} \frac{1}{4\pi R_{RH}} \frac{1}{4\pi R_{SH}} \int_{-\infty}^{+\infty} \dot{u}^{obs}(\mathbf{x}_R, t + T_{RH}) \ddot{w}(t - T_{SH}) dt \tag{13.45}$$

Finally, we consider the case where the observed field equals the reference field; that is $u^{obs}(\mathbf{x}, t) = u_0(\mathbf{x}, t)$. The "observations" align perfectly with the reference field, so $\tau_0 = 0$. The effect of a small heterogeneity is to perturb the travel time by an amount:

$$\frac{\partial}{\partial m} \tau^{(i)} \Big|_0 = \frac{-2}{D(\mathbf{x}_R)} \frac{1}{4\pi R_{SH}} \frac{1}{4\pi R_{RH}} \frac{1}{4\pi R_{SR}} \int_{-\infty}^{+\infty} \dot{w}^{obs}(\mathbf{x}_R, t - T_{SR} + T_{RH}) \ddot{w}(t - T_{SH}) dt$$

$$= \frac{-2}{D(\mathbf{x}_R)} \frac{1}{4\pi R_{SH}} \frac{1}{4\pi R_{RH}} \frac{1}{4\pi R_{SR}} \int \dot{w}(t' - \Delta T) \ddot{w}(t') dt' \quad \text{where} \tag{13.46}$$

$$\Delta T = T_{SR} - T_{RH} - T_{SH} = T_{SR} - (T_{SH} + T_{HR})$$

The last step uses the transformation of variables $t' = t - T_{SH}$. The quantity ΔT represents the difference in travel time between the direct (S-R) and scattered (S-H-R) paths. The quantify $D(\mathbf{x}_R)$ is given by:

$$D(\mathbf{x}_R) = \int \ddot{u}^{obs}(\mathbf{x}_R, t) u^0(\mathbf{x}_R, t) dt = \frac{1}{4\pi R_{SR}} \frac{1}{4\pi R_{SR}} \int \ddot{w}(t - T_{SR}) w(t - T_{SR}) dt$$

$$= \frac{1}{4\pi R_{SR}} \frac{1}{4\pi R_{SR}} \int \ddot{w}(t') w(t') dt' \quad \text{with} \quad t' = t - T_{SR} \tag{13.47}$$

The derivative $\partial \tau / \partial m$ is axially symmetric about the (S-R) line, since R_{SH} and R_{RH} depend only on the perpendicular distance r of $\mathbf{x}_H$ from the line. Sliced perpendicular to the line, $\partial \tau / \partial m$ is *doughnut-shaped*. The derivative $\partial \tau / \partial m = 0$ whenever $\Delta T = 0$. This behavior follows from d/dt being an antiself-adjoint operator, since any quantity equal to its negative is zero:

$$\int \dot{w}(t) \ddot{w}(t) dt = - \int \ddot{w}(t) \dot{w}(t) dt = - \int \dot{w}(t) \ddot{w}(t) dt = 0 \tag{13.48}$$

The time difference ΔT is zero when the heterogeneity is between S and R and on the $(S\text{-}R)$ line, so $\partial\tau/\partial m = 0$ in this case. This zero makes the *hole* in the center of the doughnut.

At finite frequency, the travel time kernel is zero along the ray path. This result is paradoxical, for the ray theoretical kernel, which is valid in the high-frequency limit, is zero everywhere *except* on the ray path. It arises because the model perturbations have been parameterized as Dirac impulse functions, which are functions of infinitesimal size. Because it lacks a scale length, the Dirac function cannot advance or delay the scattered wave field. Advances and delays result from differences in the scattered and reference path lengths, only, and for a heterogeneity on the ray path, this difference is zero.

Now consider an oscillatory, band-limited source time function with a characteristic period P. Suppose we construct the elliptical volume surrounding the points $\mathbf{x}_S$ and $\mathbf{x}_R$ for which $\Delta T < P/2$. The time integral in Eq. (13.47) will have the same sign everywhere in this volume, as will $\partial\tau/\partial m$. This is a *banana-shaped* region enclosing the source and receiver (Fig. 13.3). The banana is thinner for short periods than for long periods. Moving away from the $(S\text{-}R)$ line along its perpendicular, the time integral, and hence the derivative, oscillates in sign, as the $f(t-\Delta T)$ and $\ddot{f}(t)$ factors beat against one another. The derivative also decreases in amplitude (since the factors R_{RH} and R_{SH} grow with distance). Consequently, the central banana is surrounded by a series of larger, but less intense, bananas of alternating sign.

On account of these behaviors, the finite-frequency travel time kernel is colloquially called a *banana-doughnut kernel*.

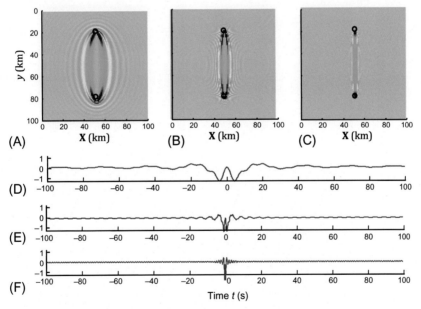

FIG. 13.3 Banana-doughnut kernels (the partial derivative $\partial\tau/\partial m$ of finite-frequency travel time τ with respect to model parameter m) for an acoustic wave propagation problem. The source $\mathbf{x}_S$ and receiver $\mathbf{x}_R$ *(back circles)* are separated by a distance of $R = 60$ km. The perturbation in acoustic velocity is $\delta v = (m/2v_0)\,\delta(x-x_h)$, where $v_0 = 1$ km/s is a constant reference velocity and $\mathbf{x}_h$ is the heterogeneity's location. (A–C) Kernels *(colors)* for a suite of $\mathbf{x}_h$'s covering the (x,y) plane, for frequency bands centers at $0.025, 0.125$, and 0.500 Hz, respectively. Note that the "banana" shape of the kernel narrows with frequency, becoming increasingly ray-like. (D–F) Corresponding band-limited source time functions used in the calculation. *MatLab* script gda13_04.m.

13.7 SEISMIC MIGRATION

Seismic *migration* is a technique for imaging the heterogeneities that cause reflected (or *scattered*) arrivals on seismic reflection profiles. In a seismic reflection experiment, a downgoing *incident* wave of known shape is sent into the Earth and up-going reflected waves (or *scattered* waves), caused by the interaction of the incident wave with heterogeneities at depth, are recorded by an array of receivers on the Earth's surface (Fig. 13.4). Seismic migration is based on the *imaging principle* that at the moment and location that the incident wave u^0 interacts with the heterogeneity to produce the scattered wave δu, the waveform of the scattered waves exactly matches $\ddot{u}^0$ (the $\partial^2/\partial t^2$ arises from the scattering interaction). This behavior suggests that the heterogeneity can be imaged by: (1) propagating the incident wave forward in time from its stating point to all possible heterogeneity locations $\mathbf{x}$; (2) propagating the scattered wave field backward in time from the receivers to these same locations; and (3) correlating (time integrating) the scattered wave field with the second derivative of the incident wavefield:

$$C(\mathbf{x}) = \int \ddot{u}^0(\mathbf{x}, t)\,\delta u(\mathbf{x}, t)\,\mathrm{d}t \qquad (13.49)$$

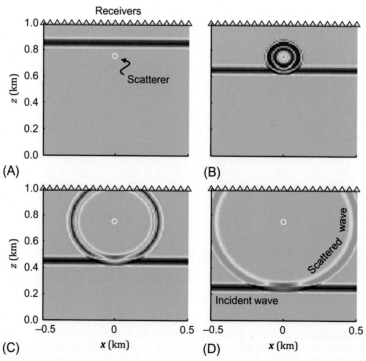

FIG. 13.4 (A–D) Sequence of four time snapshots of a downgoing incident plane wave *(red band)* interacting with a point-like heterogeneity *(white circle)* and generating a scattered spherical wave. The wave field is observed by an array of receivers *(triangles)* on the Earth's surface. *MatLab* script `gda_13.05`.

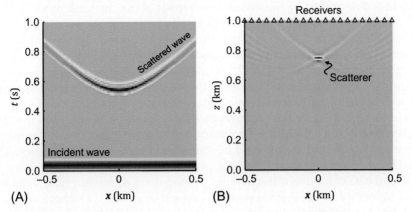

FIG. 13.5 (A) Seismic refection data recorded by the array of receivers in Fig. 13.4, containing incident and scattered waves. (B) The data from (A) is migrated to produce the correlation $C(x,z)$, which is a proxy for the estimated heterogeneity $m^{est}(x,z)$. The correlation is at the location of the true heterogeneity (compare with Fig. 13.4). *MatLab* script `gda_13.05`.

The correlation $C(x)$ will be peaked at the location of the heterogeneity (or, if several heterogeneities are present, at all their locations) and can be used as a proxy for the heterogeneous structure of the Earth (Fig. 13.5).

If this process of propagating some fields forward in time, others backward in time, and then correlating sounds familiar—it should! It is exactly the same as the process used to compute the error derivative $\partial E / \partial m$ (see Eq. 13.33). The imaging principle of seismic migration is equivalent to the idea that the error derivative is a proxy for the estimated heterogeneities $m^{est}(x)$. This equivalence is explored further by Zhu et al. (2009).

13.8 VELOCITY STRUCTURE FROM FREE OSCILLATIONS AND SEISMIC SURFACE WAVES

The earth, being a finite body, can oscillate only with certain characteristic patterns of motion (eigenfunctions) at certain corresponding discrete frequencies ω_{knm} (eigenfrequencies), where the indices (k,n,m) increase with the complexity of the motion. Most commonly, the frequencies of vibration are measured from the spectra of seismograms of very large earthquakes, and the corresponding indices are estimated by comparing the observed frequencies of vibration to those predicted using a reference model of the earth's structure. The inverse problem is to use small differences between the measured and predicted frequencies to improve the reference model.

The most fundamental part of this inverse problem is the Fréchet derivatives relating small changes in the earth's material properties to the resultant changes in the frequencies of vibration. These derivatives are computed through the pertubative techniques described in Section 12.10 and are three-dimensional analogs of Eq. (12.44) (Takeuchi and Saito, 1972). All involve integrals of the perturbation in earth structure with the eigenfunctions of the reference model, so the problem is one in linearized continuous inverse theory. The complexity of this problem is determined by the assumptions made about the earth model—whether it is isotropic or anisotropic, radially stratified or laterally heterogeneous—and how it is param-eterized. A radially stratified inversion that incorporates both body wave travel times and eigenfrequencies and that has become a standard in the seismological community is the Preliminary Reference Earth Model of Dziewonski and Anderson (1981).

Seismic *surface waves*, such as the horizontally polarized *Love* wave and the vertically po-larized *Rayleigh* wave, occur when seismic energy is trapped near the earth's surface by the rapid increase in seismic velocities with depth. The velocity of these waves is very strongly frequency dependent, since the lower-frequency waves extend deeper into the earth than the higher-frequency waves and are affected by correspondingly higher velocities. The surface wave caused by an impulsive source, such as an earthquake, is not itself impulsive, but rather is dispersed, since its component frequencies propagate at different phase velocities.

The phase velocity c of the surface wave is defined as $c = \omega(k)/k$, where ω is angular fre-quency and k is wavenumber (wavenumber is 2π divided by wavelength). The function $c(\omega)$ (or alternatively, $c(k)$) is called the *dispersion function* (or, sometimes, the *dispersion curve*) of the surface wave. It is easily measured from observations of earthquake wave fields and easily predicted for vertically stratified earth models. The inverse problem is to infer the earth's material properties from observations of the dispersion function.

As in the case of free oscillations, Fréchet derivatives can be calculated using a pertubative approach. For Love waves, the perturbation in phase velocity c due to a perturbation in shear modulus μ and density ρ is (Aki and Richards, 2002, their Eq. 7.71):

$$\left(\frac{\delta c}{c}\right)_\omega = \frac{\int_0^\infty \left[k^2 l_1^2 + \left(\frac{dl_1}{dz}\right)^2\right] \delta\mu dz - \int_0^\infty \left[\omega^2 l_1^2\right] \delta\rho dz}{2k^2 \int_0^\infty \mu_0 l_1^2 \, dz} \tag{13.50}$$

Here, shear modulus $\mu(z)$ and density $\rho(z)$ are assumed to be vertically stratified in depth z, and $u_y = l_1(\omega,k,z) \exp(ikx - i\omega t)$ is the horizontal displacement of the Love wave. The corresponding formula for the Rayleigh wave (Aki and Richards, 2002, their Eq. 7.78) is sim-ilar in form but involves perturbations in three material parameters, Lamé parameter λ, shear modulus μ, and density ρ.

When the earth model contains three-dimensional heterogeneities, the inverse problem of determining earth structure from surface waves is just a type of wavefield tomography (when seismograms are being matched) or finite-frequency tomography (when dispersion functions are being matched). The dispersion function is now understood to encode the frequency-dependent time shift between an observed and predicted surface wave field. Adjoint methods can be used to compute its Fréchet derivatives (Sieminski et al., 2007).

Ray-theoretical ideas can also be applied to surface waves, which at sufficiently short pe-riods can be approximated as propagating along a horizontal ray path connecting source and receiver (Wang and Dahlen, 1995). Each point on the earth's surface is presumed to have its own dispersion function $c(\omega,x,y)$. The dispersion function $c^{(i)}(\omega)$ observed for a particular source-observer path i is then calculated using the *pure path approximation*

$$\frac{1}{c^{(i)}(\omega)} = \frac{1}{S_i} \int_{\text{path } i} \frac{d\ell}{c(\omega, x, y)} \tag{13.51}$$

Here, S_i is the length of path i. The path is often approximated as the great circle connecting source and receiver. This is a two-dimensional tomography problem relating line integrals of $c(\omega,x,y)$ to observations of $c^{(i)}(\omega)$—hence the term *surface wave tomography*. The $c(\omega,x,y)$ then can be inverted for estimates of the vertical structure using the vertically stratified method described earlier (e.g., Boschi and Ekström, 2002). This two-step process is not as accurate as full waveform tomography but is computationally much faster.

13.9 SEISMIC ATTENUATION

Internal friction within the earth causes seismic waves to lose amplitude as they propagate. The amplitude loss can be characterized by the *attenuation factor* $\alpha(\omega,\mathbf{x})$, which can vary with angular frequency ω and position $\mathbf{x}$. In a regime where ray theory is accurate, the fractional loss of amplitude A of the seismic wave is

$$\frac{A}{A_0} = \exp\left\{ -\int_{\text{ray}} \alpha(\omega, \mathbf{x})d\ell \right\} \tag{13.52}$$

Here, A_0 is the initial amplitude. After linearization through taking its logarithm, this equation is a standard tomography problem (identical in form to the X-ray problem described in Section 1.3.4)—hence the term *attenuation tomography*. Jacobson et al. (1981) discuss its solution when the attenuation is assumed to vary only with depth. Dalton and Ekström (2006) perform a global two-dimensional tomographic inversion using surface waves. Many authors have performed three-dimensional inversions (see review by Romanowicz, 1998).

A common problem in attenuation tomography is that the initial amplitude A_0 (or equivalently, the CMT of the seismic source) is unknown (or poorly known). Menke et al. (2006) show that including A_0 as an unknown parameter in the inversion introduces a nonuniqueness that is mathematically identical to the one encountered when an unknown origin time is introduced into the velocity tomography problem.

Attenuation also has an effect on the free oscillations of the earth, causing a widening in the spectral peaks associated with the eigenfrequencies ω_{knm}. The amount of widening depends on the relationship between the spatial variation of the attenuation and the spatial dependence of the eigenfunction and can be characterized by a quality factor Q_{knm} (i.e., a fractional loss of amplitude per oscillation) for that mode. Pertubative methods are used to calculate the Fréchet derivatives of Q_{knm} with the attenuation factor. One example of such an inversion is that of Masters and Gilbert (1983).

13.10 SIGNAL CORRELATION

Geologic processes record signals only imperfectly. For example, while variations in oxygen isotopic ratios $r(t)$ through geologic time t are recorded in oxygen-bearing sediments as they are deposited, the sedimentation rate is itself a function of time. Measurements of isotopic ratio $r(z)$ as a function of depth z cannot be converted to the variation of $r(t)$ without knowledge of the sedimentation function $z(t)$ (or equivalently $t(z)$).

Under certain circumstances, the function $r(t)$ is known prior information (for instance, oxygen isotopic ratio correlates with temperature, which can be estimated independently). In these instances, it is possible to use the observed $r^{obs}(z)$ and the predicted $r^{pre}(t)$ to invert for the function $t(z)$. This is essentially a problem in signal correlation: distinctive features that can be correlated between $r^{obs}(z)$ and $r^{pre}(t)$ establish the function $t(z)$. The inverse problem

$$r^{obs}(z) = r^{pre}[t(z)] \tag{13.53}$$

is therefore a problem in nonlinear continuous inverse theory. The unknown function $t(z)$—often called the *mapping function*—must increase monotonically with z. The solution of this problem is discussed by Martinson et al. (1982) and Shure and Chave (1984).

13.11 TECTONIC PLATE MOTIONS

The motion of the earth's rigid tectonic plates can be described by an *Euler vector* ω, whose orientation gives the pole of rotation and whose magnitude gives its rate. Euler vectors can be used to represent relative rotations, that is, the rotation of one plate relative to another, or absolute motion, that is, motion relative to the earth's mantle. If we denote the relative rotation of plate A with respect to plate B as ω_{AB}, then the Euler vectors of three plates A–C satisfy the relationship $\omega_{AB} + \omega_{BC} + \omega_{CA} = 0$. Once the Euler vectors for plate motion are known, the relative velocity between two plates at any point on their boundary can easily be calculated from trigonometric formulas.

Several geologic features provide information on the relative rotation between plates, including the faulting directions of earthquakes at plate boundaries and the orientation of transform faults, which constrain the direction of the relative velocity vectors, and spreading rate estimates based on magnetic lineations at ridges, which constrain the magnitude of the relative velocity vectors.

These data can be used in an inverse problem to determine the Euler vectors (Chase and Stuart, 1972; DeMets et al., 2010; Minster et al., 1974). The main differences between various authors' approaches are in the manner in which the Euler vectors are parameterized and the types of data that are used. Some authors use Cartesian components of the Euler vectors; others use magnitude, azimuth, and inclination. The two inversions behave somewhat differently in the presence of noise; Cartesian parameterizations seem to be more stable. Data include seafloor spreading rates, azimuths of strike-slip plate boundaries, earthquake moment tensors, and Global Positioning System strain measurements.

13.12 GRAVITY AND GEOMAGNETISM

Inverse theory plays an important role in creating representations of the earth's gravity and magnetic fields. Field measurements made at many points about the earth need to be combined into a smooth representation of the field, a problem which is mainly one of interpolation in the presence of noise and incomplete data (see Section 3.9.3). Both spherical harmonic expansions and various kinds of spline functions (Sandwell, 1987; Shure et al., 1982, 1985; Wessel and Becker, 2008) have been used in the representations. In either case, the trade-off of resolution and variance is very important.

Studies of the earth's core and geodynamo require that measurements of the magnetic field at the earth's surface be extrapolated to the core-mantle boundary. This is an inverse problem of considerable complexity, since the short-wave-length components of the field that are most important at the core-mantle boundary are very poorly measured at the earth's surface. Furthermore, the rate of change of the magnetic field with time (called "secular variation") is of critical interest. This quantity must be determined from fragmentary historical measurements, including measurements of compass deviation recorded in old ship logs. Consequently, these inversions introduce substantial prior constraints on the behavior of the field near the core, including the assumption that the root-mean-square time rate of change of the field is minimized and the total energy dissipation in the core is minimized (Bloxam, 1987). Once the magnetic field and its time derivative at the core-mantle boundary have been determined, they can be used in inversions for the fluid velocity near the surface of the outer core (Bloxam, 1992).

The earth's gravity field $\mathbf{g}(\mathbf{x})$ is determined by its density structure $\rho(\mathbf{x})$. In parts of the earth where electric currents and magnetic induction are unimportant, the magnetic field $\mathbf{H}(\mathbf{x})$ is caused by the magnetization $\mathbf{M}(\mathbf{x})$ of the rocks. These quantities are related by

$$g_i(\mathbf{x}) = \int_V G_i^g(\mathbf{x}, \mathbf{x}_0)\rho(\mathbf{x}_0)dV_0 \quad \text{and} \quad H_i(\mathbf{x}) = \int_V G_{ij}^H(\mathbf{x}, \mathbf{x}_0)M_j(\mathbf{x}_0)dV_0$$

$$G_i^g(\mathbf{x}) = -\frac{\gamma(x_i - x_{0i})}{|\mathbf{x} - \mathbf{x}_0|^3} \tag{13.54}$$

$$G_{ij}^H(\mathbf{x}) = \frac{1}{4\pi\mu_0}\left[\frac{\delta_{ij}}{|\mathbf{x} - \mathbf{x}_0|^3} - \frac{3(x_i - x_{0i})(x_j - x_{0j})}{|\mathbf{x} - \mathbf{x}_0|^5}\right]$$

Here, γ is the gravitational constant and μ_0 is the magnetic permeability of the vacuum. Note that in both cases, the fields are linearly related to the sources (density and magnetization), so these are linear, continuous inverse problems. Nevertheless, inverse theory has proved to have little application to these problems, owing to their inherent underdetermined nature. In both cases, it is possible to show (e.g., Menke and Abbott, 1989, their Section 5.14) that the field outside a finite body can be generated by an infinitesimally thick spherical shell of mass or magnetization surrounding the body and below the level of the measurements. The null space of these problems is so large that it is generally impossible to formulate any useful solution (as we encountered in Fig. 7.7), except when an enormous amount of prior information is added to the problem.

Some progress has been made in special cases where prior constraints can be sensibly stated. For instance, the magnetization of sea mounts can be computed from their magnetic anomaly, assuming that their magnetization vector is everywhere parallel (or nearly parallel) to some known direction and that the shape of the magnetized region closely corresponds to the observed bathymetric expression of the sea mount (Grossling, 1970; Parker, 1988; Parker et al., 1987).

13.13 ELECTROMAGNETIC INDUCTION AND THE MAGNETOTELLURIC METHOD

An electromagnetic field is *induced* within a conducting medium when it is illuminated by a plane electromagnetic wave. In a homogeneous medium, the field decays exponentially with depth, since energy is dissipated by electric currents induced in the conductive material. In a heterogeneous medium, the behavior of the field is more complicated and depends on the details of the electrical conductivity $\sigma(\mathbf{x})$.

Consider the special case of a vertically stratified earth illuminated by a vertically incident electromagnetic wave (e.g., from space). The plane wave consists of a horizontal electric field E_x and a horizontal magnetic field H_y, with their ratio on the earth's surface (at $z=0$) being determined by the conductivity profile $\sigma(z)$. The quantity

$$Z(\omega) = \frac{E_x(\omega, z=0)}{i\omega H_y(\omega, z=0)} \tag{13.55}$$

that relates the two fields is called the *impedance*. It is frequency dependent, because the depth of penetration of an electromagnetic wave into a conductive medium depends upon its frequency, so different frequencies sense different parts of the conductivity structure. The *magnetotelluric* problem is to determine the conductivity $\sigma(z)$ from measurements of the impedance $Z(\omega)$ at a suite of angular frequencies ω. This one-dimensional nonlinear inverse problem is relatively well understood. For instance, Fréchet derivatives are relatively straightforward to compute (Parker, 1970, 1977). Oldenburg (1983) discusses the discretization of the problem and the application of extremal inversion methods (see Chapter 6).

In three-dimensional problems, all the components of magnetic $\mathbf{H}$ and electric $\mathbf{E}$ fields are now relevant, and the impedance matrix $\mathbf{Z}(\omega,\mathbf{x})$ that connects them via $\mathbf{E} = \mathbf{Z}\mathbf{H}$ constitutes the data. The inverse problem of determining $\sigma(\mathbf{x})$ from observations of $\mathbf{Z}(\omega,\mathbf{x})$ at a variety of frequencies and positions is much more difficult than the one-dimensional version (Chave et al., 2012), especially in the presence of large heterogeneities in electoral resistivity caused, for instance, by topography (e.g., Baba and Chave, 2005).

13.14 PROBLEMS

13.1 Show that the differential operator in Eq. (13.12) is self-adjoint. Hint: the adjoint of $(\partial/\partial x)$ $a(\mathbf{x})(\partial/\partial y)$ is $(\partial/\partial y)a(\mathbf{x})(\partial/\partial x)$.

13.2 13.2. Modify the seismic migration example (`gda13_05.m`) to include $N_H=5$ heterogeneities at locations of your choosing.

References

Aki, K., Richards, P.G., 2002. Quantitative Seismology, second ed. University Science Books, Sausalito, CA. 700 pp.

Aki, K., Christoffersson, A., Husebye, E., 1976. Three-dimensional seismic structure under the Montana LASA. Bull. Seismol. Soc. Am. 66, 501–524.

Baba, K., Chave, A., 2005. Correction of seafloor magnetotelluric data for topographic effects during inversion. J. Geophys. Res.. 110,B12105. https://doi.org/10.1029/2004JB003463. 16 pp.

Bloxam, J., 1987. Simultaneous inversion for geomagnetic main field and secular variation, 1. A large scale inversion problem. J. Geophys. Res. 92, 11597–11608.

Bloxam, J., 1992. The steady part of the secular variation of the earth's magnetic field. J. Geophys. Res. 97, 19565–19579.

Boschi, J., Ekström, G., 2002. New images of the Earth's upper mantle from measurements of surface wave phase velocity anomalies. J. Geophys. Res. 107, 2059.

Cerveny, V., 2001. Seismic Ray Theory. Cambridge University Press, New York. 607 pp.

Chase, E.P., Stuart, G.S., 1972. The N plate problem of plate tectonics. Geophys. J. R. Astron. Soc. 29, 117–122.

Chave, A., Jones, A., Mackie, R., Rodi, W., 2012. The Magnetotelluric Method, Theory and Practice. Cambridge University Press, New York. 590 pp.

Crosson, R.S., 1976. Crustal structure modeling of earthquake data: 1. Simultaneous least squares estimation of hypocenter and velocity parameters. J. Geophys. Res. 81, 3036–3046.

Dalton, C.A., Ekström, G., 2006. Global models of surface wave attenuation. J. Geophys. Res. 111. B05317.

DeMets, C., Gordon, R.G., Argus, D.F., 2010. Geologically current plate motions. Geophys. J. Int. 181, 1–80.

Dziewonski, A.M., Anderson, D.L., 1981. Preliminary reference earth model. Phys. Earth Planet. Inter. 25, 297–358.

Dziewonski, A.M., Woodhouse, J.H., 1983. An experiment in systematic study of global seismicity: centroid-moment tensor solutions for 201 moderate and large earthquakes of 1981. J. Geophys. Res. 88, 3247–3271.

Dziewonski, A.M., Chou, T.-A., Woodhouse, J.H., 1981. Determination of earthquake source parameters from waveform data for studies of global and regional seismicity. J. Geophys. Res. 86, 2825–2852.

Geiger, L., 1912. Probability method for the determination of earthquake epicenters from the arrival time only (translated from Geiger's 1910 German article). Bull. St. Louis Univ. 8, 56–71.

Grossling, B.F., 1970. Seamount magnetism. In: Maxwell, A.E. (Ed.), The Sea. In: vol. 4. Wiley, New York, pp. 129–156.

Jacobson, R.S., Shor, G.G., Dorman, L.M., 1981. Linear inversion of body wave data—part 2: attenuation vs. depth using spectral ratios. Geophysics 46, 152–162.

Kim, Y., Liu, Q., Tromp, J., 2011. Adjoint centroid-moment tensor inversions. Geophys. J. Int. 186, 264–278.

Klein, F.W., 1985. User's Guide to HYPOINVERSE, a Program for VAX and PC350 Computers to Solve for Earthquake Locations. U.S. Geologic Survey Open File Report 85–515, 24 pp.

Marquering, H., Dahlen, F.A., Nolet, G., 1999. Three-dimensional sensitivity kernels for finite frequency traveltimes: the banana-doughnut paradox. Geophys. J. Int. 137, 805–815.

Martinson, D.G., Menke, W., Stoffa, P., 1982. An inverse approach to signal correlation. J. Geophys. Res. 87, 4807–4818.

Masters, G., Gilbert, F., 1983. Attenuation in the earth at low frequencies. Philos. Trans. R. Soc. Lond. Ser. A 388, 479–522.

Menke, W., 2005. Case studies of seismic tomography and earthquake location in a regional context. In: Levander, A., Nolet, G. (Eds.), Seismic Earth: Array Analysis of Broadband Seismograms. In: Geophysical Monograph Series, vol. 157. American Geophysical Union, Washington, DC, pp. 7–36.

Menke, W., Abbott, D., 1989. Geophysical Theory (Textbook). Columbia University Press, New York. 458 pp.

Menke, W., Schaff, D., 2004. Absolute earthquake location with differential data. Bull. Seismol. Soc. Am. 94, 2254–2264.

Menke, W., Holmes, R.C., Xie, J., 2006. On the nonuniqueness of the coupled origin time- velocity tomography problem. Bull. Seismol. Soc. Am. 96, 1131–1139.

Menke, W., Menke, H., 2015. Improved precision of delay times determined through cross correlation achieved by out-member averaging. Jokull 64, 15–22.

Minster, J.B., Jordan, T.H., Molnar, P., Haines, E., 1974. Numerical modeling of instantaneous plate tectonics. Geophys. J. R. Astron. Soc. 36, 541–576.

Oldenburg, D.W., 1983. Funnel functions in linear and nonlinear appraisal. J. Geophys. Res. 88, 7387–7398.

Parker, R.L., 1970. The inverse problem of the electrical conductivity of the mantle. Geophys. J. R. Astron. Soc. 22, 121–138.

Parker, R.L., 1977. The Fréchet derivative for the one dimensional electromagnetic induction problem. Geophys. J. R. Astron. Soc. 39, 543–547.

Parker, R.L., 1988. A statistical theory of seamount magnetism. J. Geophys. Res. 93, 3105–3115.

Parker, R.L., Shure, L., Hildebrand, J., 1987. An application of inverse theory to seamount magnetism. Rev. Geophys. 25, 17–40.

Pavlis, G.L., Booker, J.R., 1980. The mixed discrete-continuous inverse problem: application to the simultaneous determination of earthquake hypocenters and velocity structure. J. Geophys. Res. 85, 4801–4809.

Romanowicz, B., 1998. Attenuation tomography of the earth's mantle: a review of current status. Pure Appl. Geophys. 153, 257–272.

Sandwell, D.T., 1987. Biharmonic spline interpolation of GEOS-3 and SEASAT altimeter data. Geophys. Res. Lett. 14, 139–142.

Shure, L., Chave, A.D., 1984. Comments on "An inverse approach to signal correlation" J. Geophys. Res. 89, 2497–2500.

Shure, L., Parker, R.L., Backus, G.E., 1982. Harmonic splines for geomagnetic modeling. Phys. Earth Planet. Inter. 28, 215–229.

Shure, L., Parker, R.L., Langel, R.A., 1985. A preliminary harmonic spline model for MAGSAT data. J. Geophys. Res. 90, 11505–11512.

Sieminski, A., Liu, Q., Trampert, J., Tromp, J., 2007. Finite-frequency sensitivity of surface waves to anisotropy based upon adjoint methods. Geophys. J. Int. 168, 1153–1174.

Slunga, R., Rognvaldsson, S.T., Bodvarsson, B., 1995. Absolute and relative locations of similar events with application to microearthquakes in southern Iceland. Geophys. J. Int. 123, 409–419.

Spence, W., Alexander, S., 1968. A method of determining the relative location of seismic events. Earthquake Notes 39, 13.

Takeuchi, H., Saito, M., 1972. Seismic surface waves. In: Bolt, B.A. (Ed.), Methods of Computational Physics. In: vol. 11. Academic Press, New York, pp. 217–295.

Vasco, D.W., 1986. Extremal inversion of travel time residuals. Bull. Seismol. Soc. Am. 76, 1323–1345.

Vidale, J., 1990. Finite difference calculation of travel times in 3-D. Geophysics 55, 521–526.

Waldhauser, F., Ellsworth, W., 2000. A double-difference earthquake location algorithm; method and application to the northern Hayward Fault, California. Bull. Seismol. Soc. Am. 90, 1353–1368.

Wang, Z., Dahlen, F.A., 1995. Validity of surface-wave ray theory on a laterally heterogeneous earth. Geophys. J. Int. 123, 757–773.

Wessel, P., Becker, J., 2008. Gridding of spherical data using a Green's function for splines in tension. Geophys. J. Int. 174, 21–28.

Zelt, C.A., 1998. FAST: 3-D First Arrival Seismic Tomography Programs. http://terra.rice.edu/department/faculty/zelt/fast.html.

Zelt, C.A., Barton, P.J., 1998. 3D seismic refraction tomography: a comparison of two methods applied to data from the Faeroe Basin. J. Geophys. Res. 103, 7187–7210.

Zhang, H., Thurber, C.H., 2003. Double-difference tomography; the method and its application to the Hayward Fault, California. Bull. Seismol. Soc. Am. 93, 1875–1889.

Zhu, H., Luo, Y., Nissen-Meyer, T., Morency, C., Tromp, J., 2009. Elastic imaging and time-lapse migration based on adjoint methods. Geophysics 74, WCA167–WCA177.

Further Reading

Dahlen, F.A., Hung, S.-H., Nolet, G., 2000. Fréchet kernels for finite frequency traveltimes—I. Theory. Geophys. J. Int. 141, 157–174.

Tromp, J., Tape, C., Liu, Q., 2005. Seismic tomography, adjoint methods, time reversal and banana-doughnut kernels. Geophys. J. Int. 160, 195–216.

14

Appendices

14.1 IMPLEMENTING CONSTRAINTS WITH LAGRANGE MULTIPLIERS

Consider the problem of minimizing a function of two variables, say, $E(x, y)$, with respect to x and y, subject to the constraint that $C(x, y) = 0$. One way to solve this problem is to first use $C(x, y) = 0$ to write y as a function of x and then substitute this function into $E(x, y)$. The resulting function of a single variable $E(x, y(x))$ can now be minimized by setting $dE/dx = 0$. The constraint equation is used to explicitly reduce the number of independent variables.

One problem with this method is that it is rarely possible to solve $C(x, y)$ explicitly for either $y(x)$ or $x(y)$. The method of Lagrange multipliers provides a method of dealing with the constraints in their implicit form.

The constraint equation $C(x, y) = 0$ defines a curve in (x, y) on which the solution must lie (Fig. 14.1). As a point is moved along this curve, the value of E changes, achieving a minimum at (x_0, y_0). This point is not the global minimum of E, which lies, in general, off the curve. However, it must be at a point at which ∇E is perpendicular to the tangent $\hat{t}$ to the curve, or else the point could be moved along the curve to further minimize E. The tangent is perpendicular to ∇C, so the condition that $\nabla E \perp \hat{t}$ is equivalent to $\nabla E \propto \nabla C$. Writing the proportionality factor as $-\lambda$, we have $\nabla E + \lambda \nabla C = 0$ or $\nabla(E + \lambda C) = 0$. Thus, the constrained minimization of E subject to $C = 0$ is equivalent to the unconstrained minimization of $\Phi = E + \lambda C$, except that a new variable λ is introduced that needs to be determined as part of the problem. The two equations $\nabla \Phi = 0$ and $C = 0$ must be solved simultaneously to yield (x_0, y_0) and λ.

In N-dimensions and with $L < N$ constraint equations, each constraint $C_i = 0$ describes an $N-1$ dimensional surface. Their intersection is an $N-L$ dimensional surface. We treat the $L = N - 1$ case here, where the intersection is a curve (the argument for the other cases being similar). As in the two-dimensional case, the condition that a point (x_0, y_0) is at a minimum of E on this curve is that ∇E is perpendicular to the tangent $\hat{t}$ to the curve. The curve necessarily lies within all of the intersecting surfaces, and so $\hat{t}$ is perpendicular to any surface normal ∇C_i. Thus, the condition that $\nabla E \perp \hat{t}$ is equivalent to the requirement that ∇E be constructed from a linear combination of surface normals, ∇C_i. Calling the coefficients of the linear combination $-\lambda_i$, we have

FIG. 14.1   Graphical interpretation of the method of Lagrange multipliers, in which the function $E(x, y)$ is minimized subject to the constraint that $C(x, y)=0$. The solution *(bold dot)* occurs at the point (x_0, y_0) on the curve $C(x, y)=0$, where the perpendicular direction *(gray arrows)* is parallel to the gradient $\nabla E(x, y)$ *(white arrows)*. At this point, E can only be further minimized by moving the point (x_0, y_0) off of the curve, which is disallowed by the constraint. *MatLab* script gda14_01.

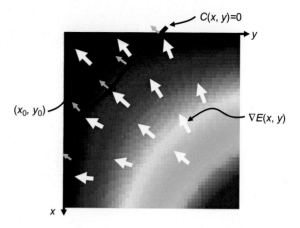

$$\nabla E = -\sum_{i=1}^{L} \lambda_i \nabla C_i \ \text{ or } \ \nabla E + \sum_{i=1}^{L} \lambda_i \nabla C_i = 0 \ \text{ or } \ \nabla \Phi = 0 \ \text{ with } \ \Phi = E + \sum_{i=1}^{L} \lambda_i C_i \qquad (14.1)$$

As stated earlier, the constrained minimization of E subject to $C_i=0$ is equivalent to the unconstrained minimization of Φ, but now Φ contains one Lagrange multiplier λ_i for each of the L constraints.

14.2 L_2 INVERSE THEORY WITH COMPLEX QUANTITIES

Some inverse problems (especially those that deal with data that have been operated on by Fourier or other transforms) involve complex quantities that are considered random variables. A complex random variable, say $z=x+iy$, has real and imaginary parts, x and y, that are themselves real-valued random variables. The probability density function $p(z)$ gives the probability that the real part of z is between x and $x+\Delta x$ and an imaginary part is in between y and $y+\Delta y$. The probability density function is normalized so that its integral over the complex plane is unity:

$$\int_{-\infty}^{+\infty}\int_{-\infty}^{+\infty} p(z)\mathrm{d}x\mathrm{d}y = 1 \qquad (14.2)$$

The mean or expected value of the z is then the obvious generalization of the real-valued case

$$\langle z \rangle = \int_{-\infty}^{+\infty}\int_{-\infty}^{+\infty} zp(z)\mathrm{d}x\mathrm{d}y \qquad (14.3)$$

In general, the real and imaginary parts of z can be of unequal variance and be correlated. However, the uncorrelated, equal variance case, which is called a *circular random variable*, suffices to describe the behavior of many systems. Its variance is defined as

$$\sigma = \int_{-\infty}^{+\infty} \int_{-\infty}^{+\infty} (z - \langle z \rangle)^*(z - \langle z \rangle)p(z)dxdy = \int_{-\infty}^{+\infty} \int_{-\infty}^{+\infty} |z - \langle z \rangle|^2 p(z)dxdy \qquad (14.4)$$

Here, the asterisk signifies complex conjugation. A vector $\mathbf{z}$ of N circular random variables, with probability density function $p(\mathbf{z})$, has mean and covariance given by

$$\langle \mathbf{z} \rangle_i = \int_{-\infty}^{+\infty} \int_{-\infty}^{+\infty} z_i p(\mathbf{z}) d^N x d^N$$
$$[\text{cov } \mathbf{z}] = \int_{-\infty}^{+\infty} \int_{-\infty}^{+\infty} (z_i - \langle z_i \rangle)^* (z_j - \langle z_j \rangle) p(\mathbf{z}) d^N x d^N y \qquad (14.5)$$

Note that $[\text{cov } \mathbf{z}]$ is a Hermitian matrix, that is, a matrix whose transform is its complex conjugate.

The definition of the L_2 norm must be changed to accommodate the complex nature of quantities. The appropriate change is to define the squared length of a vector $\mathbf{v}$ to be

$$\|\mathbf{v}\|_2^2 = \mathbf{v}^H \mathbf{v} \qquad (14.6)$$

where $\mathbf{v}^H$ is the *Hermitian transpose*, that is, the transpose of the complex conjugate of the vector $\mathbf{v}$. This choice ensures that the norm is a nonnegative real number. When the results of L_2 inverse theory are rederived for circular complex vectors, the results are very similar to those derived previously; the only difference is that all the ordinary transposes are replaced by Hermitian transposes. For instance, the least squares solution is

$$\mathbf{m}^{\text{est}} = \left[\mathbf{G}^H \mathbf{G}\right]^{-1} \mathbf{G}^H \mathbf{d} \qquad (14.7)$$

Note that all the square symmetric matrices of real inverse theory now become square Hermitian matrices:

$$[\text{cov } \mathbf{d}], \ [\text{cov } \mathbf{m}], \ \left[\mathbf{G}^H \mathbf{G}\right], \ \left[\mathbf{G}\mathbf{G}^H\right], \ \text{etc.} \qquad (14.8)$$

Hermitian matrices have real eigenvalues, so no special problems arise when deriving eigenvalues or singular-value decompositions.

The modification made to Householder transformations is also very simple. The requirement that the Hermitian length of a vector be invariant under transformation implies that a unitary transformation must satisfy $\mathbf{T}^H \mathbf{T} = \mathbf{I}$. The most general unitary transformation is therefore $\mathbf{T} = \mathbf{I} - 2\mathbf{v}\mathbf{v}^H / \mathbf{v}^H \mathbf{v}$, where $\mathbf{v}$ is any complex vector. The Householder transformation that annihilates the jth column of $\mathbf{G}$ below its main diagonal then becomes

$$\mathbf{T}_j = \mathbf{I} - \frac{1}{|\alpha|(|\alpha| + |G_{j,j}|)} \begin{bmatrix} 0 \\ \vdots \\ 0 \\ (G_{j,j} - \alpha) \\ G_{j+1,j} \\ \vdots \\ G_{N,j} \end{bmatrix} \begin{bmatrix} 0 & \cdots & 0 & \left(G_{j,j}^* - \alpha^*\right) & G_{j+1,j}^* & \cdots & G_{N,j}^* \end{bmatrix} \quad \text{where } |\alpha| = \sqrt{\sum_{i=j}^{N} \left|G_{i,j}^2\right|}$$

$$(14.9)$$

The phase of α is chosen to be π away from the phase of $G_{j,j}$. This choice guarantees that the transformation is in fact a unitary transformation and that the denominator is not zero.

14.3 METHOD SUMMARIES

Method Summary 1, Least Squares

Step 1: State the problem in words

How are the data related to the model?

Step 2: Organize the problem in standard form

Identify the data $\mathbf{d}$ (length N) and the model parameters $\mathbf{m}$ (length M). Define the data kernel $\mathbf{G}$ so that $\mathbf{Gm} = \mathbf{d}^{\text{obs}}$.

Step 3: Examine the data

Make plots of the data and look for outliers and other problems.

Step 4: Establish the accuracy of the data

State a prior covariance matrix $[\text{cov}\,\mathbf{d}]$ based on accuracy of the measurement technique. For uncorrelated data with uniform variance $[\text{cov}\,\mathbf{d}] = \sigma_d^2 \mathbf{I}$.

Step 5: State the prior information in words, for example:

The model parameters are close to a known values, $\mathbf{h}^{\text{pri}}$.
The mean of the model parameters is close to a known value.
The model parameters vary smoothly with space and/or time.

Step 6: Organize the prior information in standard form:

$\mathbf{Hm} = \mathbf{h}^{\text{pri}}$ where $\mathbf{h}^{\text{pri}}$ is of length K

Step 7: Establish the accuracy of the prior information

State a prior covariance matrix $[\text{cov}\,\mathbf{h}]$ based on accuracy of the prior information. For uncorrelated information of uniform variance $[\text{cov}\,\mathbf{h}] = \sigma_h^2 \mathbf{I}$.

Step 8: Estimate model parameter $\mathbf{m}^{\text{est}}$ and their covariance $[\text{cov}\,\mathbf{m}]$

$$\mathbf{m}^{\text{est}} = \left[\mathbf{F}^{\text{T}}\mathbf{F}\right]^{-1}\mathbf{F}^{\text{T}}\mathbf{f}^{\text{obs}} \text{ and } [\text{cov}\,\mathbf{m}] = \left[\mathbf{F}^{\text{T}}\mathbf{F}\right]^{-1}$$

$$\text{with}\,\mathbf{F} = \begin{bmatrix} [\text{cov}\,\mathbf{d}]^{-\frac{1}{2}}\,\mathbf{G} \\ [\text{cov}\,\mathbf{h}]^{-\frac{1}{2}}\,\mathbf{H} \end{bmatrix} \text{ and } \mathbf{f}^{\text{obs}} = \begin{bmatrix} [\text{cov}\,\mathbf{d}]^{-\frac{1}{2}}\,\mathbf{d}^{\text{obs}} \\ [\text{cov}\,\mathbf{h}]^{-\frac{1}{2}}\,\mathbf{h}^{\text{pri}} \end{bmatrix}$$

For the uniform, uncorrelated case:

$$\mathbf{F} = \begin{bmatrix} \sigma_d^{-1}\,\mathbf{G} \\ \sigma_h^{-1}\,\mathbf{H} \end{bmatrix} \text{ and } \mathbf{f}^{\text{obs}} = \begin{bmatrix} \sigma_d^{-1}\,\mathbf{d}^{\text{obs}} \\ \sigma_h^{-1}\,\mathbf{h}^{\text{pri}} \end{bmatrix}$$

Step 9: Examine the model resolution

$$\mathbf{R}^{G} = \mathbf{G}^{-g}\mathbf{G} \text{ with } \mathbf{G}^{-g} = \left[\mathbf{F}^{\text{T}}\mathbf{F}\right]^{-1}\mathbf{G}^{\text{T}}[\text{cov } \mathbf{d}]^{-1}$$

Any departure of $\mathbf{R}^{G}$ from a diagonal matrix indicates that only averages of the model parameters can be resolved.

Are these averages localized?

Step 10: State estimates and their 95% confidence intervals

$$m_i^{\text{true}} = m_i^{\text{est}} \pm 2\sigma_{mi} \ (95\%) \text{ with } \sigma_{mi} = \sqrt{[\text{cov } \mathbf{m}]_{ii}}$$

Step 11: Examine the individual errors

$$\mathbf{d}^{\text{pre}} = \mathbf{Gm}^{\text{est}} \text{ and } \mathbf{e} = [\text{cov } \mathbf{d}]^{-\frac{1}{2}}\left(\mathbf{d}^{\text{obs}} - \mathbf{d}^{\text{pre}}\right)$$

$$\mathbf{h}^{\text{pre}} = \mathbf{Hm}^{\text{est}} \text{ and } \boldsymbol{\ell} = [\text{cov } \mathbf{h}]^{-\frac{1}{2}}\left(\mathbf{h}^{\text{pri}} - \mathbf{h}^{\text{pre}}\right)$$

Make plots of e_i vs. i and ℓ_i vs. i.
Make scatter plots of d_i^{pre} vs. d_i^{obs} and scatter plot of h_i^{pre} vs. h_i^{pre}.
Any unusually large errors?

Step 12: Examine the total error Φ^{est}

$$\Phi^{\text{est}} = E^{\text{est}} + L^{\text{est}} \text{ with } E^{\text{est}} = \mathbf{e}^{T}\mathbf{e} \text{ and } L^{\text{est}} = \boldsymbol{\ell}^{T}\boldsymbol{\ell}$$

Φ^{est} is chi-squared distributed with $\nu = N + K - M$ degrees of freedom.
Use a chi-squared test to assess the likelihood of the null hypothesis that Φ^{est} is different than the expected vale of ν only due to random variation.

The 95% confidence interval is

$$\nu - 2(2\nu)^{\frac{1}{2}} \le \Phi \le \nu + 2(2\nu)^{\frac{1}{2}}$$

Step 13: Two different models, A and B?

Use an F-test on

$$F = \left(\Phi_A^{\text{est}}/\nu_B\right)/\left(\Phi_B^{\text{est}}/\nu_A\right)$$

to assess the likelihood of the null hypothesis that F is different from unity only due to random variation.

See exemplary *MatLab* script `gda14_02`

Method Summary 2, Nonlinear Least Squares

Step 1: State the problem in words

How are the data related to the model?
What prior information is applicable?

Step 2: Organize the problem in standard form

N data, M model parameters, K pieces of information.
Data equation $\mathbf{g(m)} = \mathbf{d}^{\mathrm{obs}}$ and prior information equation $\mathbf{Hm} = \mathbf{h}^{\mathrm{pri}}$.

Step 3: Decide upon a reasonable trial solution

Trial solution $\mathbf{m}^{(k)}$ for $k = 1$

Step 4: Linearize the data equation

$\mathbf{G}^{(k)} \Delta \mathbf{m}^{(k)} = \Delta \mathbf{d}^{(k)}$ and $\mathbf{H} \Delta \mathbf{m}^{(k)} = \Delta \mathbf{h}^{(k)}$ with $G_{ij}^{(k)} = \left. \dfrac{\partial g_i}{\partial m_j} \right|_{\mathbf{m}^{(k)}}$ and

$\Delta \mathbf{d}^{(k)} = \mathbf{d}^{\mathrm{obs}} - \mathbf{g}(\mathbf{m}^{(k)})$ and $\Delta \mathbf{h}^{(k)} = \mathbf{h}^{\mathrm{pri}} - \mathbf{H} \mathbf{m}^{(k)}$

Step 5: Form the combined equation

$\mathbf{F}^{(k)} \Delta \mathbf{m}^{(k)} = \mathbf{f}^{(k)}$ with $\mathbf{F}^{(k)} = \begin{bmatrix} [\operatorname{cov} \mathbf{d}]^{-\frac{1}{2}} \mathbf{G}^{(k)} \\ [\operatorname{cov} \mathbf{h}]^{-\frac{1}{2}} \mathbf{H} \end{bmatrix}$ and $\mathbf{f}^{(k)} = \begin{bmatrix} [\operatorname{cov} \mathbf{d}]^{-\frac{1}{2}} \Delta \mathbf{d}^{(k)} \\ [\operatorname{cov} \mathbf{h}]^{-\frac{1}{2}} \Delta \mathbf{h}^{(k)} \end{bmatrix}$

Step 6: Iteratively improve the solution

Solution: $\Delta \mathbf{m}^{(k)} = [\mathbf{F}^{(k)\mathbf{T}} \mathbf{F}^{(k)}]^{-1} \mathbf{F}^{(k)\mathbf{T}} \mathbf{f}^{(k)}$
Update rule: $\mathbf{m}^{(k+1)} = \mathbf{m}^{(k)} + \Delta \mathbf{m}^{(k)}$
Recompute $\mathbf{G}^{(k)}$ and $\mathbf{F}^{(k)}$ between each iteration.

Step 7: Stop iterating when

$\Delta \mathbf{m}^{(k)} \approx 0$ or $\Phi^{\mathrm{est}} \approx N + K - M$.
Solution $\mathbf{m}^{\mathrm{est}}$ is $\mathbf{m}^{(k)}$ of final iteration
Covariance $[\operatorname{cov} \mathbf{m}]$ is $[\mathbf{F}^{(k)\mathbf{T}} \mathbf{F}^{(k)}]^{-1}$ of final iteration.

Steps 8–12: Examine the results

Use the $\mathbf{G}^{(k)}$ and $\mathbf{F}^{(k)}$ from the last iteration to:
Examine the model resolution;
State 95% confidence intervals for $\mathbf{m}^{\mathrm{est}}$;
Examine the individual errors;
Examine the total error Φ^{est}; and
Test the difference between two different models
by following the procedures outlined in Method Summary 1 (but recognizing that these results are only approximate).
See exemplary *MatLab* script gda14_03

Method Summary 3, The Grid Search

Step 1: Which model parameters are to be searched?

Grid searches over more than a few parameters are impractical.

Dividing model parameters into two groups, one found through a search and the other by least squares, may be possible.

Step 2: Identify the range and sampling of the search

The range must bracket the solution.
The sampling of grid nodes $\mathbf{m}^{(k)}$ must not be so coarse that the point of minimum error is missed.

Step 3: Determine the error $E(\mathbf{m}^{(k)})$ at each grid point $\mathbf{m}^{(k)}$

Predict the data $\mathbf{d}^{\text{pre}} = \mathbf{g}(\mathbf{m}^{(k)})$
Compute the least squared error $E(\mathbf{m}^{(k)}) = \mathbf{e}^{\text{T}}\mathbf{e}$ with $\mathbf{e} = \mathbf{d}^{\text{obs}} - \mathbf{d}^{\text{pre}}$

Step 4: Find the grid node that minimizes the error

$\mathbf{m}^{\text{min}}$ is the $\mathbf{m}^{(k)}$ that minimizes $E(\mathbf{m}^{(k)})$
$\mathbf{m}^{\text{min}}$ is a preliminary estimate of $\mathbf{m}^{\text{est}}$

Step 5: Refine the solution using a quadratic approximation

Determine E_0, $\mathbf{b}$, and $\mathbf{B}$ in $E(\mathbf{m}) = E_0 + \mathbf{b}^{\text{T}}\Delta\mathbf{m} + \frac{1}{2}(\Delta\mathbf{m})^{\text{T}}\mathbf{B}\,\Delta\mathbf{m}$ with $\Delta\mathbf{m} = \mathbf{m} - \mathbf{m}^{\text{min}}$ using least square on a few grid nodes surrounding $\mathbf{m}^{\text{min}}$ then $\mathbf{m}^{\text{est}} = \mathbf{m}^{\text{min}} + \Delta\mathbf{m}$ with $\Delta\mathbf{m} = -\mathbf{B}^{-1}\mathbf{b}$

Step 6: Estimate the covariance of the solution

$[\text{cov }\mathbf{m}] = \sigma_d^2[\frac{1}{2}\mathbf{B}]^{-1}$ with $\sigma_d^2 = E(\mathbf{m}^{\text{est}})/(N-M)$
See exemplary *MatLab* script `gda14_04`

Method Summary 4, Factor Analysis

Step 1: State the problem in words

What are samples, elements, and factors in this problem?
Is the idea that samples are a linear mixture of factors appropriate?
Are there natural orderings in space, time, etc.?

Step 2: Organize the data as a sample matrix S

$$\mathbf{S}^{\text{obs}} = \begin{bmatrix} \text{element 1 in sample 1} & \cdots & \text{element } M \text{ in sample 1} \\ \cdots & \cdots & \cdots \\ \text{element 1 in sample } N & \cdots & \text{element } M \text{ in sample } N \end{bmatrix}$$

Step 3: Establish weights that reflect the importance of the elements

w_i for $i = 1, \ldots, M$

One possibility: $w_i = 1/\sigma_i$ with σ_i^2 the prior variance of element i.

Step 4: Perform singular-value decomposition

$\mathbf{U}\mathbf{\Sigma}\mathbf{V}^{\text{T}} = \mathbf{S}^{\text{obs}}\mathbf{W}$ with $W_{ij} = \delta_{ij}w_i$

Step 5: Determine the number P of important factors

Plot the diagonal of Σ as a function of row index i and choose P to include all rows with "large" Σ_{ii}.

Step 6: Reduce the number of factors from M to P and form the factor matrix $\mathbf{F_P}$ and loading matrix $\mathbf{C_P}$

$$\mathbf{F_P} = \mathbf{V_P^T W_P^{-1}} \text{ and } \mathbf{C_P} = \mathbf{U_P \Sigma_P}$$

$$\mathbf{F_P} = \begin{bmatrix} \text{element 1 in factor 1} & \cdots & \text{element } M \text{ in factor 1} \\ \cdots & \cdots & \cdots \\ \text{element 1 in factor } P & \cdots & \text{element } M \text{ in factor } P \end{bmatrix}$$

$$\mathbf{C_P} = \begin{bmatrix} \text{factor 1 in sample 1} & \cdots & \text{factor } P \text{ in sample 1} \\ \cdots & \cdots & \cdots \\ \text{factor 1 in sample } N & \cdots & \text{factor } P \text{ in sample } N \end{bmatrix}$$

Step 7: Predict the data and examine the error

$$\mathbf{S^{pre}} = \mathbf{C_P F_P} \quad \text{and} \quad E = \mathbf{S^{obs}} - \mathbf{S^{pre}}$$

Step 7: Interpret the factors and their loadings

Pattern of elements in the factors?
Spatial/temporal pattern of the loading?
See exemplary *MatLab* script `gda14_05`

Method Summary 5, Bootstrap Confidence Intervals

Step 1: Identify the quantity q for which confidence intervals are needed

The quantity q might be a model parameter in an inverse problem or a function of several model parameters.
Is a quicker method available for determining the confidence of q?
If so, consider using it in preference to bootstrapping.

Step 2: Identify the data and a method of estimating q from the data

The data $\mathbf{d^{obs}}$ are of length N.
The method usually involves solving an inverse problem.

Step 3: For reference, estimate q^{est} from the observed data $\mathbf{d^{obs}}$.

Step 4: Resample the data and recompute q

Let $\mathbf{j}$ be a list N of random integers, each in the range 1 to N.
The resampled data are $d_i^{rs} = d_{j_i}^{obs}$.
Determine q^{rs} from $\mathbf{d^{rs}}$ using the method in Step 2.

Step 5: Repeat Step 4 many times

Repeat, say, $N_r \approx 1000$ times.
Tabulate the results in a vector $\mathbf{q}^{rs}$ of length N_r.

Step 6: Construct an empirical p.d.f. for q

Decide upon a reasonable number N_q and range (q^{min}, q^{max}) of histogram bins by examining the range of values in $\mathbf{q}^{rs}$.
Define a vector $\mathbf{q}$ of histogram bin locations
$q_i = q^{min} + (i-1)\Delta q$ with $\Delta q = (q^{max} - q^{min})/(N_q - 1)$.
Compute a histogram $\mathbf{h}$ of the values in $\mathbf{q}^{rs}$ using this binning.
Normalize the histogram to unit area so it is an empirical p.d.f.

$$p(q_i) = h_i/A \text{ with } A = \Delta q \sum_{j=1}^{N_q} h_j$$

Integrate $p(q)$ into the cumulative probability $P(q)$

$$P(q_i) = \Delta q \sum_{j=1}^{i} p(q_j).$$

Step 7: Determine 95% confidence limits for q from $P(q)$

The left bound q_L is the q for which $P(q) \approx 0.025$.
The right bound q_R is the q for which $P(q) \approx 0.975$.
Plot $p(q)$, q_L, q_R, and q^{est} and interpret results.
The 95% confidence interval for q^{true} is approximately $q_L < q^{true} < q_R$.
See exemplary *MatLab* script `gda14_06`

Index

Note: Page numbers followed by *f* indicate figures, and *t* indicate tables.

Printed in the United States
By Bookmasters